Please Return this book
to "T" Roger

Seed Savers Exchange: The First Ten Years

Edited by
Kent Whealy and Arllys Adelmann

Summer 1993

DEDICATED

To every gardener
who has ever given
seeds to a friend.

ISBN: 0-9613977-2-1 Paperback
Library of Congress Catalogue Card Number: 86-061687

Seed Saver Publications

Decorah, Iowa

Printed in the United States of America

Cover Graphics by *Rohander*

INTRODUCTION

During 1975 a half dozen gardeners began corresponding. The only thing they had in common, besides their love for gardening, was that each of them was nurturing and protecting seeds that had been passed down from generation to generation within their families. They hoped that, by exchanging these rare seeds with each other, there would be less chance that their seeds would die out. And they wondered how many other gardeners in this nation of immigrants were also keeping seeds that their families had brought with them from the old countries. That small group has grown into the Seed Savers Exchange.

Ten years later the Seed Savers Exchange (SSE) has evolved into a national organization of 630 concerned gardeners who exchange seeds each winter through the SSE's publications. Collectively this group now offers nearly 4,000 rare and unique garden plants each year to other gardeners who pledge to grow and reoffer their seeds. During these last ten years our members have sent out enough seed samples to make an estimated 300,000 plantings of varieties that were often not in any seed catalog and in some cases were nearly extinct.

The SSE has assembled a Central Seed Collection of nearly 5,000 endangered garden varieties which it is maintaining and protecting on a permanent basis. This collection is now being stored in Decorah, Iowa (it was in Princeton, Missouri until the Whealys moved). These seeds are available through the SSE's Growers Network to gardeners who agree to "adopt" a variety, protect it on a permanent basis, offer it to other gardeners through the SSE's publications, and turn it in for redistribution when it can no longer be maintained. The SSE has also started growing a Preservation Garden each summer of up to five acres and up to 2,000 of these varieties in order to renew seeds of the central collection and put these unique varieties on display for the gardening public.

The SSE's publications have grown since 1976 from a six-page xeroxed newsletter into two semi-annual books of 256 pages each. We constantly receive requests for back-issues of our publications, but most are sold out soon after they are printed. So we decided to publish this book of the best articles from our previous issues. We have made every effort to update this information (seed catalog prices, availability of books, addresses, reports, etc.), but please realize that a small portion is probably still incorrect. This book is intended to be a look back at the growth of the SSE, not a current resource guide.

The SSE has experienced phenomenal growth during these last few years. It is having quite an impact on gardening publications and the seed industry, and is making its presence felt in plant breeding programs, government seed storage facilities, and even the gourmet restaurant trade. I certainly don't intend to take the credit for this, because I've always seen my role as just a "focus" for our members' energies. As the SSE's Director, I have tried to never impose any personal preconceptions of what I think the organization should be--I've let the SSE find its own shape. Because of this the SSE continues to evolve into an ever more powerful genetic preservation project. It will be fascinating for all of us to watch the SSE grow and mature during its second ten years.

— Kent Whealy (July 20, 1986)

CONTENTS

Interviews with Collectors

Corn Painting of the Great Horned Owl by Ernest Strubbe
(Cover of The 1982 Fall Harvest Edition)

ROBERT LOBITZ - POTATO WORKER

KENT: When did you start collecting potatoes?

ROBERT: I started in 1967, but it's been since I joined the Seed Savers Exchange that I started corresponding with universities and really built up my collection.

KENT: When you first wrote me in 1979, I think you had about 50 varieties and were just starting to work on some of your own. You've put together a tremendous collection since then which includes quite a few older named varieties. Where were you able to find them?

ROBERT: I got a few from SSE members. Others I got from a USDA collection they maintain in Maine at an Agricultural Experiment Station. A few of the first ones I got from Jim Johnson in Wisconsin. I got the Early Rose from him and that's the only source I ever found for it. The Blue Christie I got from a man in New York. And a few of the others were still sold locally some time ago.

KENT: What is the quality like on some of the older potatoes?

ROBERT: They all have a pretty good flavor and in general they are good keepers, better than most of the newer ones. Nobody really cares anymore how well a potato is going to keep because they've got modern ways of storing them--controlled environment as they call it. Everything is controlled by machine: humidity, temperature, even the atmosphere in some cases. Modern storage. It doesn't matter to most people because they have gotten used to buying a little bag of potatoes at the store every other week. But if they'd try storing some of them in a basement like ours for the winter, they'd sprout like crazy. Most of these old ones don't sprout that bad at all. That characteristic is valuable to rural folks who grow potatoes, store them over the winter and then plant them each spring. Really good keepers are fairly rare.

KENT: I know that you especially prize a potato named Garnet Chili. Could you tell me about it?

ROBERT: It's the oldest potato in my collection and dates back to 1853. I got it through the Potato Introduction Station in Wisconsin. In 1851 a Rev. Goodrich from New York received a small quantity of potatoes from the American Consul to Panama. One of the varieties he called Rough Purple Chili. In 1853 he grew seedlings of Rough Purple Chili and out of those he selected one which he called Garnet Chili which he distributed to other growers. About 170 varieties can be traced to Garnet Chili including many of the old varieties of commerce. I would guess that Early Ohio and Early Rose are both seedling varieties of Garnet Chili, because they are quite similar to it. I also have an idea that Blue Christie is too, because the leaves on the seedling potatoes are quite similar.

KENT: I've gotten numerous letters from gardeners who were worried that two potatoes they were growing would cross because they flowered at the same time. Why don't you clear that up for everyone and then tell me how you get your crosses.

ROBERT: Potatoes produce asexually as far as tuber production is concerned. Home production of tubers is actually a cloning process and the flowers don't affect the tubers at all....potato flowers are self-pollinating, but can be crossed by insects or by hand. Occasionally the flowers will produce a seed ball which, depending on its size, may contain from 12-250 seeds which are extremely toxic. I put the seeds through a fermentation process for several days and then dry and store them. The next spring I plant those seeds under hotcaps in the garden. Each seed in each seed ball is a different variety. If there is a broad enough genetic base, from either those that crossed or even inherent in one that's just self-pollinated, you can get a whole array of colors....I have potatoes that have white, yellow, red or purple flesh. And I now have skin colors of red, white, russet, purple, lavender, pink, brown, buff, purple & white, purple & red, and black (the black-skinned types have very dark purple flesh)....Also each plant can be a different shade of green. They have different shaped leaves and the flower colors can be different once they start to bloom. You just never know what you are going to get. That's really what I find so fascinating about all of this....

KENT: What do you look for in seedlings when developing new varieties?

ROBERT: I look for disease-resistant plants that will stay healthy. I also look for good production and decent size. Some will set a lot of potatoes, but there's not much size to them. So I select for disease resistance, good-sized potatoes, high yields, shallow eyes, good coloring, how well they respond to stress and how well they store. Characteristics like those.

KENT: You said this year you're growing 650 varieties that were seedlings last year?

ROBERT: Yes, but those were selected from approximately 1,300. And I'd estimate that out of those 650 I will probably only keep 200-300 over for another year. This is a continuing process, of course. I've got 250 hotcaps of seedlings out there right now. If everything went perfectly, each one could produce 20 small potatoes. So at best I could end up with 5,000 new varieties, or at least 1,000 even if the season goes poorly....

KENT: How many named varieties are you also growing?

ROBERT: Probably about 300....

KENT:....and you mentioned some wild species....

ROBERT:....about a half dozen species of wild potatoes that are said to be native to Arizona and Mexico. And then I'm also growing probably about a thousand foreign varieties which I grew last year from seed that I got from the Potato Introduction Station. Some of them are from Mexico and Guatemala. Most of them are from Colombia, Argentina, Chile, Peru and Bolivia.

KENT: You've told me that you plant whole potatoes?

ROBERT: I've found that by planting whole small potatoes I can sometimes get yields that are double that of cut potatoes, at least on most varieties....I select the potatoes I intend to keep for seed right when I'm digging them. I usually select ones that are about the size of a hen's egg or a little larger. Smaller potatoes usually don't have as many sprouts in the spring as the larger ones. Depending on the variety, they will usually have a 2" or longer sprout when I plant them. If there are too many sprouts, I just break off all but two to four. Always be sure to leave the main sprout on the eye, because it will produce the most vigorous plant....I wouldn't be at all surprised if studies showed that yields could be increased by planting sprouted potatoes. The plants emerge sometimes in only a few days and as much as two weeks ahead of non-sprouted potatoes....As I said, the sprout on the eye end of the potato is by far the strongest. Lots of times the other sprouts or eyes won't even develop. Many people really make a mistake by cutting their seed potatoes too small, especially if it's not a large potato. With some varieties every piece will grow a nice plant. With some others the eye end of the potato is all that grows. And you never know from one year to the next how they are going to react in that way. So if your potatoes are coming up spotty, that may be part of your problem.

KENT: How much of each variety do you plant? Obviously it can't be much with the number you're keeping.

ROBERT: That varies on what I happen to have for seed. I try to plant a half a dozen hills of each kind. But sometimes I have as few as one. The more rare ones I try to plant 15 to 20 hills....Usually I plant my potatoes in rows that are 500 ft. long. I could just about plant one of those in a day. But this year I lengthened the rows on out to 700 ft. I have seven full rows this year, plus two other rows of foreign species and wild species.

KENT: With such huge plantings, do you ever have problems with disease?

ROBERT: Not really. Oh, I have had varieties that were diseased for from one to several years. But I just continue to plant them year after year and eventually, for whatever reason, one year they just won't have the disease. Or one year some plants would be free of the disease and I would save seed from those hills, thus gradually developing a healthier strain.

KENT: When do you start digging your potatoes?

ROBERT: Some of them start dying back by the middle of July, so that's about when I start digging them. Last year I finished on the 20th of September. My dad and I are truck gardeners, so that's really a busy time for other crops too. I try to do some digging each day during the first of that period. Of course as more and more of the potatoes start to mature it gets increasingly hectic. There are times when I've been doing nothing but dig potatoes for over two weeks straight. The soil will have dried out and I feel like I'm going at a cement road with a table fork. My arms and back will be so sore that I can hardly stand it. It gets to the place that I'm almost tempted to start praying for rain....

Robert Lobitz
And Some of His Potatoes

KENT: Why can't people find a source for Green Mountain? From what I've been told, until just recently it was fairly common in the Northeast.

ROBERT: It was very common. But there are a lot of these varieties that were common as recently as five years ago that you can hardly find being grown anymore. Like everything else, it's getting down to a half dozen to a dozen standard varieties....One variety that I'm afraid is going to be dropped soon is Early Ohio. It's a really fine-flavored eating potato and quite a potato for frying. It's such an early potato and yet it's a really good keeper too. Just a few years ago they used to grow better than 2,000 acres of certified Early Ohio seed. Many of the older gardeners in this state and the surrounding ones used to grow it. But now many of them have quit gardening and there is less of a demand. I don't think we can even get them through the store here now. I'd really hate to see it die out.

KENT: Somebody in the SSE this year was talking about a Canadian potato that was banana-shaped. They said it grew in a cluster and almost grew out of the ground.

ROBERT: Yeah, I've got that one, but I never had too much luck with it. They set like mad last year, but most of them didn't get very big. They claim that up there it's really a nice big potato. But they just don't seem to do too good here.

KENT: Are there any specific varieties that you are trying to find?

ROBERT: In general I would like any named variety I don't have. Specifically I'm looking for Aroostook Wonder, Sutton's Flourball, Sequoia, and a yellow-fleshed Irish Cobbler.

KENT: We're growing Sequoia. Amy Rogers from Kentucky sent it to me this year. I can send you some of it.

ROBERT: That's great. I've been looking for it for some time. It's claimed to be quite insect resistant, but I've never been able to get ahold of it. The USDA doesn't have it and neither do any of the breeding stations I've tried.

KENT: Somebody wrote that they used to have a variety named Cowhorn and they remembered kids walking in from the garden with an armload of potatoes that were really long. They said they looked like they were carrying in a load of stovewood. Have you ever seen a potato like that?

ROBERT: The Cowhorns I have grown were only about three-four inches long and about the size of your finger. Maybe I'm too far north for them.

KENT: Almost sounds like they were talking about a sweet potato, doesn't it?

ROBERT: Maybe so....There is a guy in Georgia that has a variety called White Elephant, a huge white potato. He wrote me this spring and said he's going to try to get some to me this fall. I sure hope that he does. He said it has been in their family for almost 70 years....Jim Johnson from Wisconsin wrote that he had a variety called Snowflake 1873. The one I had was from 1961. Same variety? So we both exchanged and grew them. I could tell right away that they weren't the same potato. I wrote to the University of North Dakota and they said that they had released a variety named Snowflake in 1961. But my dad remembers another from many years back. So there is one from 1873 and one from 1961. They are both white potatoes, a little different in maturity, and the older one has a bigger plant and is a much better yielder.

KENT: I'm surprised they'll let them use an old name like that.

ROBERT: They probably figured the other one was long gone. So many are.

KENT: What are some of your most productive varieties?

ROBERT: Hudson is very productive. Last year I had a hill of La Salle that produced 15 3/4 lbs. That was my best one by weight. The Viking (purple) and Royal Viking (a mutation of Viking that's red) are both very productive. One year I had one hill of Viking that produced twelve potatoes that totaled 11 lbs. Also the two Rural New Yorkers are both quite productive. They are very similar (white potatoes, large heavy production), but the eyes are a little different. The Russet Rural is actually a mutation of Rural New Yorker. It has more of a brown skin.

KENT: Did you find these mutations?

ROBERT: Yes. Mutations usually occur on only one eye of a potato, usually resulting in a different skin color. A patch of a new color will appear on a potato, but it must completely surround the eye in order for a new variety to develop.

KENT: You mentioned that so many of the European potatoes are yellow-fleshed. They are in general better flavored, aren't they?

ROBERT: Generally they are. They have a higher dry-matter content. They aren't as watery as a lot of the white varieties. Just a better quality flesh. Nutritionally, they aren't very different from the whites. Size-wise, in general, they are not nearly as big.

KENT: I have always wondered why there weren't more yellow-fleshed potatoes grown here.

ROBERT: They got started many years ago pushing just the white-fleshed ones (generally with red, white or russet skin colors) and that was it. Now everybody is pretty well used to that. They figure if it isn't that color, it can't be eaten. When you talk about yellow flesh, they might try it. But when you talk about blue or purple flesh, they think, "What kind of poison are you trying to feed me?" I consider most of the purple-fleshed ones to be the best flavored of any of them. There are, of course, really good-flavored ones in the whites too. There is one white-fleshed variety that I started myself, a seedling of Chieftain, that I call White Chief. It is quite similar with white skin and just a little bit of red by the eye. I've sent it to people who live where it rains quite a bit and it does really well even in wet, heavy soils. It is a round, slightly oval potato. I dug some last year that were almost 8" in diameter. They are only an average keeper, but for eating and yield they're great. Even better quality than the Chieftain as far as taste is concerned. We had a few of them left and we finished them up late this spring. Even that late they were still nice and mealy with a real good flavor. The red-fleshed one I developed has quite a good flavor too. They have the brightest red skin. You've never seen such a potato ....Sometimes at a meal you'll end up with purple and red and yellow and white potatoes all on your plate there together. Mash those up and add a little butter and you've got some really good eating. And that's something not many people have ever seen.

KENT: Several years ago there was a lady in the SSE who was working on a potato project back East. As I remember it she just requested correspondence from other potato workers. Later a garden writer who had been interviewing collectors contacted her. It seems that for some reason her root cellar froze. She was just sick. She lost the complete collection that she had built up over the years....Now that I've seen the real scope of what you are doing, it really scares me to think of such an incredible amount of material being stored in only one location. Do you ever worry about losing any of your collection?

ROBERT: To tell you the truth, I never think about it. Our basement has always been excellent for storage, but I guess freezing isn't the only thing that could happen to them.

KENT: I've been seriously thinking about setting up a Maintenance Program within the SSE in which people would be specifically responsible for maintaining certain varieties. I'd keep track of who was maintaining what and if a person could no longer maintain the materials, they would turn them back in for reassignment. I wouldn't ever send anyone to them unless I had a request for that variety and it wasn't available through the Yearbook or the Growers Network. I think it would especially work well for things like potatoes and perennials. There's quite a difference between sending out a packet of seeds and sending out a box of potatoes. Do you think that would be valuable?

ROBERT: I think it is really a good idea. We could start with the ones in my collection that are either historic, colored, highly productive or good keepers. In general, the ones that are good home garden types. Maybe several people could maintain each variety, or at least the older ones....I've become so busy that I really can't answer requests any more. So I'm really not sure they will get out unless you are able to get people in your Growers Network to maintain them. Even in my limited fashion of developing these new varieties, there are quite a number of really fine ones that I feel should be kept. Of course, I'm the only one that has them and it would be a shame if they were not increased.

(The 1982 Fall Harvest Edition, pages 25-31)

(From earlier correspondence with Robert Lobitz) -- Your best, richest, well-drained soil should go to your potatoes; do not have excessive nitrogen for potatoes. The most important stage of growth when plenty of moisture is essential is from the time the flower buds form until two weeks after they are through blooming....The universities of many states have potato breeding programs in which the potatoes are grown in greenhouses the first year. After that they are planted in the field where they are tested for yield, insect and disease resistance and other qualities. These universities may start from 30,000 to 70,000 new varieties each year, but as the years go by very few are kept and released as new named varieties....As for storage, they keep best the colder the better, 34-40 degrees F. Some varieties start to sprout and grow even at this low a temperature, others show no sign of a sprout or very little. As for darkening during cooking because of the colder storage temperature, this varies widely depending on the variety. Some varieties can be stored only a month or two to make a good eating potato....I have stored both washed and unwashed potatoes for seed or eating and have found no difference in their keeping qualities....After they are dried off, either washed or unwashed, drying in the shade for only a few hours, they are ready for storage....

(The 1980 Seed Savers Exchange, pages 25-26)

SSE POTATO MAINTENANCE PROJECT

Robert Lobitz has sent me 54 varieties of potatoes (eight tubers each). I want to send each to two different growers (four potatoes). I've got a list worked up of all their characteristics and their date and place of introduction. If you are interested in maintaining at least two varieties, send a self-addressed stamped envelope with two stamps to: Potato Maintenance Project, c/o Seed Savers Exchange, RFD 2, Princeton, MO 64673. Please don't apply unless: 1) You are already growing and storing your own seed potatoes each year; 2) You're sure you won't be moving within the next several years; and 3) You are located in the upper Midwest. (This is Robert's suggestion because most of these varieties are either adapted to or developed in Minnesota or Wisconsin and the states surrounding them, plus slightly eastward. He also said that since we intend to maintain these over a long period, being kept in these cool climates would minimize insect and disease problems.) Also, if you run a business that uses small boxes (about a 6" cube), I could sure use about a hundred of them. Call me collect at 816-748-3785 and we'll work something out.

(The 1982 Fall Harvest Edition, page 31)

BEN QUISENBERRY - "BIG TOMATO GARDENS"

(Over the years many of you have bought tomato seeds from Ben Quisenberry's "Big Tomato Gardens." Ben turned 95 last August, making him probably the oldest active seedsman in the country....I want to thank John Gorman for the pictures and tape he sent me of a recent visit with Ben. Some of you tomato enthusiasts may wish to participate in John's "Biggest Tomato Contest." Thanks again for sharing with us, John.)

BEN: Let me show you through my garden....These are Big Ben. Those took a notion to get natural size. They're so heavy they're breaking the vines....Those are Long Toms. They are one of my best. The Long Tom is a dandy salad tomato....This is Ruby Gold....That one I call the Red Cup. I think the right name is Tomango. It's a stuffing tomato....This yellow tomato is Golden Sunray. The neighbor lady makes the best yellow tomato juice out of it....Here's Evergreen. They're green inside, too. They usually get pretty large. That's the one that got me started. I started out with Evergreen and Ruby Gold. That's the two I started with. Then good people just kept sending me their favorites....

JOHN: What is this little tomato?

BEN: This is my Large Cherry tomato. They're just loaded. They yield real heavy and bear up to frost....This is the Golden Sunray. I want you to cut that and see how nice it is inside....And here are some Mortgage Lifters.

JOHN: It's sort of a rough tomato, isn't it?

BEN: Yes it is. Rather rough, but they're quite productive. They yield pretty heavy and get really large. The name Mortgage Lifter tells you they're a good size, doesn't it?

JOHN: Do you like to eat tomatoes as much as I do?

BEN: I sure do. I'm very fond of tomatoes.

JOHN: What is your favorite for taste?

BEN: Brandywine. It has kind of a sharp taste to it. It's a potato-leaf. You can spot it just by the leaves. Potato-leaf and long-jointed. Brandywine is as fine a tasting tomato as ever was raised. You know, the lady that sent me the seed told me they had been in her family for over a hundred years.

JOHN: Do you ever sell tomatoes or do you just sell seeds?

BEN: I give people tomatoes and I sell the seeds. No, I never do sell any tomatoes....

I'll show you how I clean my seed. First I cut the tomatoes in half. Then I work my finger around in the cavities, just push them out, and let them drop in a bucket of water. That way you don't have much of the pulp to deal with. Then I take a strainer and put it over another bucket and pour the juice, the water, the seed and everything into the strainer. I work it like this for a while and just push the pulp on through the strainer. That pulp is pretty well worked loose now. Then I dump the strainer right out on a newspaper. Take a knife and spread them out over the paper. Aim to spread them out even in a single layer and kind of press them down with the knife against the paper. It's surprising how much the newspaper will absorb the moisture. Now I take this and set it back here in the sun and forget it for about five minutes or a little more. Then I take a dry piece of newspaper and move them off that wet one onto it. Someone asked how do you keep them from sticking to the paper. I keep them moved from the wet to the dry places. I'll put them right in the window here and let the sunshine dry them out again. The third time I move these, I'll leave them loose in little piles and I don't press them down.

JOHN: When they get really dry, how do you store them? Do you put them in the refrigerator?

BEN: No, I put them in an airtight container. If I have just a few seed, I take tinfoil and fold it tight around them and put it in an envelope with their name on it. Tomato seeds deteriorate awful fast if you just throw them around any old way.

JOHN: How do you start your plants?

BEN: To grow good tomato plants, you want a shallow box, maybe 12" x 18". And you want some light fluffy soil. Maybe you have to go to the woods and get woods dirt—around a rotten stump would be ideal. Or you can go to a farm and get dry chicken mix or a professional mix which is a light fluffy soil. I plant them out across the box with a row for each different variety I've got. Sow the seed very sparingly in the rows. Drop them along about a half-inch apart. You get a better, stronger plant if you give them more room....See that little shelf on the wall back behind the stove? I put that box on there so they get a little bit of the bottom heat from the stove. Just a little bit. That'll make them come quicker. You won't need to put them in a bright window until about 6-7 days when they'll be coming up. Then you want to put them in full sunlight. Never let them get crowded in the box. Water between the rows. When they get 2" x 3" tall and get their permanent leaves, then you move them into peat pots if you want. Then when the weather is favorable, you can move them out into your garden. Put them out in rows about two feet apart and about a foot apart in the rows. I think the plant should be set pretty deep. Put some good soil down in the bottom. If it is hot dry weather, a mulch counts for a lot. A mulch of hay or straw, grass cuttings, or if you have access to a stable, get barnyard litter. When you mulch you keep the soil at a more even temperature, cooler in the daytime and warmer at night. And it discourages the weeds....I stake all my plants....

Once a fellow from a television station asked me how I accounted for keeping active 95 years. I said to keep happy and to keep content, do something that's worthwhile. If you have a hobby and you can make your hobby your business, you're all the better off. Do something good....I do marvel at myself sometimes the way I keep going. Indeed I do. I marvel at myself. I'm surprised sometimes how I can keep going from 7 o'clock until dark. About dark, I'm like the chickens. I hunt for my roost....Aging in rural America, the elderly man is very fortunate if he is in a rural district. Out in nature. Out where he can work. Out where he can fasten his hands onto the end of a hoe handle and make things grow. He's with nature and when he's with nature he's close to God. God is nature and nature is God. So you're in good company when you are out in a rural district.

(The 1982 Fall Harvest Edition, pages 41-43)

ERNEST STRUBBE - CORN BREEDER

(I had been driving since early morning. Late in the afternoon I arrived at Ernest Strubbe's farm in western Minnesota just ahead of a huge black prairie thunderstorm that was rolling in from the west. Ernest met me at the door and it cut loose raining just as we went inside....There was an old wood stove over by a desk where he had been doing some paintings and matting some photos of birds and wild flowers. The archway into the next room was just solid with ears of corn. Back through it I could see a window lined with corn and a rod high across the room that had corn hanging all over it. Shelves on the wall in the front room were lined with ears of corn hanging on display....It was every color of the rainbow--reddish blacks to pure whites, emerald greens, turquoise blues, light blues, purples, pinks, browns, calico blues, calico pinks, white-capped reds, oranges, yellows....He said the sun had faded most of it, but it was still dazzling. He took me into a back room and showed me a bunch of it that the light hadn't got to. Then he brought out two display cases that he had made....)

KENT: These are the most beautiful dent corns I've ever seen. How did this whole process start?

ERNEST: In 1929 I crossed some yellow field corn with some black popcorn, Black Beauty which Shumway dropped a few years ago. What I got was a dark-colored black/bronze dented ear. At first the color was real spotty with bronzes and brownish/purples. I worked with it for quite awhile and finally got the color really even and uniform. I was pretty proud of that....Then I crossed some of that original dark-colored black/bronze with some Rustler's White Dent, which was an old field corn that farmers around here used to raise in the 1920s. A lot of interesting things started cropping up right then. That year I got one ear that was the start of my Blue/Gray Dent. When I planted seed from that the next year, you should have seen the firing and albinoism. When it started coming up it looked just like somebody had sprinkled salt and pepper all over the field. About 25% of the seedlings were pure white--no chlorophyll. They grew until they were about 3-4" tall, until they used up the food in the kernel, and then they died. In lessening degrees it showed up the following three or four years until it weeded itself out.

KENT:....and you said firing?....

ERNEST: Yes, that appeared in about 10% of the other plants. When they were about ready to tassel suddenly the bottom leaves, and sometimes all of them, just shriveled up and fell off. But that weeded itself out too, since they died and didn't produce seed....When I finally thought I had all of that cleared up and had my Blue/Gray Dent stabilized, some pink suddenly appeared in it. So I picked out these pink colors and isolated them. And when I grew them I had the albinoism and firing to deal with all over again....Those first few years would have been enough to discourage most breeders.

KENT: Were you picking out your best ear to save or just picking kernels?

ERNEST: That stuff came so fast that within just a few years I had solid-colored ears and could just save the ears. But right at first I had to pick the kernels for most of the colors....The hull color comes from the female parent (the seed that you plant). The inner colors come from the male parent (from the pollen). So you can actually develop anything you want if you try hard enough. It's fascinating....

KENT: The one that always catches my eye is your Green Dent because it's such an unusual color for corn. How did you come up with it?

ERNEST: That came from the original bronze/black corn that I got from crossing the black popcorn and the yellow field corn. I kept a little patch of it by itself just to see what developed. One fall I happened to be looking at it in the sun to see how many different colors I could get to reflect. There were some blues, some purples, some brownish-bronze and the like. And one time here was this bright green reflection. You had to hold it just right. That fired my imagination! I wondered if it would ever be possible to produce a green color. I could see the possibility of producing most of these others, but green? I hunted and hunted through that stuff and every time I would find a kernel that had the faintest suggestion of a green reflection I picked it out and grew that in isolation....For the first ten years or so it seemed quite hopeless, just plain frustrating. Because only occasionally would I find a kernel that was really green. But all at once it started to get greener and greener and greener and greener. Every year it looked a little more promising, so I kept going. After about 40 years, lo and behold one fall when I was picking I found a couple of ears that were really green....But there's still a lot of work to be done on it yet. Even with the ear here in the case, if I were to plant that I would throw out all of the lighter green kernels. And a few of the kernels might have just a little too much of a purple cast to them. Throw them out because they are just going to pollute whatever you're after....One year Gurney Seed and Nursery wrote to me and wanted to grow 50 bushels of my Green Dent. They'd heard about me through different corn breeders I was corresponding with. That struck me kind of funny because all of the green corn I had right then I could hold in two hands. And they wanted 50 bushels!....I think they did get ahold of some of it in the late 1960s, probably from someone I'd traded seeds with. One year in their catalog I noticed what looked like my Green Dent, so I wrote them a really pointed letter asking them where they got that green corn. They immediately took it out of their catalog. They even refunded money to some of my friends who had ordered it. I wasn't really mad because they were offering it, but more because it wasn't perfected yet....

KENT: Where did the pure black and also the turquoise blue come from?

ERNEST: I found blue and black kernels both that same year. I don't think I've ever found any before or since. Just that one year, and I isolated them and eventually got the Blue Dent and the Black Dent....That Blue Dent is really the prettiest one of the whole bunch to my eye....I found them both in my Green Dent back in the very early 1950s. I was out there looking through my green corn patch. It still had a lot of yellow speckled through it. In fact, the little kids called it John Deere corn....I always like to look at the corn very closely. I think every husk had been opened, even when it wasn't ripe enough yet to pick. I'd usually already been through every ear kernel by kernel, so I'd know who was out there.

KENT: Where did the Brown Dent and the Purple Dent come from?

ERNEST: When I had most of these colors at least partially isolated, I'd take the purest seeds for my pure patches. The odds and ends I'd throw in a big washtub and plant it for pheasant food. That produced some interesting results. Especially after I got it mixed up with this Purple Husked corn. But the Brown Dent and the Purple Dent showed up in that mix. So I picked them out and isolated them. It didn't take very long to get them, either. After the first couple of years, I was getting quite a a few ears that were solid purple. But that was really exceptionally fast....Usually when I isolated a new color, the first few years I was getting popcorn and speckled and everything else. But there might be a little nubbin or small ear that would be perfectly dented and uniformly colored a bright rose-pink. That would be my seed from then on. Then every year it would get a little more and more like I wanted it.

KENT: Did you ever do any hand-pollination during any of this process?

ERNEST: No. I let the wind do my work for me. You see, the prevailing winds around here are always from the northwest. And I've got a long, really narrow patch of ground up on the other side of the road that runs north and south. So after I got my colors up to the point where I had enough seed, I used my four-row planter to plant them instead of planting isolated patches. So I'd put one color in and plant a 30-50 ft. section that was four rows wide. Then I'd clean out the planter real good and plant the next one....So the wind would blow across this long north/south row from the northwest. What little mixing I got was right where two patches joined, but you could easily see that in the ears. So I just kept seed from what I could see was improving. And year after year it improved....Eventually I was planting those patches 20 rows wide.

KENT: So over the years you've selected all of these to the point that they are really stable.

ERNEST: Quite stable! But by no near means do I consider any of these a finished variety. But they are to the point now that if you plant blue seed you will get all blue, but you may get some slightly different shadings of blue....I just wish I had 20 years more to work on them, mainly to get the ear sizes up some more. I was afraid I was going to have to give up working with them completely. I've got real bad feet and can't stand to be on them for more than a half hour. But this spring I bought a six-horse Troy-Bilt.

KENT: Yeah, we've got the seven-horse Troy-Bilt with the Kohler motor. They're really incredible machines.

ERNEST: My, the way they work! My garden lay idle for four years and was full of box elder sprouts, some of them up to five feet tall. I figured I'd have to chop all of them out before I'd even dare to work it up, and I wasn't sure I could do it. So I thought, heck, I'll just see what this Troy-Bilt is made of. The first time through I got about half the trees. The second pass took about half of what was left and after the third pass there were hardly any left. Even the big ones were just gone! Mine came with a Tecumseh motor and I've never heard it labor really. It just pops away and the little dinger keeps right on going. They're so smooth to operate, too. I'm really happy with it. And this year I'm growing some of my Green Dent in the garden.

KENT: So most of your corns were last grown in 1975?

ERNEST: Yeah. I do have some blue that I raised in 1979.

KENT: Just how long will corn keep?

ERNEST: Dent corn will keep quite a while. The corn I sent Carl Barnes I delayed sending for awhile because I wanted to check the germination. By testing it here myself, I found out that the germination was still over 92%.

KENT: Since 1975?

ERNEST: Most of that was from 1973. A lot depends on how it's stored. Some that tested the best I just had in metal cream cans downstairs. But most of it I've got in glass honey jars back here in a closet. I must have about a hundred of them....Over the years I've never made a practice of planting all of my seed. When I'd sort out my green or my blue or whatever, if I had just a little handful I'd just plant about a third of it a year.

KENT: I really try to tell the folks in the Seed Exchange to never plant all of any seed. If you do, you're just asking to lose it.

ERNEST: Yes, you are. One year I had a nice patch of the green corn planted in the bottom in the richest soil you'd ever want to see. But it rained a lot that summer and drowned it all out. It's a good thing I had some in reserve.

ERNEST: (Gets out another display case) - Here's brown pods, purple pods and calico pods in two colors.

KENT: I've only seen this white husked pod corn.

ERNEST: There's also this purple, a brown in addition, and the white. Once in a while for some reason this will throw a regular ear and when it does it has a brown cob. Down inside there's not much cob at all, it's more like a big stem.

KENT: People have written me about a red-kerneled pod corn and a yellow-kerneled pod corn.

ERNEST: You can get that any color you want. One year I picked a lot of this that had green kernels in it. You can cross it with anything. I think most of this was more or less a red popcorn inside when I got it. The stalk was purple.

KENT: These on the right are all popcorns, aren't they?

ERNEST: Yes. This one is Golden Australian-Hulless. It's one of the prettiest corns I've ever raised and is really a good popcorn too. This is the Black Beauty popcorn that I started all my breeding with. This probably is a cross between Red Beauty and Garnet. And this pinkish one, that's Pinkie Popcorn. That came from the old Oscar H. Will Co. over in Bismarck, North Dakota. It seemed like about half their catalog was just to satisfy people's curiosity. They had all these odd vegetables. They had a black sweet corn that was excellent with a fine flavor on a par with Golden Bantam. It was supposed to be ready to pick 60 days after planting. The ears on it were right down next to the ground and the whole stalk wasn't very big. They also had another excellent red Indian sweet corn called Nuetta. This Northwestern Dent, the white-capped red dent here in the display case, that's an Oscar H. Will corn too. At one time that was about the earliest dent corn in the country. I'm not sure, but I think that came from one of the North Dakota Indian tribes. The Oscar H. Will Company got most of their stuff from the Indians. I used to raise their Hidatsa Red Bean and another they called Arikara Yellow. I haven't seen either of those for years and years....

KENT: I dealt out both of them through our Growers Network this spring.

ERNEST: You did? I wouldn't have thought anybody in the whole country still had them. I like what you folks are doing....

KENT: This Rustlers White, you said that's an old field corn that farmers used to raise?

ERNEST: Yes. A lot of farmers raised that here in the 1920s. A lot of them raised this Calico Dent, too. I guess before my time a lot of farmers raised Bloody Butcher, too.

KENT: Did you introduce Bloody Butcher into your blend to get any of your colors?

ERNEST: None of them except this Orange Dent. There's kind of a story behind that, too. Once in a while in a patch of this Bloody Butcher red corn you'll get a few yellow ears. Once in a great while you'll find an orange one. So I'd plant the orange hoping to get orange. I never did. All I'd get was either the Yellow Dent or Bloody Butcher. I didn't know what to do about it.... One year I was planting my regular field corn and I ran out of seed. I only needed just a little bit to finish up a strip. I had a sack full of Yellow that I had saved out of the Bloody Butcher, so I used that. I figured it would probably produce mostly yellow. And you know that darn stuff came about one-third of each color—yellow or orange or red. I got more orange in that than I ever did trying to plant from the orange. Now after saving orange ears from that, I can get orange ears every year.

KENT: What's this Purple Husked corn?

ERNEST: Many seed companies use that purple husked stuff to separate or mark seed plots. I've seen it around experiment stations a lot in the fall....This pretty striped one here is Calico Dent. I've used it to originate a lot of corn of my own. Cross the Calico with the Blue Dent and you get Blue Calico. Here's a cross between the Pink Dent and the Calico with the red stripes on the pink kernels. Besides the red stripes on blue and the red on pink, I've seen red on green and red on brown....You know, some of the prettiest corns that you can get--I've seen a few ears that were accidents in my mixed pheasant plots--are a solid black ear with either bright blue kernels speckled through it, or bright green kernels speckled through it, or a bright rose-pink. I think you can get any combination you want, really. You won't get 100% for a while, but as long as you get just a few ears you can always start in from there and go on....

There was a kid up there by Elbow Lake who used to peddle fish out here. He heard about my corns and got interested in raising some. Well, he had a dollar sign in each eye, thanks to the coaching he got from his relatives. He grew a wonderful crop of that corn and then asked me if I had any more seed he could get next year. I said, "What, didn't you save any?" He said, "Nope, I sold it all"....Make money. That's the object of the game today. Over the years it's really bothered me that nobody kept any interest in anything like this corn, or the wildflowers, or the birds. Nobody was interested because it didn't make them any money. And all the while, life was getting so monotonous. Everything is the same. There is no variety anymore. Back in the old days, you had variety in the birds and the flowers and in the garden. You'd go to the county fair and people would exhibit all different kinds of garden vegetables, all different kinds of corns. Today, you go down there and one or two hybrid companies show off some of their yellow corns, brag about their yields, and that's your corn exhibit....Well, variety is not only the spice of life, but it's essential if our civilization is even to survive!

Ernest Strubbe and Some of His Corn Paintings

(The 1982 Fall Harvest Edition, pages 32-37)

THE BEAN MEN

I was finally going to meet John Withee. We were both in Tucson to speak at the Seed Banks Serving People workshop. I'd had a display case built so John could display his collection at the workshop. I was going to send it to him to fill in Massachusetts, but time had gotten too short and he had brought samples of about 800 different heirloom beans to Tucson. It was the day before the Conference and Aaron and I were taking the display case to his motel to help him fill it.

John had asked me to take over the distribution of the Wanigan Associates bean collection late last fall, almost a year ago. But the transfer had never taken place because an avalanche of publicity had rolled over John like a snowball rolling downhill. For almost a year he struggled trying to get the collection straightened out enough to turn it over, but it had been a losing battle.

I had mentioned John's problems to Dr. Richard Harwood of the Organic Gardening and Farming Research Center. Within a couple of weeks, the Rodale organization had sent a team of their people up to John's house. They stayed for a week, threshed beans, answered all of John's correspondence, and attempted to get the collection in a manageable order. Less than a month after I had mentioned the problem, UPS delivered three boxes to my house that contained 1,185 samples of beans and the accession list for the entire collection. Samples of the entire collection are also being kept at the Organic Gardening and Farming Center in Pennsylvania. Anyhow, on the afternoon before the workshop started, Aaron and I met John in Room 23 of the Motel-6 and it wasn't long before we were all busy filling the bean case. The following are excerpts from a cassette recording of that afternoon. John was telling about the Rodale people coming up—

John Withee

JOHN: Four young girls knocked on the door Sunday night. They were in a big Ford van. They brought a winnower, but I convinced them that the one I had built myself was better, so they used mine. They divided up the chores and all worked very well together. One girl was the typist. She brought up an electric typewriter and table and reams of paper and was really good at it. Two of the girls worked at thrashing....They accepted my method of thrashing, because what I had were dry pods. Almost all of them were pole beans, and I pick the dry pods as soon as they form on the plants over a period of weeks, ripening low on the vines and progressively upwards....The girls thrashed these on the big picnic table I had out back and then put them through my winnower. I was putting together a lot of metal shelves which they had brought up and given to me. Peggy meanwhile was working on all of the stuff that I had in my bean building, which was in a variety of containers. She had five-pound paper bags, one for each variety that I had. She had those laid out all over the front lawn and they even filled the whole 20-foot-wide carport....

KENT: Sounds like you were hoping for good weather.

JOHN: Fortunately we only had one light shower while they were there and that was during the night. They were all wonderful....Skip Kauffman came down on Wednesday. The determination he made was that it would be best for them to ignore the age of the seeds. So everything is there in the bags on the new shelves regardless of the year. I still have a tremendous amount of work to do, even with all of the consolidation that they did....And I still have the problem of not having them in good containers....

KENT: Ralph Stevenson keeps his in baby juice jars. Do you want to see some pictures?....There's Ralph....and there's his setup....The juice jars are air-tight and about twice the size of baby food jars.

JOHN: Now that's the kind of storage I'd like. You need to be able to see through your storage containers.

KENT: Yeah, Russell Crow had his in pint and quart green plastic containers, but after he saw Ralph's collection, he went home and bought a whole bunch of canning jars.

JOHN: Those are good photos.

KENT: These two were taken by Craig Hopps, a young friend of Ralph's. I took a bunch at both of their gardens this last summer just about blossom time ....Aaron, what does that one look like?

AARON: We've grown that one. That's Cliff Dweller. But the packet says "No Name"....

JOHN: There are a lot of "No Names." I don't have access to any authority like Prof. Hepler who can identify. I was told early in the game that the only true way to identify them is by growout tests.

KENT: Even though this one came to us as Cliff Dweller, isn't its correct name Florida Speckled Lima?

JOHN: Yes, it's been called that commercially, but it'll come to you under Speckled Lima or Florida Lima....

I had difficulty in obtaining space to grow out. I had a good garden spot up in Danvers which is about eight miles away. It was on state hospital property. But eight miles is too far. The gas crunch came and then there was some vandalism to the place up there....

KENT:....to your garden?....

JOHN: Yes. It was a sort of community operation, you see, with the state taking care of the plowing....The trouble I have now is with rotation. There's two theories about it. If you are in a professional situation, you're not going to plant beans on the same plot more than once, or certainly not more than every third year. On the other hand, one of the big beliefs that I have, and I think I am correct in it, is that in most home garden situations, even in farm gardens, they still plant their garden in a plot. They tend to repeat....I feel that a plant growing in the conditions of home garden, which pays no attention to rotation, generally, or not that much, or if they do rotate in a home garden, they are not rotating all that great a distance. A good home gardener might have 100 feet on a side and he would consciously move his crops from here to here to here, but with the tilling and all that, he's not getting all the cleanliness that he needs. But in the saving of seed such as beans or other home-saved seed and planting them over and over again, I believe we are getting a built-in strength into those plants. I think that those plants, if they are planted year after year and they maintain themselves, you're keeping only the best and your plant is developing a resistance to whatever might be present....So you have a pattern where the self-saved seed is developing a strength to cope with whatever is in that neighborhood and it exhibits it year after year....Gee, this case is the most fabulous thing. Look at it building up its good looks already.

KENT: It's going to be a knockout. And we haven't even finished one page yet....

Ralph is really wanting to grow as many of Burt Berrier's varieties as possible. But on your accession list that the Rodale people sent me, there are no sources. Will Burt's beans be 100 or so varieties between this and that accession number?

JOHN: Not really, because I was circulating with Burt before he died. There was some mixup right after his death because I had been dealing with him anyway....After Burt passed away the National Seed Storage Lab, at his previous request, picked up his collection. But because the sample sizes were so small they decided to offer them to me. When Dr. Bass offered them to me, I assumed that they were also offering them to others. So when I made my selection from the Lab's printout sheet, I avoided taking any of Burt's that I already had--either ones that he and I both had or any of his that I had gotten earlier. I didn't want to be hoggish about it....But somewhere in my file, I think I have the printout sheet that the Lab sent me that lists his on there and I can make a copy of that and send it to Ralph.

KENT: That's good, because I would especially like to get those to Ralph.... And Russell is simply fascinated with crosses....

JOHN: Oh really? That's the way Ernest Dana is up in New Hampshire.... He's got one that he calls Dana's Cross and another that is a Soldier that is crossed with something else. But when I've spoken with Prof. Meador about crosses, he just sort of grins....The common Phaseolus just resists crossing.... How are you going to prove it's a cross? I guess I just haven't paid that much attention to them, having had the word from the expert (Prof. Meador) that it ....

KENT:....That it doesn't happen?....

JOHN: Oh absolutely, it happens. But the way they like to describe it is that it more often is a genetic shift than it is a real cross. The weak genes in the plant succumb of increase according to conditions or soil pH or whatever....that sort of a shift is actually what happens more often than actual crossing due to insects or weather....In all of the times that I have tried them, and every year I have tried some of these outcrosses or mixups or whatever you want to call them, I come up with all sorts of variable results.

KENT: Have you ever become so interested in one that you tried to stabilize it?

JOHN: Oh yes, and I have done it on one occasion when I gave a variety a name--Lynnfield, after my town. It was out of a pole bean that Ernest Dana sent me. When I grew it I saw a difference in some of the plants and in their earliness. So I marked those, separated them and found that the difference showed up in the seed....Prof. Meador had told me that if it would maintain that characteristic through three plantings, three growouts, then it could be called a new variety....And that is what happened and I gave it the name Lynnfield. But I would not be so bold as to say that definitively, because as you know, grown somewhere else, it might revert back to the other....

I think we should have a contest and tell people--In this collection of over 800 different beans are three look-alikes. If you can find them....

KENT:....we'll give you a packet of beans that needs multiplying....

JOHN: Right. That'll shake them up....

Years ago I remember getting my first Wild Goose bean. It was a pole variety from Rhode Island. Then I read this article by Dixon, I think it was, up in New York who was describing his attempt at collecting heirlooms. They had a regular program for it at one time at the University. They went heavily into West Virginia. He described a person who had given him some Wild Goose beans and he described the same story as the woman had described who gave me the ones from Rhode Island. It turned out when we got together on it that the two beans with the same name were altogether different in appearance. And over the years I have gotten a number of Wild Goose names with different seeds....They've come from different states, they all have the same story--from the crop of a wild goose, and they are all different beans. Now I have gotten one that has the name Turkey Craw, the same story only with a wild turkey. I think that's just cute as the devil. Makes a good story.

KENT: I got a Wild Goose bean this last year that was supposedly taken from the crop of a wild goose that had been shot by a slave.

JOHN: They really went back on that one, didn't they?....

Isn't that the most beautiful display. I just hope somebody can get a picture of it.

KENT: I'll be taking as many as I can.

JOHN: I'm going to be allowed to take it home?

KENT: Oh, certainly. But I have to list it as an asset of the SSE....

JOHN: Then let's just call it a long-term loan....

KENT: Exactly.

JOHN: Right after I get home I get to speak to the Historical Society in my town and this will be the perfect thing....it's heavenly....just beautiful....

KENT: You know, a person can read about a collection of 1,200 different beans, but that's just a number. I figured that the visual impact of seeing 800 different beans displayed would be incredible....What better way to spread the word....

John Withee
With Part of His
Wanigan Associates Bean Collection

Ralph Stevenson & Russell Crow

During July of last summer, I took a trip to meet two bean collectors--Ralph Stevenson and Russell Crow. My family drove up to Diane's folks' home in Festina, Iowa, which is in the extreme northeast corner of Iowa, just 30 miles from Minnesota and 50 miles from the Mississippi. From there I drove over to Russell Crow's in Woodstock, Illinois, which is northwest of Chicago.

Russ was in his garden when I got there. I'd never seen such a garden. I thought we had a good-sized garden at home, but this one was huge and except for a couple of rows, it was all beans. The soil was black and Russ had it worked up really fine. I was really fascinated by the differences you could see in the varieties when they are grown side by side like that. There were leaves of all shapes and sizes and textures and shades of green. There were many beans with runners of different lengths that didn't climb.

Then we went inside to see his seeds. There were metal shelves filled with different containers. There were shallow flat boxes with row upon row of seed packets. The ones to be planted next season were already lined up. There were new treasures that had come too late to be planted this season. I got my first look at Butterfly Runner--the most beautiful bean I've ever seen.

Russ works night shift at a Chrysler plant and he had to go to work in the middle of the afternoon. I went through his packets and containers of beans for hours and went back out three times to walk some more in his garden. Russ got off work for the weekend late that night and we left right then for Ralph Stevenson's.

By early morning we were there, tired but excited. Ralph and his wife Clara live just outside of Tekonsha, Michigan in a lovely farmhouse surrounded by big trees and barns. Clara has flowers everywhere, really beautiful. The flower beds outside their back door contained an incredible collection of hen-and-chicks. And out behind the house was Ralph's bean garden. It was a little bigger than Russ's and neither one of their gardens had a weed in them. The following are from tapes of that day--

RALPH: This used to be a bean center here in southern Michigan. They used to ship carloads of beans out of here every fall. They had two thrashers around here all fall thrashing beans into the winter. We'd pull the plants, load them into a wagon and store them in the barn until the thrasher came. Now there isn't hardly a bean raised around here. It's all up around Bad Axe in the thumb of Michigan. The last two years up there they have been raising the Black Mexican bean and they're all being shipped to Mexico.

KENT: I've heard there's quite a bean center on the west slope of the Rockies near Grand Junction, Colorado where farmers are still planting and combining fields of dry beans. Do you think there are many dry beans that are still grown in field conditions like that?

RALPH: The Northerns are still being grown in Minnesota that way.

Ralph Stevenson
In His Bean Sorting and Storage Room

RUSS: Navys, kidneys, pinto, the common ones you see in the grocery stores are being grown that way, but I doubt that many others are....

KENT: This seed was sent to me by an 80-year-old woman in Arkansas who said that it was the really rare Ice bean, but it's nothing like the Ice bean that Ralph showed me.

RUSS: That looks like Ram's Horn.

RALPH: Sure does.

KENT: Do people tend to call anything that has a whitish pod an Ice bean? Are there different Ice beans?

RALPH: No. The true Ice bean has a greenish pod that doesn't have too much color. It's the color of ice--just a light-colored frosty-looking bean.

KENT: Are Ice bean and Icicle bean and White Cabbage bean all the same?

RALPH: I'd say so.

RUSS: In Beans of New York, the variety name was originally Cabbage bean and it's grouped with the wax pods and Ice bean was one of its synonyms.

KENT: Someone wrote me that white cabbage is the same color as the bean's pod.

RUSS: What amazes me is that this far down the road in this century, there is still even this much, that people have bothered to keep this much alive.

RALPH: The way people are getting nowadays, I'm just afraid that there will be a lot more available this century than there will be next.

RUSS: If we could just get more young people excited about gardening and helping keep older varieties alive.

KENT: I think that may already be happening. The people who are in the SSE right now are mainly either in their 20's and 30's or in their 60's or 70's. There is a tremendous amount of interest right now. So many more people are writing that they have this, or that they are looking for that, or that they are willing to grow varieties that need multiplying. And that's totally different than it was even two years ago. Maybe it's because I have finally built it to the point where I am really contacting them. Or maybe people's awareness has changed and they realize that there is a real problem and they are willing to work for it....

Tom Knoche, the squash collector, was at the Campout Convention. He told me, "I've got a real funny story to tell you. This magazine writer was interviewing me over the phone and asked me, 'Why are you growing all this?' I told her, 'I just like to see them grow.' She went on for awhile and then asked me the same question again. And again I told her, 'I just like to see them grow.' She came back and asked me that same question four different times...."

RALPH: She asked me the same thing.

KENT: What'd you tell her?

RALPH: Just like he did--I just like to see them grow....

I thought Aaron was going to come up with you.

KENT: He was going to, but his grandpa is combining oats. I wish he had come because he would have been fascinated with your collections. This next generation we're teaching is really going to be something. I think he knows more about gardening at age nine than I knew at 25. He really loves it. I'll be sitting at my desk typing and he'll come up and start tugging at my arm and say, "Come on, let's go look at the garden....come on, let's go look at the garden"....

RUSS: I've got a really good name for the next bean cross that I stabilize.

RALPH: What's that?

RUSS: I'm going to call it "Your Guess Is As Good As Mine."

RALPH: That'll about cover it....

KENT: Is there anything that either of you are trying to find?

RALPH: I'm looking for a large red lima bean the size of Christmas Lima. And of course any out-of-this-world bean is always welcome.

RUSS: I'd like to find a lot of those that are in Beans of New York.

KENT: How big a list of those did you send to Dr. Bass and how many did he have?

RUSS: Probably 80, and he sent me six of them.

CLARA STEVENSON: How do you find time to keep a garden that size clean and work full-time too?

RUSS: I work at night and spend three hours or more a day in the garden. I work at it in sections.

CLARA: Ralph just about lives in the garden.

RUSS: One time this spring I had some really heavy weeding to do and I just disappeared into the garden for the whole weekend. But now the rows have closed in so much that I can't till anymore.

RALPH: When they start blooming and setting pods, you want to stop your cultivating. Oh, maybe a little weeding and hoeing. But mainly you want to leave them alone.

KENT: You don't fertilize, or you didn't this year?

RUSS: No. Beans aren't heavy feeders.

RALPH: I tried some 5-10-5 on about a third of a row of the navy beans, but I can't tell where I started and where I left off. What kind of tiller have you got?

RUSS: It's a Troy-Bilt. I've found that you've got to till about 6" deep and soft and fluffy if you want to get things off to a good growing start. I had a patch in a garden that was rough one year and the beans were stunted and didn't grow right.

KENT: I've heard people talk about planting some beans eye down. Do you ever do that?

RALPH: It's good to plant all of your larger beans, especially the limas, with the eye down. The root comes out of the eye of the bean and starts pushing the bean up out of the ground. In some cases if the ground gets hard, if you can't keep it loose or wet, a large bean planted eye up will be in such a bind that it will actually break itself off. With the little beans, it doesn't matter....

Ralph Stevenson In His Bean Garden
Just As It Was Starting To Blossom

HOW TO PLANT AND TEND 250 VARIETIES OF BEANS

Ralph fall plows and works his soil up in the spring. He lays open 2" deep rows with a large hoe. Then he spaces and drops the seeds with one of those hand-pushed seeding machines, putting in numbered metal tags on metal stakes as he goes. Then he covers the seeds with about an inch of dirt and firms it. He has a little garden tractor with a spring tooth cultivator bolted on its back end. It has seven teeth and he removes the middle one. He cultivates right over the top of the plants two different times until they are too tall to pass under the tractor. Then he switches to a couple of contraptions like I had never seen before. They were really old manufactured jobs that he'd fixed up. They reminded me of push cultivators with the big front metal tires that many people use in their gardens. The teeth and the handles were the same, but they each had a small rubber tire up front and a gasoline motor. One was larger than the other and he used it until the rows closed in so much that he had to switch to the smaller one. When the rows closed in so much that he couldn't use the smaller one, he went to his hoe.

(The 1981 Fall Harvest Edition, pages 14-20)

****

(From earlier correspondence with Russ Crow) -- I am a bean collector whose ambition at one time was to start a small bean seed company. I have since decided that collecting sales taxes for the state, maintaining sales and expense records for tax purposes, sending to the state seed samples for germination tests and maintaining a seed file in compliance with the seed trade laws, etc., is not for me. I feel the Seed Savers Exchange is a more enjoyable and better way for me to help other gardeners....

Beans and peas like open air storage and will remain viable for about four years ....About two months after harvesting bean seed, when the moisture content has dropped, I put all of my beans in a tightly closed jar and freeze them for a day. This is to kill weevil eggs which are almost always present under the seedcoat, which hatch after about a three or four month dormancy period and ruin the seed....

It has been quite an exciting year collecting beans. Between my membership with the Seed Savers Exchange and Wanigan Associates, I added about 200 varieties to my collection....my collection now contains over 310 varieties.... This year I managed to get nearly 180 varieties planted. Have been shelling the new seed crop for about two weeks now (mid-October) and have about a quarter of them shelled and sorted....

I have been thinking there might be gardeners who would be interested in seeing a complete listing of my collection, mainly a seed description of all my varieties....So I have been working on a catalog and hope to have it finished by late November 1979....Everyone who has traded beans with me this past season will get a free copy. Just a little something to thank my bean friends for their generosity, and they have been truly generous....As for a price on the booklet, the printer said he would call me and give me a unit price, but so far has not. I intend to mail these booklets in a manila-colored envelope by First Class mail. I'm going to take a guess and say the cost of the booklet including mailing will be 50 cents for members....(later letter)....The printer just called and when I heard the price my mouth literally dropped open. For the 100 copies I ordered the price is $2.70 per booklet. This is my first experience with having anything printed and I had no idea of costs....I'm simply flabbergasted. (Printing costs have skyrocketed in the last two years. This is a continuing problem for me. I suggest that if you do not receive a copy of the booklet from Russell, the price will be $3.50 postpaid - Kent.)....I have enjoyed my first year with the Seed Savers Exchange tremendously and I want to thank you, Kent, for creating the Seed Savers Exchange....

(The 1980 Seed Savers Exchange, pages 25-26)

This past spring I had the good fortune of receiving a copy of Beans of New York from Dorotha Shortridge. I began a search for some of the many old varieties described in the book. I wrote to Edward Wisk who informed me that Agriculture Experiment Stations probably do not maintain old vegetable varieties. He referred me to the U. S. Dept. of Agriculture Research Center.... Dr. J. R. Stavely at the USDA Research Center, Beltsville, MD, wrote that some of the bean varieties on the list I sent to him were at the National Seed Storage Laboratory at Fort Collins, CO....I decided to take a chance and sent Dr. Louis Bass, Director of the NSSL, a very extensive list from Beans of New York. I had picked those varieties which had the best useful characteristics and those which seemed to have the most interesting seedcoat colors. Dr. Bass sent me 50 seeds each of the following varieties: Kentucky Wonder Wax, Corbett Refugee, Tiny Green, Seminole, Rival, Uniwaled Wax and Full Measure. Dr. Bass states because of the good supply of high germinating seed they have on hand of these varieties, he will not even require me to increase and return seed of them. The other varieties I asked about, they do not have. But these seven kinds are just a tremendous thrill to have found....I also had the pleasure of doing some bean trading with the Henry Doubleday Research Association in England--a friendship that I hope will continue....Among about 15 varieties received from them this year was the Black Runner described in Beans of New York. The Black Runner goes under a variety name of Black Knight....Dorotha Shortridge has coined a phrase: "The Big Bean Hunt Is On!"....

(The 1981 Winter Yearbook, page 35)

People should always wait until the weather and soil are both warm when planting beans. Don't try to rush it. Beans planted two or three weeks later will often do better and mature just as fast as those that have been struggling during that period to overcome cold weather....The biggest problem for the bean grower is wet weather during harvest. If it is raining when the pods should be drying, the beans just sprout in the pods. It can really be a mess. A week of rain at the wrong time can take you from having a good crop to just hoping you get seed back....

(The 1981 Fall Harvest Edition, page 21)

Late this last summer when my beans were entering their peak drying period, we got eight days of nearly solid rain. I worked like crazy trying to save what I could. I did manage to do quite well and harvested nearly 20 gallons of good seed....I've just finished making a huge bean display (plaque) of almost all the varieties I grew this year. Nearly 300 samples of beans are sealed in Deco Pour Polymer Resin. Sealed off from the air this way they will never change color. I plan to bring this display with me to the next Campout Convention. I think it is quite impressive-looking and hope it will spark some enthusiasm among gardeners who view it....I wish I could figure out a way to collect and grow beans full-time. I know I would love it....

Russ Crow and Some of His Pole Beans

(The 1982 Winter Yearbook, page 17)

****

A Message from John Withee

I appreciate this opportunity to thank Kent publicly for taking on the load of work that the Wanigan Associates collection represents. I am sure that the combination of his energy and experience along with the proven value of the Seed Savers program will guarantee success....My remaining stock of seeds, which consists of the full collection in mini amounts, will again become a part-time hobby. I plan to maintain it with the smallest increase possible--one plant each of 500 different varieties each year. This will provide a three-year renewal. The risk of failure is evident, but now the back-up of other Seed Savers reduces the odds on complete loss. Except for a few very loyal bean-growing friends, my surplus will be available only to Kent and to the Rodale Press group....

The cultivation method I plan to try this year might make an agronomist smile in disbelief, but it's my garden! I'll select the oldest stock and also scarce and special varieties. Each packet of four or five seeds will carry its accession number. Brief notes about its origin, etc. will be on separate punched sheets which will become pages in a field notes binder. After the garden is prepared and soil tests are made (I try to stay above 5.6 pH), the wait for warm soil begins. Ideally, it will be 55-60 degrees F. at 3" depth. There will be no attempts at rushing the season with tunnels or tents this year. Early starts with long-season varieties seem to be a waste of effort, because so many of them need the shorter days of August to trigger pod production. I'll try to protect against fall frosts instead of spring....This year all seeds will be pre-sprouted at an optimum temperature of 70-75 degrees F. Depending on conditions at the time, sprouts will be either: planted outside at once; held in a gel (plain gelatin made at 1/4 strength) in the refrigerator; or planted in 2" peat pots in the greenhouse. Outside planting will be done just prior to a natural warm rain. This method will help guarantee a good start and will enable me to re-sprout any having defects, or which failed to sprout on the first try....

I calculate that 400 feet of single rows 40" apart will accommodate up to 500 plants (the limas and runner beans will be kept distant from one another, as they seem to cross too readily). Overhead wire on poles will support dropped binder cord to each climbing plant as it develops. A plot plan will enable me to locate any one of the varieties without too much searching for the individual labels. The notes to be made will be the weak part of this growing experiment, in a scientific sense, because of the lack of quantity, replication and weather control. It will, however, keep me in heirloom beans along with the rest of you.

(The 1982 Winter Yearbook, page 87)

BURT F. BERRIER - BEAN COLLECTOR

You asked how I got into this bean collecting. I spent some 30 years as a sales and serviceman for farm machinery manufacturers--John Deere, Allis Chalmers and Case Co. Part of my work was to show new owners of threshers and combines how to thresh grain, beans and other seeds. To thresh beans without splitting them was some job in the early days. Split beans have no value. I found how to do this to as little as 3%, so each fall I was sent all over the West where beans were grown. Each grower would give me a sample to take home, and I soon had quite a collection....Some 40 or more years ago I was near Shiprock, New Mexico and I met an old Navajo woman who had beans growing and had to haul water many miles to them. She gave me a pod of ripe beans and told me these were grown by the Cliff Dwellers....A lady sent me the kind of beans Brigham Young brought with him and his followers to Utah....Science magazine says that beans, corn, squash and potatoes all originated in Peru or Equador....The University of Colorado has built a building to house, catalog and collect all kinds of seeds, and they have space for a million kinds. They tell me so many plants are being lost, told me they were going to see me to get beans....

You asked me how I keep them pure. I've never had any that crossed. Have had only one person send me a cross. A seed company in Minnesota sent me instructions on how to make a cross—has to be done by hand by taking the pollen from one to another and has to be done just at the right time....You have in your Seed Savers Exchange the names of all the bean people I know, as well as the collectors....

Our humidity here is from zero to 25%, have no mold. All of my seeds are in glass, and the temperature is never more than 50 degrees....I plant as many varieties of beans each year as I can take care of. Planting, irrigating, weeding, picking and shelling each bean by hand takes so much time. This year I planted 125 kinds. I'll have a mess to sort and get them back in my file....I just finished my files and have 448 varieties....One thing about collecting beans, each has a life in it; it's not dead as collecting clocks, dolls, guns, etc....There is no end to the subject of beans. I was told as a boy I didn't know beans, and I find that now at nearly 85 it's still the same. I work some 8-12 hours a day just to give them away. I got in the same mail as your letter a package of 48 kinds of beans, and all were really large samples. It's been like a snowball rolling downhill. I'm swamped....

(The 1978 Seed Savers Exchange, page 18)

CARL BARNES - "CORNS"

KENT: How long have you been working with corn?

CARL: 1946 is when I really got in it. My grandfather came here in 1902 and he planted corn that year. Some of our immediate family has grown corn here every year since. Even the years that I've been off going to school or somewhere, I'd come back and we always had some corn and roasting ears. I still have some of that Cherokee corn and some of the old Red Dent field corns that my granddaddy had years ago. He came from southeastern Kansas to this area.

KENT: Did you ever get any of Dr. Wyche's Cherokee corn that came over the Trail of Tears in 1837?

CARL: Yes. I've got three different strains of Cherokee corn and they are all a little bit different. But they are grown in different parts of the country from Indiana to southeastern Oklahoma to the Carolinas. That Cherokee corn is unique. Its kernel has got kind of a rough edge on the top of it. I've had old people tell me that the Choctaw had a corn that was very similar to the Cherokee.

KENT: Almost all of what you've shown me are traditional and ceremonial corns, most of it from the Southwest. Do you work with sweet corns at all?

CARL: I won't plant sweet corn here because of pollination problems. The sweet corn pollen seems to just float. One year I had to do away with quite a bit of our Indian corns because they cross-pollinated with sweet corn that was a long distance away. Now I don't take that chance. The other pollens are more granular and heavy. They usually drift 15-20 feet and then hit the ground. The sweet corn pollen, seems like it will sail forever. There's enough other people in the SSE interested in the sweet corns. However, I would like a couple of representative ears of any sweet corns that are not available commercially for my collection.

KENT: I have growers I can distribute sweet corns to. You're working a lot with crosses of the old Hickory King.

CARL: Yes, those have always fascinated me.

KENT: You said the best roasting ears in your whole collection are crosses of it, aren't they?

CARL: Yes. That's a good eating corn and it's an all-purpose corn. You can do anything with it. You can make hominy out of it, good for roasting ears, and for meal it's hard to beat.

KENT: Now those are the ones that have the big wide square Hickory King kernel and are orangish with the red stripes?

CARL: Yes. Those are Gila crosses....When I say cross, realize that I don't mean hybrids, as most people think of them. These are selected blends. This process is continually ongoing.

KENT: In many cases aren't you picking out things that are similar and planting them back together to let them cross?

CARL: You just keep doing this. I think most Indians did this, too. They'd move to a different location farther north or to a different elevation. They'd be planting in a different soil, different mineral, and a certain color would come out a little more intense. They'd have all these different colors of reds and suddenly this black comes out, and it worked at their new elevation. So they kept that separate. Same way with the blues. So in different generations and under different conditions it eventually kicks out what all's there. It's all back there from the original. And this is just grandpa and grandma coming out.... And then they continued to trade these selected inbred lines and blend them back together occasionally to renew their vigor.

KENT: Mr. Strubbe had his seed in jars, that weren't airtight, in an inside closet where it was sort of cool. The seed he gave me was from 1973 and 1975. I know one person who grew some of it last year who said the germination was over 90% and it was incredibly vigorous. How long will corn keep?

CARL: Once I had some corn stored up at the folks' in an old granary. The rats and the mice worked on it, the dirt blew in on it, it rained in on it and the snow came in and covered it up each winter. I was looking for a certain kind of seed, so I went up there and found some ears. I said, well, I'd better try to germinate this because it's nine years old. I only got 97% germination!....That depends on your conditions, though. You drive 150 miles from here and you've got a lot of insects and the humidity is a lot higher. Whereas I just don't have those problems....Sweet corns won't store that good though....

KENT: You told me a story one time about how you got your friend Lynn Hagaman started working with corns.

CARL: A few years ago when I was getting ready to plant, I picked a handful of blue kernels out of just about every kind of corn that I had which contained any blue. I took it to Lynn and told him, "I finally got some 'pure Hopi Blue' and thought you might want to try some." Well, that year he got the most incredible rainbow of corn he'd ever seen. I just shook my head and marveled right along with him. We discussed and worked with that rainbow for a couple of years before I told him what I'd done. By that time he was hooked. We still laugh about that "pure Hopi Blue...."

Carl Barnes Displaying Corns (Note the Crookneck Milo)

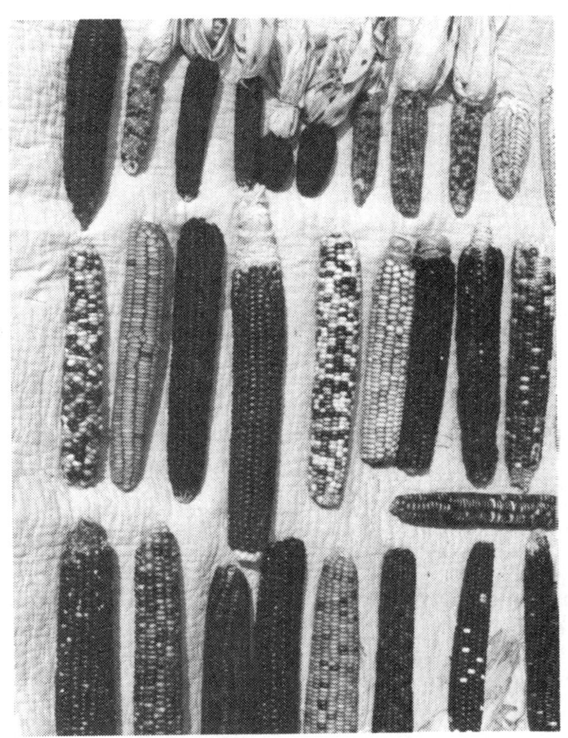

A Small Part of the Carl Barnes Collection

(The 1982 Fall Harvest Edition, pages 38-41)

O. J. LOUGHEED

I think we need to identify what it is we really want to do with storage. Are the Seed Savers and different people who are saving seeds wanting to become long-term repositories or are they just trying to save seed for five to ten years, in which case some pretty simple systems will suffice. I think we get too carried away with fancy frozen storage facilities and sealing materials in nitrogen-filled tin cans. I'm not sure that's actually our purpose in all of this....

We've even gotten to a point where science no longer refers to these endangered materials as seeds. These seeds....these little dormant plants just waiting to live again....they refer to them as germplasm....The fact of the matter is that our civilization didn't come up with any of these plants. It was peasants and cave dwellers and people thousands of years ago who came up with all of these things. And we haven't really done a whole lot of work or improvement with them since. I mean, we think we do, but we're just sort of messing things up as we go along....Primitive people, who for thousands of years selected all these food crops, know a whole lot more about them than we do, or at least they feel a lot more about them than we do....

So here we are down to the point where we call them germplasm and want to stuff them in nitrogen-filled tin cans....We really need to be growing them! Figure out how to use them as subsistence crops for those people who would like to get by in a more healthy, more balanced sort of life. And not get carried away with a lot of high-tech solutions where there may not even really be a problem. How did all these seeds get to us over thousands of years without nuclear-powered freezers? If all those people back through 10,000 years had to depend on all this technology, we wouldn't have the seeds now. We would have lost them a long time ago to power outages....

With living, working collections of seeds, long-term storage shouldn't even enter into the picture. If it's stored reasonably well, it should probably be good for five years. Even my collections, I plan on growing everything <u>at least</u> every five years, just because after five years one forgets what it looked like and how it grew. I mean after all, the seed is setting there and it wants to grow. It wants to take part in life just as everything else does. And to deprive it of that, to stuff it in this tin can for the next 60 years, seems to be the height of human arrogance. Let it grow!....That's what always struck me in your different articles about the collectors in the SSE. When people asked them, "Why do you do this?" the answer always came back, "Well, I just like to see it grow." What better reason! If there were more people around like that, we wouldn't be concerned with storing seeds for 30 or 60 years. Everybody would be out there growing them....

(<u>The 1982 Fall Harvest Edition</u>, page 44)

(O. J. is establishing a Seed Library high in the mountains of New Mexico for "Traditional, Subsistence and Folk Agri-culture Crops." With five acres of flat bottom land at 7,500 feet, only the hardier, cool-night tolerant and short-season varieties of many crops can be grown. The emphasis will be on "Mountain" varieties including hulless barley and oats, dryland wheat, millets, soybeans, potatoes and forage crops, as well as the cole crops, oriental vegetables and other short season crops - Kent.)

(<u>The 1983 Winter Yearbook</u>, page 112)

The Annual Campout Conventions

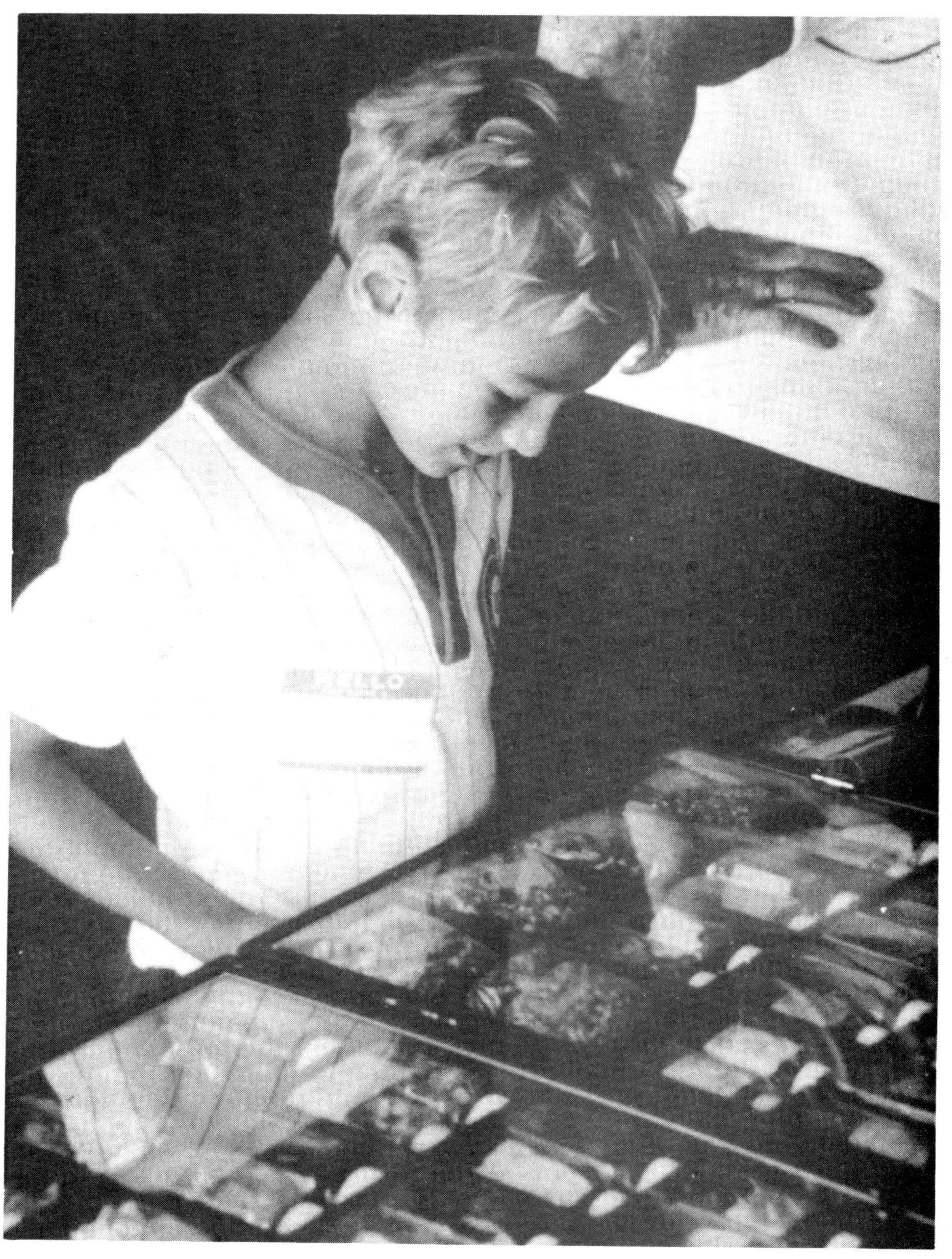

Planting the Seeds of Fascination and Wonder
Nathan Goc Admires a Display at the Campout

THE FIRST ANNUAL CAMPOUT CONVENTION

Some of those Seed Savers were coming for a visit and there was a lot to do to get ready. Our house is out on sort of a point above a broad valley, so I spent over a week building a huge white oak deck around two sides of the house for us all to get acquainted on. Diane planned out all the meals and got all the fixings. We sent out maps of how to get to our place because we're way back in the woods. The kids helped me get the gardens to looking better than they've ever looked. Finally everything was ready and not a minute too soon because it was the morning of June 13. And that was the first day of The First Annual Campout Convention of the Seed Savers Exchange.

Then the "seedy characters" started arriving. First on the scene was Russell Crow, a bean collector from Woodstock, Illinois. Next came Faxon and Mary Stinnett from Vian, Oklahoma. We sat around the picnic table talking. Then Virgil and Hazel Johnson from Liberty, Missouri arrived. Then in fast succession: Auburn and Clarice Cooper from Overland Park, Kansas; Tom Knoche, the squash collector, and his wife Sue from Sardinia, Ohio; Dale Anderson and his nephew Rex Anderson from Anderson, Indiana; and Al and Mary Razor from Collins, Iowa (who brought us some of the best comb honey I've ever eaten).

We shared a wonderful dinner and then all settled back and started talking.... and never quit. No agenda, no schedule, and the group was small enough that

Left to right: Mary Stinnett, Clarice Cooper, Hazel Johnson, Diane Whealy and Mary Razor. Not pictured—Sue Knoche

most often everyone was in on the same conversation. It was a warm, sunny day, but it was blowing a gale outside. Somebody said they didn't know if it was more windy outside or inside.

Mary Stinnett discovered that we hadn't had time to pick our peas yet, and in no time at all that was remedied. The women decided that they might as well work up the peas as they talked. They soon discovered that they all had a different term for what they were doing. One was shucking the peas, another was shelling them, another was podding them and still another was hulling them.

That night we barbecued chicken on the deck and relaxed out there until the mosquitos got too bad. Then we all moved inside and talked until after midnight. Everyone stayed at our place except Tom and Sue Knoche who had gotten a motel room that morning on the way to our place. We were sorry they hadn't waited, because they could have stayed with us. Everybody else either pitched tents in the yard or stayed in their campers. Aaron and Amy caught the campout fever and I pitched their pup tent in the yard too. It was really blowing and we fully expected to hear Amy at the front door in the middle of the night, but we didn't.

Bright and early the next morning everyone was up for coffee and rolls. That morning we did our seed trading. Faxon had brought the most, including a large collection of peppers. Everybody had fun looking through the box full of seeds that I had received in the mail during the previous few months. That was also just after the series of articles on seed saving and the SSE that were published in the June issue of Organic Gardening. Russell had sent me small handpicked samples of 60 different varieties of beans which I held in my cupped hands for the cover photo. (David Dinges wrote and said, "I see you whacked your thumb.") Anyhow, because of that beautiful photo I had just received a tremendous number of letters with small samples of beans that people wanted me to identify. Watching and listening to Russell Crow and Virgil Johnson going through all of them was one of the most interesting parts of the convention for me.

We had a delicious Sunday dinner together and everybody said they didn't know how Diane had done it. The women said that next year they wanted to help her out more by bringing their favorite dishes or making them in their campers. We all had a fantastic time those two days, and all too quickly they were over. Everyone really had a hard time saying goodby. Some tears were shed. It's amazing how close you can get to people in only a couple of days. But gardeners are special people and seed savers are special gardeners.

Everyone wrote us such wonderful thank you letters. The one from Dale Anderson made Diane and me feel so good that I want to share it with you. He wrote, "This can't be much of a letter as last weekend seems like a dream and I can't seem to get my thoughts together or my feet on the ground. Rex and I had such a wonderful time and want to thank you for your friendly hospitality. Thinking it over, I don't think there is anything you two can't do. I must have talked 10 hours non-stop telling my family about the Whealys and my experience...." Then Dale wrote something that I think pretty well sums up the way we were all feeling about each other when we had to leave: "I consider it my lucky day when I got to know you...."

People have been asking if we are going to make this an annual event. Nothing could stop us! I'll print the notice for The Second Annual Campout Convention in the "Winter Yearbook." We held it in mid-June this year because we thought we ought to have it when everyone's kids were out of school. But there weren't any kids here, much to Aaron and Amy's dismay. You know, our kids ended up having a ball anyway.

Before Christmas, if you are thinking about coming next year, drop us a line and tell us when would be the best time for you. We are all busy in the spring with planting, and many of our people are farmers and once harvest starts in the fall, that leaves them out. We want to have it when the most of us can get together....I just want to tell all of you that were here, my family and I will always remember those two days with a lot of warmth, and we feel blessed to have met you.

Clockwise: Amy Whealy, Al Razor, Rex Anderson, Auburn Cooper, Virgil Johnson, Faxon Stinnett, Russ Crow, Kent Whealy, Aaron Whealy holding Tracy, and Dale Anderson. Not pictured—Tom Knoche

(The 1981 Fall Harvest Edition, pages 42-43)

THE 1982 SSE CAMPOUT CONVENTION

It's almost impossible to describe the Campout Convention....it's fun....it's intense....it's too short. About 35 people attended this year. One couple came all the way from Florida. Some folks rolled in Friday evening and set up their tents and we stayed up well past midnight talking. Just couldn't wait.

People kept arriving all the next morning and by dinnertime everyone was here. Beth, a retired lady who used to run a restaurant in Princeton, cooked and even delivered all of the meals, and they were great! Because of that, Diane and I were able to just relax with everyone and visit. And that's really what the Campout is all about.

If anybody came here expecting a scheduled agenda and speakers, they would have been disappointed. We had 35 speakers and they were all speaking all of the time. It was hard to tell whether it was windier outside the house or inside.

I had hoped I'd have a half dozen pages for this article, because I sure had the photos to fill them. But, as usual, I'm running out of space. So I'll just show you a few of my favorites and hope that it will give you a slight idea of what those two (too) short days were like.

Carl Barnes and his wife Karen came to the Campout this year. It was the first time I had seen them since the Workshop in Tucson. Carl brought a fantastic display of some of his corns with him. Most of them were traditional and ceremonial corns from the Southwest and some were very beautiful. He had a pink Hopi ceremonial corn that I had never seen, and I especially remember a Hickory King X Gila cross--huge square kernels that were orange with red stripes and only six rows.

He also had two types of crookneck milo which I'd seen for the first time in Tucson. And he had some small samples of all of Ernest Strubbe's dent corns, and they were real eye-catchers, especially the green and the turquoise. We certainly all enjoy it when someone displays a prized collection.

I got tickled at Carl. Folks were constantly asking him, "What's this?....what's this?....what's this?...." Carl got to the point that he'd just grin that ornery grin and say, "Oh, that's just a heavenly mutation...."

Russ Crow is a bean collector who has been with the SSE almost since it was founded. He offers a huge selection of beans each year and may have distributed more samples than any other SSE member.

In 1981 Russ and I drove to Michigan to meet Ralph Stevenson, another bean collector. We talked about the problems of putting together a display of beans. Most beans darken with age, so you only have an accurate visible display when the seed is new.

Russ had some small bean plaques he had made several years earlier. The beans were sealed in clear plastic (hobby store stuff). We compared them to new seed and they hadn't changed at all. We imagined building a large display.

Russ Crow
With His Bean Display

Left to right: Aaron Whealy,
Mark Knoche and Howard Cooper

Well, Russ did it, and he brought it to the Campout. It was fantastic--really beautiful.

We all talked until way late Saturday night, too. Before sunrise the next morning, several of us, both in the house and out in the tents, were awakened by "....Howard....Come on Howard, wake up!....Howard!...." It was Aaron and Mark Knoche trying to wake Howard Cooper, so they could go fishing. They finally succeeded.

There is a big new pond about half a mile from our house that we stocked with bass about a year earlier. A couple of hours later, they came swaggering down the road with a nice stringer of fish. (I got them to pose in front of Paul and Donna Kline's tipi.) They cleaned them all too.

(The 1983 Winter Yearbook, pages 219-220)

"Seedy Characters" Who Attended The 1982 Campout Convention (Photo by Mark Fox)

THE 1983 CAMPOUT CONVENTION

The Campout Convention has quickly turned into an annual family reunion. This was our third year and more than 50 "seedy characters" found their way to our home in these wooded hills of northern Missouri. The food was great, the company was even better and the conversation never stopped. It was a very high-powered weekend and my entire family thoroughly enjoyed it. It always really charges me up with renewed energy and determination.

Each year I have plenty of photos to share with all of you, but never enough room to print them. But there was a lot of empty space scattered through this issue, so here are some candid pictures from the Campout. I am very grateful to Mark Fox who sent me most of these Campout photos.

Left to Right: Mark Fox, Jack Rice, Carl Barnes

(The 1983 Fall Harvest Edition, page 23)

The 1983 Campout Convention at Princeton, Missouri (Photo by Mark Fox)

FOURTH ANNUAL CAMPOUT CONVENTION

The Fourth Annual Campout Convention of Seed Savers Exchange was held on June 9 and 10, 1984 at the Pinebluff 4-H Camp. The Camp includes 115 acres of pine-covered bluffs, beautiful trails and a suspension bridge over the Upper Iowa River. The magic of the Campout is that it brings together a diverse group of people who share the common bond of working to preserve our seed heritage.

By Friday evening the Camp was already starting to fill up with "seedy characters." Over 30 of us had supper together at a local restaurant. With that many old friends talking up a storm, it was by far the noisiest group in the place.

Saturday morning was reserved for registration and by the time the dinner was served, over 60 people had gathered. The year before, John Amory had finally talked me into getting up to speak, so I decided to turn the tables on everyone. Without any warning, I asked all of them to get up and give short speeches about their preservation projects. It turned out to be one of the most fascinating afternoons I have ever spent. Portions of those speeches have been published here, so that all of you can enjoy them too.

That evening we gathered around the fireplace for our Saturday Night Brainstorming Session. These hard-core seed savers are exactly the kind of folks that I like to bounce ideas off of and ask for suggestions. Many are either collectors or are working with various maintenance projects, so we are all dealing with essentially the same problems. Most of us are building up huge collections of seeds, so we talked a lot about documenting characteristics and what sort of data we should be observing and recording during grow-outs. We talked about ways to standardize the photographic records of fruits which many of us are beginning to take. How to improve the SSE was discussed at length along with various schemes to minimize the shock of its rapid growth. We discussed how to improve the Growers Network, which resulted in the changes you see in this issue. And we dreamed about establishing a Preservation Farm where we could grow out the Growers Network collection, where young people could come to study as apprentices, and where the work of the SSE would always be carried on.

Then, while all the kids were roasting marshmallows, Mark Widrlechner and Sherry Dragula showed a fascinating series of slides on their work with ornamental corns. Everybody sure enjoyed it, and evening slide shows are sure to become a regular feature of the Campout.

The next morning everyone got down to some serious seed trading. Many folks are keeping huge collections of seeds, so that was really something to see. There was too little time to enjoy the camp, but many folks did walk across the suspension bridge and explore the beautiful trails on the bluffs across the river. All too soon it was time to go. Those good-byes are getting harder every year. But as busy as we all are, we knew it wouldn't be that long until the next Campout.

****

Tom Woods

I'm manager of the Oliver H. Kelley Farm, which is a living history farm near Elk River, Minnesota about 40 miles northwest of Minneapolis....A living history farm is a museum organization that picks a particular date in history and then tries to replicate farming activities of that period by: reconstructing, or restoring, barns and houses and all the other outbuildings; using draft animals like horses and oxen; and trying to locate sheep and cattle and other animals that would have been typical on a farm during that time period. One thing that such farms try to do--or at least the best ones try to do--is to locate the same breeds of animals that were raised in the period that the farm museum has been selected to portray....At the Oliver Kelley Farm we portray the change in agriculture in Minnesota from 1850 to 1870. We also have to think about the ethnic composition of the farm, too. Oliver Kelley was a Yankee, and so we portray Yankee farming. We try to get the kind of animals that Kelley would have had, like Shorthorn oxen and Shorthorn cows and Southdown sheep, and the poultry that he had like Dominiques and Dorkings, and also Berkshire hogs....

A good living history farm should also be presenting the same kinds of crops and even the same varieties that farmers--in that period and in that region and of that ethnicity--were growing. And so the best living history farms will do their research to find out what varieties were being grown at the time and place that they are presenting, and then try to find those varieties. We started that research about four or five years ago. I don't remember how I ran across the Seed Savers Exchange, but I did, luckily. We found several seeds that we were looking for through the Seed Savers Exchange and also just found support--that there are other people who are engaged in the same task, other people who are also looking for old varieties of seeds, and preserving them....

As a farm museum we have two missions--to preserve and to interpret to the public. Most living history farms interpret, but they're not preserving vegetables and field crops because they don't use period varieties. They use whatever they can get their hands on, because they say there's no difference. The more I work with old varieties, the more I find out that there is a great deal of difference. Each variety of oats is different from another variety. You can see it. It is apparent! They are so different from modern varieties....We started working with the Department of Agriculture's Plant Introduction Stations. We used the one in Ames, Iowa, to obtain our corns and we used the station in Maryland to get some frozen storage wheat. We got our oats from the Department of Agriculture in Canada....We've gotten some vegetable varieties from members of the Seed Savers Exchange. In return we introduced the Champion of England pea into the Seed Savers network, which we got from the Department of Agriculture Station in New York. So we've been really active in Seed Savers in the last few years....

We have also developed various program using old varieties of vegetables. We are trying to get people in our area interested in saving seeds. We have a seed exchange of our own every March. We get interested people together and we talk to them about what's happening with seeds. Why are all the old standard varieties disappearing? We talk about all the things that Pat Roy Mooney talks about and all the things that Kent has written about in the Yearbooks. We give them seeds to grow out and then we have an agricultural fair in the fall. We try to encourage people to bring those crops back that they've grown. And then we display them like an agricultural fair of the 19th century, so people can look at the various vegetables and crops and see how they are different....we're a very new organization. We've only really been in operation for three years, so it's a developing project. We've got a long way to go yet, but I also think we've come quite a ways in the three years we have been operating....

The press loves the idea of old varieties of plants. They love the idea of seed savers....Museums are always subjects of publicity and lots of people coming wanting to write articles. I start talking to these reporters and magazine writers about period varieties of crops and they get all excited right away. They forget about things like Oliver Kelley was the founder of the Grange and that's why they came out there....(laughter)....You start talking about what you're doing with historic seeds and that's what they write the article about....

I've also been working hard to get other living history farms and open-air museums interested in saving old varieties of seed and interpreting them to the public. We have a great potential for popularizing the saving of heirloom varieties because so many thousands of people come through the door to the farm museums throughout the country. And Kent helped by initiating his Historic Gardens program in The 1982 Fall Harvest Edition. He sent out free copies of that issue to nearly 500 organizations and individuals who are working with historic garden projects. Unfortunately, I don't think he got a great initial response from museums, but we all realize you have to be patient. You plant the seed first, and then it grows. I hope we'll come to fruition in a couple of years. And I think it will because it's a growing interest among living history farms and open-air museums now. People are starting to catch onto it. It's a topic that's coming up again and again and again. And I think that there can be a really good rapport built up between organizations like the Seed Savers Exchange and museums which try to interpret plants and animals.

There are basically two organizations that I'm working with to get people interested in the seed saving. One is the Midwest Open Air Museums Coordinating Council, of which I'm the president right now, and Darrell Henning, sitting over here, from the Norwegian-American Museum in Decorah, is vice-president. And the other is the Association of Living Historical Farms and Agricultural Museums, which is an international organization of living history farms. There is interest growing. Everybody isn't jumping into it right away because they see all kinds of logistical problems about saving seeds, such as just having the staff to do it....and many museums seem to think that they are interpreting technology. They save the McCormick reaper and they save the threshing machine, but they don't save the wheat or the oats that went through those machines. But why not? Certainly they are as important as something like a reaper. They are part of our heritage and should be interpreted....

I visited one of the largest living history farms in the country a couple of years ago, when we were moving back from the East Coast after we had bought our oxen. It is supposed to be the model of living history farms. I started talking to them about crops. I said, "What do you do for oats?" And he said, "Oats

The 1984 Campout Convention near Decorah, Iowa (Photo by Mark Fox)

are oats. They haven't changed at all."....He is so wrong! But that just shows you that even what is called the best of the living history farms has not yet awakened to the fact that plants and crops are really essential in interpreting agriculture. Most modern oats and wheat have a very short stalk because it's easier to harvest with modern harvesting equipment. But 19th century oats and wheat were extremely tall because, of course, you had to bind them. You can't even bind modern oats and wheat effectively. And they don't stand up in the shock either....they shouldn't be growing modern oats and wheat on a historical farm any more than they should be using a modern combine....

The one big problem that keeps museums from accepting this whole idea and using historic varieties is that these institutions would have to become really committed to saving seeds on a long-term basis. And that's tough, because generally that means growing them out every year and being careful not to let them cross. You can throw a reaper into a darkened room and forget about it for years and years and years....(laughter)....and you may be able to freeze seeds and save them years and years too, but ideally they're going to be grown out every year. And most museums just have not accepted that as part of their responsibilities at this point. It is a whole new area and it's going to take quite some time for that to become institutionalized....

There's a great potential here. If members of this organization would work together with living history farms and open-air museums, the future is bright. It's just getting the people together and sharing ideas that is necessary. It'll come, but it'll take time.

Mark Widrlechner

I have an interest both in maintaining varieties and in selection....If you're interested in maintaining varieties, you should try to keep your prejudices out of it and that's not always such an easy thing to do....My interest in the long term has mostly been in selecting, so in essence I guess I'm a plant breeder as opposed to somebody who is mostly interested in maintaining heirlooms. However, last November I accepted a new job. After I finished my studies in breeding at the University of Minnesota, I was hired as the horticulturalist at the USDA's Plant Introduction Station at Ames, Iowa. In that capacity I am interested in locating varieties that the USDA does not have and getting them into our system so that they can be maintained over a long period of time, increased in quantity, stored, and described for their horticultural characteristics and for their disease and insect resistances. So now I guess my interest is more in maintaining these varieties than it was before.

Since 1977, I've been working with corn. I worked with Jack Harlan in the Crop Evolution Lab at the University of Illinois for two years....during that time I learned that there is a tremendous amount of diversity in corn and it's just an amazing crop to work with....I moved from Illinois to the University of Minnesota and there I was working in azalea breeding. Just for the fun of it, I ordered different kinds of ornamental corn from various seed companies and grew them out in a plot to see what they looked like. When I grew out the seed companies' idea of ornamental corn and compared it to the experiences that I'd had growing sweet corn and field corn, it just didn't look very good. The plants fell over. There was smut and rust and ear worms. So that's when I really got interested in trying to select better types of ornamental corn. My wife and I have been working on that for the last five years now. I imagine it'll be many more years before we really have something we're happy with....So my experiences working with the corn have been both as a breeder and working through the USDA maintaining lines at the Plant Introduction Station. If there's any way that I can be of help to any of you as far as advice with breeding, or with trying to explain how the USDA works and how the Plant Introduction Station works, please don't hesitate to ask....

At the Plant Introduction Station we are in charge of maintaining about 3,300 different lines of corns, about 5,000 lines of tomatoes, carrots, radishes, cucumbers and Cucurbita pepo. We have a large staff to handle that, and a lot of the work is done with controlled pollinations. It's a massive undertaking....

QUESTION (Tom Knoche): What do you think about the receptivity of the USDA's Plant Introduction Stations to requests from ordinary citizens? How do they feel about that? I've always had really good luck with their response, but....

The Plant Introduction Stations are designed to provide research material. If you have any intention of research, I don't think there are any problems with getting the material from Plant Introduction. If you are willing to report at all on the performance of this material, you won't have any problems with Plant Introduction. Plant Introduction is not a seed company to provide people with seed to just grow in their gardens and eat. But I think that any serious request from any of you for materials that you can't locate elsewhere, especially if you are interested in maintaining the varieties, such requests will be well received....

The USDA's vegetable collections are generally lacking in heirloom varieties collected in the U. S. The Ames collections include carrots and radishes and Cucurbita pepo, but the emphasis has been on germ plasm collected in other parts of the world. Another area that's been neglected is that we don't watch the seed companies to determine what they sell. If they discontinue something, we're likely to miss it. So I hope that Plant Introduction can work together with the Seed Savers Exchange to make sure that Kent's work with The Garden Seed Inventory will allow Plant Introduction to maintain lines that are being discontinued....

QUESTION (Kent Whealy): I've talked briefly with Dr. Bass at the National Seed Storage Laboratory in Fort Collins about getting our material into the federal system. Should material like that go to the Plant Introduction Stations or to the Fort Collins facility?

It's my understanding that the National Seed Storage Lab is really set up as a collection of last resort. They are not really the way to enter materials into the Plant Introduction System. Each of the regional Plant Introduction Stations sends material to Fort Collins to be entered into frozen storage. If a disease epidemic or a local or national emergency wiped us out, they would still have the material in long-term storage at Fort Collins. They are very reluctant to give out materials....There are four regional Plant Introduction Stations: one here in Ames, Iowa; another in Pullman, Washington; one in Geneva, New York; and another down in Experiment, Georgia. These are the working collections. They're the ones that generally receive new materials and send out materials to researchers. If you've got material that you want to bring into the system, I'd contact the station in charge of maintaining that crop....

QUESTION (concerning hand-pollination of corn....)

Sherry and I will be giving a slide show this evening that covers hand-pollination and shows a lot of our work with ornamental corn. But I can describe the way I do hand-pollination....There are two types of bags that we work with to do a controlled pollination. We put a white waxy paper bag on the ear, and a waterproof paper bag that looks much like a grocery sack--a little smaller, but that same kind of paper--is placed over the tassel. We tassel bag usually in the evening before, just as the anthers are coming out in decent quantity. The tassel is covered usually for at least 12 hours to prevent it from being contaminated by foreign pollen. Pollen tends to shed in the morning, so about noon we collect the pollen....We bend the tassel over and hit it. You take the bag off and it now contains a bunch of anthers and a lot of pollen.... Although it varies a lot from variety to variety, after you've worked with a certain corn for a while you can tell about a day or two before the silks are going to emerge from the husk. I use the small scissors that are on a Swiss Army knife and snip off the top of the ear so that you can see a ring of silks sort of like a bulls-eye in the center of your husk leaf. Then I put a white waxy paper bag over that ear. Every couple of days I'll check it and see how far the silks have come out. When the silks are out about one inch and look sort of like a brush, I take whatever pollen I'm using in the brown paper bag and shake on enough pollen to cover the silks. Then I put a brown waterproof paper bag over the ear and staple the bag around the stem....

QUESTION (John Hartman): Does the pollen have to be collected that day or can you use older pollen?

You can't hold your pollen for more than a day or two. That's a real problem....

QUESTION (Kent Whealy): Can you trim the tip of the ear in a way that will speed the silk emergence?

Yes, I do that to a certain extent. You can cut down a little further on the ear and you'll get your silks a day or two earlier. But you can only do that up to a certain point. There are some varieties that just don't have good synchronization. In those cases you may have to plant a larger population in order to find the earliest silking one and the latest shedding one. Then you can cross those together. That's all I can recommend....

QUESTION (Ted Gibbs): Does the pollen have to hit the tip of the silk?

No. Most times the tip of the silk is actually cut off when you snip off the end of the ear....

QUESTION (Carl Barnes): Any place that pollen hits that silk, it will grow and go into the ear....

Yes. Corn is one of the few plants that the stigma--the female surface--is receptive all the way along its length. In a lot of plants, if you clip off that tip, the pollen won't grow very well. But on corn that whole surface of the silk is receptive. Even though you've cut off the end of it, the pollen will grow....

QUESTION (Mark Fox): Have you ever tried freezing the pollen and then seeing if it's still viable?

If you try freezing corn pollen, because of the amount of water that's in it most of the grains will explode. That's not a very good way of keeping it....I have tried keeping pollen in refrigeration at very high humidity. There's an old corn geneticist by the name of Charley Burnham who's about 85 years old now. He told me that one time he accidentally left some pollen in a bag on a rainy cold day sitting on the ground. He came back three days later and under those cool, high humidity conditions, the pollen survived....

QUESTION (Mark Fox): Where can we get the bags?

Lawson Bag Co. makes the best bags for corn pollination. We use them at Plant Introduction Station. Their address is Lawson Bag Co., PO Box 8577, Northfield, IL 60093 (312/446-8812). I'm not sure what minimum quantity you will have to order....

QUESTION (Kent Whealy): Could you discuss the size of the plantings we should grow in order to maintain the diversity within populations?

Okay. That depends on the crop you're working with. If you're working with a self-pollinated crop like tomatoes or beans, your population size doesn't necessarily have to be very large....If you're going to maintain a corn variety, you should grow out <u>at least</u> a hundred plants and save ears from maybe thirty of the best plants and use as many different pollen parents as possible to pollinate those ears. That would maintain a very large percentage of the variability in that corn. Sure, five hundred plants would be better, but you're cutting down the number of varieties you can keep....But if you grow out only twenty-five or thirty plants each year, you're going to gradually reduce the amount of variability with each passing generation. Eventually you'll pick up inbreeding depression and after maybe four or five generations what you'll have left is going to be very different and usually worse than what you started with. So I'd say that you should grow at least a hundred plants.

Thane Earle

(From earlier correspondence with Thane Earle--....The Society for the Preservation of Poultry Antiquities (SPPA) was first organized by Neil Jones of Virginia. He did his best to promote the cause of perpetuating the old and vanishing breeds of poultry. But the membership was quite small so there were only

Mark Widrlechner

Thane Earle

limited funds for promotion. Finally, he decided to give it up....Loyl Stromberg heard about the plight of the SPPA and contacted several breeders of rare poultry. His idea was to try to reorganize and see what could be accomplished with better promotion. A meeting was held at the Minnesota Poultry Breeders fall show in Minneapolis. New officers were elected and the SPPA was off to a new start....The object of the SPPA is to perpetuate the vanishing breeds or varieties of domestic poultry. "Poultry" includes standard-size chickens, bantams, ducks, geese and turkeys. A list compiled by the Society is brought up to date periodically telling what is considered to be on the endangered list. A directory of the membership is published every other year. It lists what they are currently breeding and if they sell stock, eggs or chicks. It is probably the best sourcebook for many of the rare breeds. This book is included in the membership dues along with a quarterly newsletter....Each year a poultry show is chosen to host the annual meeting and competitive show. Many cash awards are offered to encourage the showing of these endangered poultry breeds. This show is rotated into different areas of the United States so, sooner or later, one will be close enough for you to attend....Poultry and gardening go together very nicely. Poultry manure is one of the best natural fertilizers. And the garden can furnish part of the food for a small flock of fowl....Why not make the poultry you raise one of the rare breeds? There are many breeds and varieties from which to choose. Just as you are trying to save many of the heirloom varieties of vegetables, there are those trying to do the same with poultry. Write to the Secretary of SPPA, Marion Nash, P. O. Box 102, Murphysboro, IL 62966, for membership information.)

****

I was involved with the Society for the Preservation of Poultry Antiquities long before I discovered the Seed Savers Exchange. And when I heard about the Seed Savers Exchange, I thought here are two groups working in different fields, but they have the same goals....When the SPPA was reorganized about ten years ago, I was approached to be its president. At that particular time I was quite involved with a lot of the rare breeds of poultry. In fact, I had 140 breeds of large fowl and 120 breeds of bantams that we were actually breeding. These were all penned separately....We have to worry about cross-pollination too—that's for sure--but not the pollen flying around in the air....(laughter)....If we just have wire between them, there's going to be no problem. So it's a little bit easier than when you're talking about pollinating corn....(loud laughter).... Well, it's the truth!....Anyhow I took over as president. But as the years passed the work got to be too much. I figured somebody else should take over, which they did.

At the present time my wife and I are the Bulletin editors. We put out a quarterly bulletin. I don't have to tell Kent how hard it is to get somebody to write for these things. Even though he only puts out two issues a year, he can't get anybody to write....I've got a long background of interest in poultry....and gardening, too. So a lot of the Bulletin is really drawn on my past knowledge and past experiences in both breeding and showing and working with 4-H kids....My wife and I used to run a show string of about 600 birds. When you're getting that many birds ready to go to a show, you can't start crating them a week ahead of time. They've got to be crated, put in the truck, and you <u>go.</u> Otherwise, you're going to lose them. When you're driving about 300 miles and unloading about 300 chickens and driving back the same day, that's a <u>long</u> day. So my wife said, "Why don't you get into something a little bit easier?" So I said, "I've always liked gardening. Why don't we increase the garden a little more?" She agreed, but didn't realize what she was saying. The other day she said, "My god, we weren't working this hard when we had the chickens!".... (laughter)....My interests <u>always</u> get out of hand. I'm growing out 125 kinds of tomatoes this year. The plants look great if the weather will only cooperate. I've got several kinds of corn and about 115 kinds of beans. Also a few peppers and eggplants and okra....My wife is also into herbs. She does the processing and I do the growing....

COMMENT (Tom Woods): You said that you've worked a lot with 4-H kids. I think that would be a really excellent avenue for the Seed Savers Exchange to pursue. This year we set up a historical potpourri of projects that 4-H kids could enter, and one of the projects is historic vegetables. They come to the Oliver Kelley Farm and we have research material and give them some seeds to start them off. And they take those seeds and it becomes an actual 4-H project that they follow throughout the summer and then display at the County Fair. We're hoping that it will catch on in some of the other county extension services and become a statewide project where they can progress up to the Minnesota State Fair. Right now, it's pretty much a county affair. And since the kids can't take it to the state level, some of them don't want to get involved in it. But I think it's got the potential for really expanding seed-saving work.

Everybody around my area has me pegged as "the poultry man." But since my interest in gardening is increasing, I'm trying to get the kids interested too....

For neary 40 years in our county, each of the banks has kicked in so much money and they award the kids cash prizes for the 20 best gardens in the county. But it's gotten to the point where they can hardly find five or six good gardens today. And it boils down to the simple fact that gardening is a lot of work....the same problems apply to poultry as well. It's easier for the kids to go to a hatchery and buy baby chicks. It's exactly the same way they go to a nursery or garden supply center and buy plants. And what do they get? In both cases they get the hybrids. It's going to be just as hard to get these kids to think about saving seeds as it is to try to keep some breeder chickens back and hatch their own. It's the same thing, really. It's just the difference between horticulture or aviculture....

QUESTION (Carl Barnes): Have you ever worked with garden clubs and do any of their programs include kids?

We've got a fellow in our county--he's a down-to-earth sort of a man--who has started an organization called the National Junior Horticultural Association. They have a national contest each year where kids are picked as state representatives—not county representatives—and they have to identify all kinds of plants. It's quite a deal for those kids. When this fellow was trying to start the program, he went to the State Horticulturist and do you suppose he would give any help? No! They wanted something that could give them more recognition. Why should we bother with these kids? That is the attitude. But the fellow didn't take "no" for an answer. And he kept at them and kept at them. This year the state has finally come around and they're going to give some help ....So, we've got some people in our county who really are trying to do something on the state level....We've got to get these kids back into gardening....

COMMENT (John Hartman): I've worked with quite a few garden clubs in Indianapolis. Most of this work was with flowers, but they're interested in hearing about the Seed Savers Exchange and about heirlooms—vegetable varieties. What I do, after I give them a talk, is give each person who is there a couple of packets of seed. And the results these last couple of years have been amazing. People will call to tell me that the squash or whatever was so wonderful and they're going to grow them again....(laughter)....So if you ever talk to your local garden club or whatever, be sure to give them some seeds....Another thing I do to get kids interested in gardening is to plant "munchables" which they can easily reach over my garden fence. That eventually gets a conversation going....A couple of years ago I grew Moon & Stars watermelon for the first time. It's one of the most beautiful garden plants I've ever seen with its yellow-spotted gray-green leaves. The neighborhood kids were puzzled with the plant and wanted to know what it was. When I said that it was a watermelon, I could see the glances being exchanged. I told them I didn't mind if they stole them, but to wait because they weren't ready yet and all that I really wanted was the seeds. After I explained how rare they were and about the Seed Savers Exchange, the kids were fascinated and seemed to come around more often asking questions....

We always raise more than we can plant ourselves, so we usually sell a few plants in the farmers markets in our area. The majority of the people come up and ask, "Do you have any Big Boys or Early Girls?" They know all of the hybrids by name. I'll say, "No, but we've got some Peron. It's a very good open-pollinated tomato." "Never heard of it! You sure it's good?" And I tell them, "It's the best we've ever grown." And usually they'll buy a couple plants. The next year those same people come back to find us and say, "You know those tomatoes you sold us last year? Best tomatoes we ever had!".... Most gardeners are so brainwashed with all this hybrid business. But that's all they ever see anymore in the catalogs....If we can just get them to try the best of our varieties, they'll switch....

John Hartman

It's really kind of funny how this all started with an interest in seed saving and ended up as the J. M. Hartman and Daughters Seed Co. I did not originally intend to start a seed company. I really didn't....(laughter)....It just happened. I got my retail experience in the seed business by purposely hiring on at a seed company in Indianapolis called Bash's Seed Co. If anyone ever gets a chance to go there, they should. The building is on the National Register of Historic Places. You walk into the place and it has this marvelous old small octagonal ceramic tile floor and huge bins of seed on the side walls. There are hundreds of drawers all the way from little tiny ones to great big drawers....Everything's measured using little brass ladles. This certain seed takes one #6 spoon, or this seed takes two #8 spoons....and that is the way the seed packages are all filled....it is actually a very profitable venture. It's right in the heart of down-town Indianapolis in the high rent district....And yet the major clientele are inner-city blacks....During a season they buy 200-300 lbs. of Seven Top turnip seed. And it's all measured out in half-ounces and quarter ounces. Lots of collards and Mississippi Silver crowders. Tremendous quantities of seed go out of there. It's all sold on an individual basis. You sit and you talk to your cus-tomers as you measure out the seeds. You find out recipes and what they like or don't like. And this is the way they run their seed business. I've never seen anything like it! They've been there since 1856. They moved once.... across the street....(laughter)....

While working at Bash's one day, an elderly black gentleman walked in. His name was Mr. Williams....I have to digress. He wasn't as elderly as one of my other customers. One of my daughters is ten years old, and I've always wanted her to meet Rev. Hunt. On June 13, Rev. Hunt will be a hundred and eleven years old and he still plants Hickory King corn every year....At any rate Mr. Williams came in and he was kinda shuffling around and looking at the water-melon seed. He said, "You don't, by any chance, have any white watermelon?" I asked, "White-seeded?" And he said, "No, the meat's white. And very sweet." I said, "No, I've never even heard of any." And then he said, "Well, I've got some old seeds at home." Of course my ears perked up. "I got them from my aunt down in Georgia and she's been raising 'em for over thirty years. But I don't think the seed'd probably be any good. It's been a couple of years since I saved 'em." "They might be good. Any chance of getting some?" I asked. "Well, if I come down this way again, I'll drop in and I might bring you some," he said as he left. I didn't know if I'd see him again. But he came back two or three days later with a Scotch snuff can full of the seeds. And he said,

"Here they are." I asked, "Can I have a few of them?" "Take the whole thing," he said. Well, I profusely thanked him. He said, "They're the best watermelon in the world, those there. They came from Georgia and they'so sweet and just as pure white ivory flesh." Then he just started out the door. And I called after him, "Well, sir, well, what do you call them?" He turned around, gave me a stern look and said, "Son, these the ol' famous Cream Watermelon."

I also lived in Europe for quite a number of years and was bringing back local varieties of seed. I discovered that many of the open-pollinated and non-protected varieties which I had been purchasing had completely disappeared within a year or two. This was right at the tail end of when they were still allowing many of the older varieties to be sold by the companies. And I had many of them. I started planting and found out that they are really excellent varieties, which are no longer being offered anywhere....About this same time I became acquainted with Seed Savers Exchange. I started offering my foreign varieties through the SSE and in return received many really excellent heirloom varieties....Each year I experiment with a large number of new varieties. Last year, for example, I grew 75 varieties of beans, 30 varieties of tomatoes, a dozen lettuces and so on. Every year I discover a few varieties that are so exceptional that I feel they should be offered commercially....

I remember going to a convention in Florida years ago. There was a display of Florida fruit tree seedlings in the airport, which were selling quite briskly. I've always remembered that and often wondered if it would be possible to promote rack sales of heirloom or regional seeds to the tourist trade....I envisioned a polished wooden stand that would hold five packets high and five packets across with old-fashioned lettering burned into the wood across the top....The packets would be the standard size and contain a generous sample of seed. They would be "buff" manila envelopes with a border that was Prussian blue baroque scroll-work. Very Victorian-looking. Across the top, using an old-fashioned print face in black, would be "J. M. Hartman and Daughters Seed Co., Antique and Heirloom Varieties of Seeds." Then a description of that variety on the back and an explanation of heirloom and open-pollinated vegetables. Each of the packets would be priced at $1.00, no matter what type of seed it contained. They would be given out on consignment at 50 cents per packet....The first place that I approached was the giftshop in the Adam's Mart Hotel. They thought it was a marvelous idea. Then I went to the Fort Wayne Zoo....(laughter)....the Tippecanoe Historical Society, the Holiday Inn giftshop, the Fort Wayne Botanical Conservatory, Stuckee's, the Marriott giftshop, Nickerson Farms, several muzzleloader reenactments such as the Feast of the Hunter's Moon, several giftshops in the state parks near Indianapolis, and many others. No one has turned me down. They want the displays before Christmas and they want me to keep them stocked right through the Indianapolis 500. The production of the seed crops that are being grown this summer are already gone....(applause)....

I intend to give myself a couple of years to get my distribution systems down. Then I'll start to think about mail-order and other methods of distribution. I don't want to jump into it with both feet without testing the water first. I think that is the wise thing to do. When I do go into mail-order, the publications I will use for promotion will also be unique. The first publication I intend to try will be Legionnaire Magazine. It's certainly not your normal outlet. I've never seen an ad for seeds in there. But it's a natural. And there are many history-oriented magazines which would be perfect places to sell heirloom seeds. I wouldn't think of trying to use Yankee Magazine or Organic Gardening. I can't compete with Jung's 10-cent packet of seeds and I don't intend to try....Why try to compete with the big seed companies? We can do things on a small scale that they can't do. We can specialize in small labor-intensive horticultural projects and make money doing it. This has certainly been proven out in Europe. I've seen people in Holland make a year's living for a family growing cucumbers for seed in an area no larger than our shelterhouse here....

I will initially be raising only about 25 varieties for commercial production. That's all I can handle. I'm sharecropping a lot of my seeds out. Most of this has been with back-to-the-land groups. Also I know a large group of people in Mensa who are doing a cooperative small farm project in southern Indiana and they're interested. Various religious groups would be interested. It's a perfect fit. They have the land and they need small labor-intensive cash crops. I have found several people who are interested in growing a variety of something for me. A 10' x 50' patch of one variety of lettuce will yield more seed than you know what to do with on a commercial venture like mine. People are very happy to grow a crop that size. We come in and do the harvesting of the seed and pay them for that seed. And it's more than they would make, certainly, putting it into beans or normal field crops....I have one fellow who's growing a half an acre of Moon and Stars watermelons for me this summer. We've got a nice barter arrangement. I take the seeds and he gets the juice. I think he intends to make quite a bit of watermelon wine....(laughter)....Most of these people have access to small isolated fields. So you guide the layout of all of it. You keep things pure through isolation and by giving different crops to different folks. You can give them instructions and help them harvest the seeds to make sure things go right....I intend to sell 40,000 packets of seeds my first year at 50 cents profit per packet. That seed is all being grown right now and so far I have less than $1,000 invested and $500 of that was for a roto-tiller....

I know that a lot of people are reluctant to start a seed company because they are scared of the regulations and taxes and paperwork. I was very apprehensive. But if you live in the state of Indiana, it's very easy to start a seed company. Your license costs you nothing. You write to the Seed Commissioner and tell him what you want to do. He sends back a form. You mail it back to him, and then you have to send in a one-sheet form on a quarterly basis, whether you sell anything or not....When I first started, I took a huge amount of Indiana seed regulations home with me. But after wading through them, it turned out that almost all of them applied to agricultural seeds. We're talking about vegetable seeds, and the regulations are totally different. They are much simpler and much cheaper. It turns out that for vegetables there's a six-dollar testing fee per variety. You don't have to have them tested by the state. You can have an independent test, but it's just as simple to send them to Purdue. They test 400 seeds and send the results back to you. And you're in business....

In Indiana there is a seed tax on sales. The tax is $1.00 per consignment of seed, which is automatically passed on to the buyer. Now it doesn't make any difference whether you sell 10,000 packages of seed or one package of seed--it's $1.00....Each time you refill a display it's another $1.00 tax, whether it's ten packets or one hundred packets. You have to report this and pay your tax to the Seed Commissioner in Indiana on a quarterly basis. That's all there is to it! Of course you have to keep a record of who you sold to, but you'd be doing that anyway for Uncle Sam. And mail-order from Indiana is the easiest thing possible. There is no tax at all because every mail-order shipment is considered an out-of-state sale, whether it's sold within the state or not. It's a very simple process. So in Indiana, it's easy. I can't speak for other states. And I understand talking to other people that it is different in each state.

John Hartman

Paul Griepentrog

Paul Griepentrog
(Turtle Enterprises, Rt. 1, Box 242, Dalton, WI 53926)

Wisconsin's seed laws are a little bit different. If John Hartman sells his vegetable seeds in Wisconsin, they better meet minimum germination requirements or they're going to call him up and say, "You've got a $100 fine here waiting for you."....(laughter)....Sorghum seed falls under the Agricultural Seed Act and I've got to run the germination tests and the purity tests for noxious weeds and weed seed and inert matter. All that's got to be stated on the label....I have to buy a Seed License that costs $10. That allows me to buy seed in bulk, and then break it down into little packages and sell it again, and it allows me to sell the seed I produce myself. The license fee is determined by how much you sell each year. The first $20,000 costs $10. I don't think I'll ever get over that ....(laughter)....

I started out with two or three varieties of sorghum and right now I'm at forty. I've also got a twenty-eight page computer list that's twenty lines to a page of sorghum varieties that are available to me through the Georgia Plant Introduction Station. This last year I had five varieties commercially available. One of my last orders went out to a Mennonite fellow in New York. One of his friends wanted seed of an early sorghum because they were having a lot of cool, wet weather. So I had a couple pounds of Blackstrap which is about an 85-day sorghum. The rest of the ten-pound order I filled with Honey Drip....I don't know what he's going to do with ten pounds of sorghum seed because that's enough to plant ten acres....(laughter)....And that would be more than all of us here could strip in a month and process....

There's only a few varieties of sweet sorghum that are still available commercially and two of them are inbred lines. Most of the commercial varieties seem to no longer be pure. They're not at all the same as what I obtained from the Introduction Station which would have been foundation seeds....the Sugar Drip seed I've bought commercially is only about 4-5' tall. The Sugar Drip I got from the Introduction Station will grow 8-10' tall....The old stories say you should get between 100-200 gallons per acre. Last year I got 35 gallons off a half an acre. But as I bring in more varieties and continue to work with them and breed them up, I'm pretty sure I'll be able to see at least that 100 gallon figure again. Especially with a couple that I obtained through the Plant Introduction Station at Ames....

My interest in sorghum started because of an old sorghum mill near where I live. It's owned and operated by this old couple. He's blind in one eye and going bad in the other one. She's got about eight or ten things wrong with her. She can hardly walk around, but she's the boss. There are six people that work there. The cane is brought in on pickup trucks or trailers and it's stacked up in bundles. They haul the bundles with a wagon and bring them and put them through a press which squeezes the juice out of it. And from there it gravity-flows down into a tank. As it cooks, you turn a spigot and draw off five to seven gallons at a time. And while you're cooking it, you constantly skim off the green foam that comes to the top. The first set of pans is about 5' x 10' and the last set of pans is about 4' x 8'....It takes about 20-40 gallons of juice to make a gallon of syrup....They use slab lumber for the heat. It's an art in itself just to fire the thing. You have to keep everything boiling the whole time....If it's a nice, clear day and it's cool out, there's no problem. But if it's a rainy day, you've got to fire it harder in order to keep it cooking. The day that you cook it determines a lot of the quality in the finished product....

There are a lot of tourists that come by to watch and usually one or two newspaper people. Every couple of years a TV crew comes down and does a little bit on it. But when they try to show the inside of the building, all you can see is steam. You can't see anybody. Just three dark figures standing in the mist. We always tell the boss, "We're just lost in the fog."....(laughter)....They've all worked there for about fifteen years. It's one of the last true mills in the state of Wisconsin. They even bring cane in from the outside and have it done. There are a couple of privately owned mills, but they've got to have a food processor's license similar to what a dairy farmer would have. They have to be inspected and meet the standards and everything else. It sounds like as soon as the old woman passes away, they'll probably close the mill. But as long as she can hobble from the house to the sorghum mill, they'll keep it open....

Carl Barnes

(From previous correspondence -- We have formed an organization called CORNS which is dedicated to the collection, propagation, distribution and preservation of any and all open-pollinated corn varieties. These include teosinte, pod corn, flour or flint corn, sweet corn, pop corn and dent corn....We are working with a small growers network in this area, but are willing to work with those genuinely interested in following a few simple guidelines to isolate and increase varieties to be returned for distribution. Currently we have over 200 varieties of corn seed. We need all the help we can get from those who have O. P. corns and can furnish history, origins, uses and any available literature. We will try to provide seeds on the basis of climatic conditions in the area of the country where the grower lives. When sufficient seed is grown, it will be stored and offered through the SSE and Growers Network....Carl & Karen Barnes, CORNS, Route 1, Box 32, Turpin, OK 73950.)

****

I could spend three or four hours or the next three or four days talking about corn....(laughter)....But I'm going to be very brief....Roy Ensley in North Carolina and I have been corresponding back and forth for quite some time. He sent me an outstanding Cherokee all-purpose corn....(holds up an ear).... There's a fellow that has a $42,000 grant to study this to see how it relates to the original eight-row that grows in Mexico and Central America. And in the meantime, before it becomes commercialized, I'd like to have this in the hands of a hundred or a thousand people and already on the people's market....You understand what I'm talking about now....(laughter)....If you can't motivate people, irritate them....(laughter)....This has a sweet gene. It's a good green corn, as the Indians call it, in the roasting ear stage. It's a good hominy corn. It's also a good flour corn and a good meal corn. I think that maybe us Indians knew what we wuz doin'....(laughter)....

There's a fellow who is working with corn out in Big Timber, Montana. He has an archeologist friend who opened up a dry cave which they carbon-dated at 850-900 years old. Some people here are growing some of the corn and beans taken from that cave. This is from the third generation....(shows another ear) ....and I got one solid ear. Larry Sallee is looking for the Blue Eagle corn. It's a beautiful blue corn with the white dot at each silk insertion point. Even in the first generation this corn has a lot of dots at the insertion point on different-colored kernels....So, with some work, maybe Larry will finally have his Blue Eagle corn....

This work has been very rewarding. There's no way that I could do this all myself. I have 160 people who are cooperating with corn, all the way from one variety to fifty or sixty. It's grown into a real maintenance program....I need some of you to help me come up with some forms that I can send out to these people to evaluate characteristics. Mark is going to help me with that and oth-

Carl Barnes

ers have offered too. So we'll start getting some of this down. That will also help us get these corns into the right regions of the United States. The Mandan corns will not mature down in my country and I have 22 varieties from Central America that need someone below Tucson or in Florida to grow them out. They all have promise. They're mostly flour-type corns. I've had an Indian friend in New Mexico in the past to grow some of these corns, but they really need to be grown farther south. My wife is working more and more with popcorns because she does some arranging work....There are some very unique corns in Arizona and New Mexico in all the little different valleys. Just some beautiful corns. Every time I go out there I find a little different place and find a little different type of corn. It just blows your mind....Don't ask any questions, because I don't have any answers....(laughter)....This is short and sweet. There's some wonderful stories that I could tell, but some of them may not be true....(laughter)....

Steve Neal

These past two years that I have been a member of the Seed Savers Exchange have been two of the most exciting years in my gardening experience. I have met a group of the finest people in the world and if that isn't enough, it turns out that they are all deeply involved in preserving a very important segment of our heritage. I am a person who likes to save everything (ask my wife) because if I don't save it, I'll need it the day after it's gone or some friend will call and ask if I know somebody who has it. Well, I save letters too, and I'll tell you I have some of the most interesting ones from Seed Savers. It is great to hear from people with plants or information to share or those looking for these things. I am really impressed with the attitudes, ideas, desires and dedication of the SSE people. I would like to express a special gratitude to Kent and Diane for devoting so much to SSE.

Obviously, my greatest personal interest and most of my work with SSE has revolved around potatoes. The letters with all their questions and remarks, and the concern expressed by other people, have resulted in my establishing the organization called "TUBERS." This is an organization dedicated to the collection, preservation, propagation and distribution of any and all potatoes, both heirloom and modern varieties endemic to the United States, as well as to the development and introduction of new and desirable varieties. One of our main goals is education. By this I mean that we will try to help homestead gardeners learn about the many options available to them in their potato production besides helping them solve their problems.

Our first and current working project is to index and catalog potato varieties, listing detailed descriptions, regionalized growing and production information, storage methods and the specific qualities and uses of each variety. We call this our Potato Index Project and we have developed a record sheet that all of our growers fill out as they are growing out the varieties. If we all catalog these differences, within a few years we hope to put together a book that will describe the potatoes and also let folks know which ones will do well in different areas of the country.

Our collection of potatoes is rapidly increasing and it appears that by spring 1985, we will have over two hundred named varieties and possibly many more. We also have unnamed varieties, not all of our own development, some of which are showing promise. We are doing a little work on some foreign species (with only slight success I must admit) made available to me through the USDA.

The preservation program is made possible directly by the propagation of all these varieties, a few at a time, by individual members scattered throughout the country. These same members facilitate the distribution themselves in that each member must be willing to make samples available to fill requests as long as sufficient supply exists. This should greatly increase the offerings in next year's SSE Winter Yearbook.

Many people cannot understand why we would be too concerned about developing new varieties of potatoes. In trying to answer this, let me say that as with all our SSE friends, we see the climatic, disease and insect vector changes coming upon us as one of the greatest obstacles in the path of continued advancement of mankind. The United States is in a low third place in world potato production, since we only produce five percent of the total world potato crop. Most people in the U. S. think of potatoes as a low grade food despite the high vitamin and mineral content, despite the high quality protein that is quite likely better than soybean protein, despite the low caloric content and nearly zero fat content, along with their fiber content. Still, with all this in mind, some of the recent U. S. potato introductions have achieved a higher food value, more disease resistance and greater production levels than the older varieties. The dietary problems many of us have with potatoes should more accurately be blamed on the rich toppings such as butter, gravy and sour cream, or the grease in which we fry our potatoes, rather than the potatoes themselves. New varieties can only come from one place and that is from the genetic diversity allowed us, by an intelligence much greater than any of us can claim, in our older varieties. If through our efforts we can, in some small way, help to change any of these problems, then our goals will be met.

Once more, we are a network of lay potato workers throughout the United States and we are willing to work with anyone genuinely interested in furthering potatoes. Anyone who is currently growing potatoes and has some sort of facility for winter storage (many use a refrigerator) and is willing to follow a few simple guidelines to keeping their records and reporting their work would be welcome in our organization.

We are actively searching for potato varieties which we don't have in our collection. Should you have or know the whereabouts of any rare, unusual or not readily available varieties of potatoes, please contact me immediately. While not all varieties which we have are currently ready for distribution, I can send you a list of what we have. Please include a S.A.S.E. whenever you write. We might also note that we would welcome the chance to work with historical gardens, societies and natural history museums.

("TUBERS," Steve Neal, Rt. 4, Box 169-D, Gainesville, MO 65655)

Ted Gibbs

I grow okra, basically because I enjoy eating okra....And I like to have other people enjoy it too. I currently have 23 varieties. This year my grow-out will include 12 of those. All, so far, have been unique....I have been searching for information on okra and am trying to develop better methods for growing okra and preserving okra. I find that there has not been a lot of work done on okra in agriculture departments in recent years. There is a study starting right now with the U. S. Vegetable Laboratory in Charlotte, South Carolina....Also I've been in contact with Dr. McFerran at the University of Arkansas, offering what assistance I might give him in okra development. He does have an okra breeding program going currently....I am cataloging all of the characteristics of these various okra varieties in my collection....that information will be made available through the plant descriptions in my SSE listings so that others can obtain varieties suited to their particular area....If you're in the Midwest, as I am near St. Louis, you become concerned about the strength of the stalk when you come out after a strong windstorm and find certain varieties knocked over.... Okra, being the kind of plant it is, usually is harvested perhaps every other day. That leads to waste and I'm very interested in eliminating waste....In our world today, when there are so many people that are hungry, I think it's a terrible shame that okra is enjoyed by so few....I feel okra does have a tremendous potential, both as a food crop for all of our people and as a commercial crop....

QUESTION: Is all okra grown for the pods? Is there potential for the seed as well?

Yes. Recent research by the USDA has evaluated the oil from okra seeds, and those tests indicate that it is top quality. They feel that it would be competitive with safflower oil, for example....The seed is also made into a beverage that is used as a coffee substitute....the seeds can be ground into flour....tests show that the flour doesn't stand alone—it needs to be mixed with other grains—but certainly it has potential....Many gardeners cook the immature seeds like peas.... okra is a very nutritious low-calorie food....

QUESTION: What are some of the characteristics you are examining for the varieties you grow?

Spines, certainly....Those little stickery things that aggravate you when you pick okra....It bothers some people more than others. Should you ever allow an okra pod that's heavily spined to strike you in the eye, as I did, you will remember it and not let that happen again. There are varieties called "spineless" but they're not....One of the primary goals of past USDA breeding has been the development of okra that is spine-free (smooth)....It has been my experience that the heavier spined okras actually have the better flavor.... Another characteristic is the number of ridges on the pod....one extreme would be White Velvet, which has a pod that's essentially round in cross section. It's very hard to see any ridging on it at all. The other extreme is an okra called Star of David, which is a large okra with very pronounced, heavy ridges. As a matter of fact, a good third of the pod's diameter is ridged. It's a very good okra, by the way. Mine was 14' tall in late September, making it very difficult to harvest. And it has no side branches. Its opposite would be Candelabra which Park Seed Co. recently introduced. Unlike other okras, it branches out from the base of the plant. Star of David produces pods strictly off of the main stem and there's no branching whatsoever....Other characteristics include pod size and color, stalk strength, plant height at maturity and internode length, which seems to relate to productivity....

QUESTION: Which varieties have the most strongly pronounced, distinctive flavor? I've tried several and they all tended, in my opinion, to be on the bland side.

I think, generally, that's true....Longhorn has a very strong flavor. People that do not care for okra would really dislike Longhorn. The complete opposite of that would be Ever Tender which is the most productive okra that I have. I got it from Kent and it is very, very mild. I use Ever Tender to introduce people to okra, because they don't run and hide....(laughter)....and of course it is very tender. If you grow Ever Tender, you'll have plenty of okra. It is tremendously productive....Star of David has a distinctive flavor too, but it's not quite as strong as Longhorn. And I appreciate it. I enjoy it. But if somebody's new to okra I wouldn't start them off with Star of David or some of the stronger flavors. I'd start them off with a mild, bland type....

Ted Gibbs

Tom Knoche

Tom Knoche

I think this is my seventh year to be in the Seed Exchange. During that time I've amassed a huge collection of seed, actually much more than I've been able to grow out and test. So I developed a mini frozen seed bank. I've been processing the seeds I haven't had time to grow and storing them. I've collected 260 pumpkins and squash, 50 watermelons, 250 bush beans, 250 pole beans and so on....I've never had time to do any hand-pollinations....I work all night so I'm asleep during the mornings. The way I keep things pure is by growing eight isolated pumpkin patches....I use "pumpkin patch" because it's easy to say and people relate to it. I tried calling it a squash patch, but everybody thought it was some kind of a sport or something....those eight gardens are all separated by over 1,000 feet and with either woods or field crops in between. That allows me to raise 32 varieties of squash, 8 varieties of watermelons and 8 varieties of muskmelons....

This year I'm trying to unravel some of the mysteries involved in my collection. I have about a dozen different Sweet Potato squashes that I know aren't the same....Also there's an old Cherokee squash called Candy Roaster that was offered a few years ago in the Seed Exchange. The seed I received was dead when I went to plant it. Now I have eight different Roaster squashes and I'm growing those all out this summer. We'll find out if any of them are the true Cherokee Candy Roaster....I'm not interested in fruit shape or size or color or anything else. I have only one concern. I'm interested in flavor....you'll never hook me on growing beans for the seed coat color. I'm just not a bit interested in that. But you might hook me on size a little bit, because I like to raise the big beans....

My grandma is going on eighty-nine years old, and she's said all her life that no bean was worth eating unless you had to string it first. A lot of the old folks say that about the string beans. I'm zeroing in on flavor of these old snap beans, because there really is a big variation in the ones that I have tried. I would appreciate it if folks would send me any snap beans (green beans) that have really excellent flavor....most folks think that Lazy Wife pole bean is one of the real old pole beans. It's old, but it isn't one of the oldest. It may be one of the oldest stringless, but Scotia Cornfield bean and all the Cutshorts are much older. This year I want to plant every Cutshort that I can find. I've located about 12 different kinds so far. The ones that I have raised in the past have been very individualistic in their flavor. I will also be growing five kinds of Lazy Wife beans....a lot of people come up from the Appalachian area to Cincinnati for work and money. I've talked about growing garden to some of them at work. Different ones at different times say that when they raise a garden there's only one bean that they would think of growing and that's the Greasy bean. And so that put me on the track of the Greasy bean. I've found out there are many types of Greasy beans and I will be testing out probably a dozen kinds this summer....Greasy beans got that name because they look like they've been dipped in grease. They don't have any fuzz like most beans. I've run onto several from North Carolina, and some others were offered in the Seed Exchange. They are a string bean; they have to be strung....So this is how I've decided to go about my work. I'm going to zero in on certain areas of special concern, rather than trying to spread myself so thin....

Another good thing that I have to report is that I believe through one of my wife's aunts I've discovered the old-time black-seeded Kentucky Wonder. It fits the Beans of New York description perfectly. And if so, that would be one of the real old-timers....my wife said that during the Depression years they were raising five hundred poles of it. Can you feature that? Five hundred poles! And they were picking and taking it to Cincinnati to make the only spending money that they had during that time. The folks in town, when they saw those beautiful Kentucky Wonder beans, would buy them with what little money they had. So I may have that to offer this fall. At least I hope so....

I want a white-meated squash. If anybody happens onto a white-meated winter squash, I sure would do anything to acquire it....Well, pretty near....(laughter) ....I have a special recipe for squash that just doesn't look the same, if you don't use a white-meated squash....Also, I'd like to mention something that Kent talked about last year or the year before. It would be really great if we could work together as a group to assemble some recipes....I'd love to have twenty-five good squash recipes. There's got to be many more than just eating pumpkin pie. I don't like to waste anything. One of the things that's hurt me the last year was when I took the seeds from my squash and threw many of the fruits out on the garden. I really felt bad about that. If you want to do something great for a pumpkin raiser, send him some good recipes, some unique ways to use pumpkin that he doesn't have. And I think that would carry over into the different things. I thought of John Withee and the bean cookbook that he assembled. It'd be great if all of us could come up with some new and unique ways to use some of these good fruits that we're raising....

The squash that I've collected and that I'm the most concerned about are the large-fruited ones. These types are dying out so fast that there is no way they're going to be preserved if somebody doesn't take an interest in them. People want the little tiny handy size. Nobody wants to raise the large family-sized types anymore....If I were to take some of my precious squash to our County Fair, there'd be no place for them. They have everything so categorized that if mine isn't a Hubbard or a Butternut or a Bush Scallop, there's no place for them. That's how bad things have gotten. And how on earth are young people ever going to know that there's anything different? I went to the State Fair for several years and tried to acquire seed from some of the growers. But there's so little interest that they don't even bother anymore to put the names of the growers on the specimens at the State Fair....

Down in the South in the mountain areas probably fifty or sixty years back, they had a favorite bean that was a staple with the folks and they planted them in the field corn. They called them Fall beans. Now you might think that's a generic term, but an old Kentuckian who was in his eighties before he passed away told me that a Fall bean is a pole bean that's large and very round. They come in white, red, brown, buff, tan, speckled and striped. And as long as they bear no more than four beans to a pod and they're very round, then they are a true Fall bean. If you ever run onto any of them, let me know! Most of what people call Fall beans nowadays aren't the true Fali beans. And I want to work on finding real ones and preserving them....

KENT WHEALY:....Tom, I've got some of them....

You do? That's great! Fantastic....(loud applause)....

KENT WHEALY: There was a girl who was in the Seed Exchange for just a little bit. She went to summer school to get her teaching certificate not far from where we lived in Missouri. Her name was Amy Rogers, and she was from Kentucky. She was only in the Seed Exchange one year, but she sent me two beans that she called Fall beans and said there was a solid red one and several others. I'm just sure that they've both been multiplied and are in the Growers Network collection. They were both perfectly round. The color on one was split in half between shiny white and reddish-purple. The other was white with a few BB-sized, perfectly round reddish markings....

That sure sounds like them.

Glenn Drowns

My interest in gardening started when I was about two and a half years old. By the time I was four, I had about a 500 foot square garden that I took care of myself. When I was eight I took over the whole family's garden and started ordering my own seeds....I began saving my own seed when I was 13 and started seeing varieties disappear in about 1975. I really started saving seed after I purchased a copy of Burpee's 1888 catalog reprint and saw all of the varieties that were no longer available....I can't say for sure what first sparked my interest. I guess it is just something that I was meant to do. I became really interested in squash when I was seven years old. Every year I'd send for more squash. I'd send for all the catalogs I could find and grew all the squash they had available. But I was lucky if I could mature half of what I planted every year because our season was so short....

My involvement with the Seed Savers Exchange started when I was doing a government fruit project for my senior class in high school. We were looking up a lot of these materials and I came across this ad for the SSE....I remember the first time I wrote Kent I said I'd like to have all the squash varieties that are available....(laughter)....Well, I thought at that time I was talking of maybe 150. It was amazing! All of a sudden they started rolling in from the SSE. At last count I have somewhere over 270. Last year I planted 232 types. Six of them didn't come up and I ended up with about 220, because I lost a few to pests and things. I hand-pollinated them all. That was a struggle, but it was fun too....Along the way, I've become more and more interested in other crops. I sort of held back because I didn't know enough about how to save carrot seeds, for example....all the questions I ever asked about isolation distances, people could never answer. I'd go to the county agent's office and ask, and he'd just give me that blank look, like what do you want to know for....(laughter)....Whoever thought anybody would want to know that. They didn't know and they didn't have any idea where to find the answers....So I learned the hard way by making mistakes and practicing. I had to learn just by doing it ....I've also gotten quite interested in watermelons and cantaloupe. But I've always had a hard time hand-pollinating the cantaloupes. So I'm just going to find about fifteen that I like, and maintain them. The watermelons I've had better luck with and I'm now keeping about 115 kinds....

As far as goals, I don't know what I'd like to see the Seed Savers Exchange turn into. It's grown so much the last two years that it's hard to say. I know my goals have changed. At one time I thought that I would never have an interest in starting a seed company....But as things progress, I certainly see the possibilities....I'd love to see a seed company like the ones they used to have back in the 1800s with page after page of listings. Not like the catalogs of today, which seem to think it's one of the worst things in the world to have more than three types of carrots or four types of cucumbers. I think they are scared that people don't know which one to choose. The more the better is my philosopy....This summer I'll be working with Kent helping him finish up The Garden Seed Inventory, buying up endangered commercial varieties and processing them and getting the Growers Network collection in good shape....Next summer we hope to grow out about five acres of heirloom material. Maybe at the Campout next year, we'll all be able to look at that....

Glenn Drowns Hand-Pollinating a Watermelon to Prevent Crossing

Mark Fox

Once you get deeply involved with anything, you find out that there's a thousand times more to learn than you knew to start with....There are probably over 2,000 different named beans. Kent and everyone else need a way of eliminating the duplicates and also some sort of standard data form for gathering information about them....I want to compile a book that would sort of be a cross between The Beans of New York and the little field guide books which are used to identify trees. Kent and I have also talked at length about doing a large true-color (for a time) wall poster with life-size seed samples of several hundred of the most common heirloom beans. Although it would be beautiful, we feel that its value as an identification guide would be limited....If this book included all of the information and all of the varieties that I wish it would, an encyclopedia would look small by comparison....I envision it as an 8 1/2" x 11" book. Each page would picture a single leaf, a full pod and a cross section of that pod, a picture of a group of the seeds, and also a single seed looking at the hilum and another seed on its side. All of these would be life-size and true color....On the facing page would be the statistical information. This would include: genus and species; plant habit and height; stem characteristics; eight different leaf characteristics. Some blossoms have more than one color, so you would have to give the color for the standard and for the alae. There are nearly three dozen possible pod characteristics such as color, shape (curved, straight, recurved, etc.), cross section (double-barrelled, crease-backed, egg-shaped, oval, round, etc.) and whether the pod is fleshy, fibrous, stringless, or how much string, and whether the spur at the end of the pod is straight, curved, long, short or non-existent; and about two dozen possible characteristics concerning the shape of the seed, its color and the way the colors are patterned. A person could get a bit carried away with all this....(laughter)....A certain amount of this data should be worked up into a standard form which we could all use to record characteristics as we grow the beans out.

There is one problem area (other than flavor, which would be impossible to describe in meaningful terms) and that is trying to describe colors. We all need a standard color system, so that we can describe not only bean seeds but other fruits and plants which we are growing....There are several hundred growers working with the Growers Network collection. If one of them describes a seed as being "ox blood red," that would mean a dozen different colors to that many different growers. "Mouse gray" was the very best descriptive term I ever read in The Beans of New York, as long as we are all thinking about the same type of mouse....(laughter)....There are several standard color systems that artists use which have various types of non-fading paint chips, but most of those are really expensive. I tried to think of something that would be cheap and yet work really well. Now don't laugh, but I think we should all go down to our local Sears store and beg, borrow or steal a complete set of the little strips that they have for paint colors....(laughter)....It would work. They're the same nationwide and are really good quality because each little bar of color is actually the paint. Of course we'd all have to do it the same year, because they're liable to change the colors on us. Seriously, we really do need some standard color system. Because when you read through the Yearbook, all of the members are describing colors that don't mean beans to me....(laughter)....

DARRELL HENNING: Many of the museums use the Munsell Color Book to match colors during restoration work. It's known the world over as a standard system. They have one set of charts which is in book form that costs, if I remember right, about $250. There is another set that has color chips which you can take out and hold up against the object you are trying to match. But I heard that one was in the neighborhood of $400 to $500.

MARK WIDRLECHNER: There's also the Royal Horticultural Society's color chart which is put out in London. We're thinking about ordering one of those at the Plant Introduction Station. Before long we'll know how much that one costs.

KENT WHEALY: About a year ago the Plant Introduction Station at Pullman, Washington published a new list of the data which they are going to use to describe the characteristics of their beans. There were about 70 names included that they were going to use to describe colors. It didn't look like those colors related to a standard color system. It's surprising that the Introduction Stations and all of the university breeders didn't adopt a standard system long ago.

Mark Fox

John Amery

John Amery

I've been in the Seed Savers Exchange about four years, and for three years I've exchanged. At the same time I got into seed saving, I met Rev. Lammers, an old fellow who lives near me. His family had raised soybeans for some twenty or so years, since they'd been missionaries in China and brought out a variety of brown soybean. Every year he raised this brown variety, which we will be having for lunch tomorrow, by the way. Over the years several genetic mutations appeared: a tan-colored variety, a black and a large green. And I have all four of those varieties to pass on to you which came from Rev. Lammers....My interest in soybeans has grown and during the last few years I've taken over most of O. J. Lougheed's soybean collection. I have about 30 varieties from O. J. I obtained about 15 varieties from Will Bonsall this year, which I'll offer this fall. And I also got about 15 new varieties from Illinois University. So I've got about 60 varieties offered now, and I'll have another 30 or so this fall. My goal is to specialize in soybeans. This is a rather narrow group that I can take care of--they're pretty easy. I recommend this to anybody who feels overwhelmed with trying to collect everything. Just pick a narrow group that is especially interesting to you and specialize in that....So my goal is to collect as many north to central U. S. type soybeans that grow from the Pinto region to the Northern--the shorter season types....I've got about 30 or so varieties on display over here, if anyone is interested in obtaining some seed....

Barbara Bond

I'm Barbara Bond. Milton and I live in rural Nebraska City (RFD 1, Box 124A, Nebraska City, NE 68410). I grow approximately 100 different varieties of herbs. I love the prairie and am especially interested in prairie flowers and the medicinals and dyes used by the Indians. Milton tends our half-acre vegetable garden. He enjoys growing crops that were grown by the Indians, particularly the Mandan tribe. We're new Lifetime Members of the Seed Savers Exchange....We have decided to start "The Old Thyme Flower and Herbal Seed Exchange." It will run parallel to the Seed Savers Exchange and will concentrate on native flowers and herbs, and those that were brought to this country by our colonists and immigrants....We've written about the Seed Savers Exchange in the "Nebraska Herb Society Newsletter." Within the Society we are also starting a seed exchange....We're looking for herbs, dye herbs, any of the old-time herbs that can't be found now....When we first started growing herbs, I'd go through every catalog to see what it had to offer. Now I'm trying to locate everything that isn't offered commercially....We would like association or correspondence with anybody interested in herbs and in forming a seed exchange....

****

Everyone helps make the Campout a success, but there are a few people I want to especially thank: Steve Neal, who barbecued about 30 chickens Ozark-style on Saturday evening; Katherine Adam, who transcribed all of the tapes from the Campout; Mark Fox, who took many of the photos in this section; Paul Griepentrog, who brought crates of fresh garden produce; all of the folks who helped in the kitchen (Diane says she couldn't make it without them); and Clarice Cooper who helped with the registration. My sincere thanks to all of them and everyone else who helped make the Campout such a success. --Kent

(The 1984 Fall Harvest Edition, pages 52-88)

The 1985 Campout Convention—My, How This Reunion Has Grown
(Photo by Mark Fox)

Plant Patenting

Cary Fowler

Pat Roy Mooney

THE CONTROVERSY OVER PLANT PATENTING LEGISLATION

For several years now I have been publishing warnings from scientists about the dangers of genetic wipe-out. Last year I told you what I'd learned about plant patenting legislation in England and also the fears of Canadians that similar legislation would be attempted there. In the fall of 1979 widespread grass-roots opposition forced the Canadian government to reluctantly cease its efforts to introduce plant patenting legislation. The general consensus was that plant patenting laws were dangerous and unnecessary, such laws would greatly restrict genetic diversity, would create concentration and monopolistic conditions in the seed industry, that 95% of their plant patents had gone to foreign applicants and that publicly-funded Canadian breeding stations were already doing a fine job. Groups in Canada now hope to be able to stop future seed-patenting proposals before they get to Parliament.

In January of 1979 Sen. Frank Church (D-ID) and Rep. E. de la Garza (D-TX), acting on behalf of the American Seed Trade Association and its 600 member companies, introduced two identical proposals in the House and Senate which seek to extend and strengthen the Plant Variety Protection Act of 1970 (which already affects 224 vegetable crops). These amendments would allow companies the exclusive rights for 18 years to distribute, import, export and use for breeding purposes their "new" varieties. The amendments were unpublicized, and the USDA and the American Seed Trade Association expected no public awareness on the issue and no trouble running them through. They were wrong.

Due to the efforts of a few concerned individuals, the word quickly got out and within a week a huge flood of angry protest mail hit Washington, just as the hearings were starting. At the center of this effort stands Cary Fowler. He is Director of the Resource Center at the National Sharecroppers Fund's Frank Porter Graham Center, has been a consultant to the United Nations' Centre on Transnational Corporations, and a consultant to the Ministry of Agriculture of the Province of Saskatchewan, Canada, on the subject of plant patenting proposals in Canada. Following is part of his testimony at the hearings on July 19, 1979:

(Testimony by Cary Fowler before The House Agriculture Subcommittee on Department Investigations, Oversight and Research on HR 999: An Amendment to the Plant Variety Protection Act)--

The Crisis of Genetic Uniformity.... All the major food crops grown in North America originated elsewhere in what we call the Third World.... Thousands of years ago our Stone Age ancestors began domesticating plants, saving the best seed for replanting the next year. Human efforts and natural selection processes resulted in different varieties of food crops becoming adapted to different niches in the ecosystem. The result was thousands of varieties of wheat, rice, corn and other crops as genetically distinct as beagles and Great Danes. In diversity there was strength. As pests and diseases changed or mounted more powerful attacks, plants evolved different or better defenses....Modern agriculture is changing this natural system....With the breeding and marketing of new "improved" varieties, traditional varieties are being replaced....Field after field is planted with one variety. Where thousands of varieties of wheat once grew, only a few can now be seen. When these traditional plant varieties are lost, their genetic material is lost forever. Herein lies the danger. Each variety of wheat, for example, is genetically unique. It contains genetic "material" not found in other varieties. If, because of genetic limitations which result from inbreeding, new varieties are no longer resistant to certain insects or diseases (even insects or diseases never before known to attack wheat), then real catastrophe could strike....How serious is the situation? The concerns of the scientific community speak loud and clear:

Dr. Jack Harlan, University of Illinois, Journal of Environmental Quality -- "These resources stand between us and catastrophic starvation on a scale we cannot imagine. In a very real sense, the future of the human race rides on these materials....The Mexican wheats have washed over Asia with astonishing speed, replacing major centers of diversity almost overnight....In many areas it is already too late to salvage anything. The line between abundance and disaster is becoming thinner and thinner, and the public is unaware and unconcerned. Must we wait for disaster to be real before we are heard? Will people listen only after it is too late?"

Dr. J. A. Browning, Iowa State University, before Senate Agriculture Committee -- "The situation of genetic uniformity is like a tinder-dry prairie waiting for a spark to ignite it."

National Academy of Sciences, Genetic Vulnerability of Major Crops -- "....most crops are impressively uniform genetically and impressively vulnerable. This uniformity derives from powerful economic and legislative forces."

Dr. Garrison Wilkes, University of Massachusetts, The Bulletin of the Atomic Scientists - "The extinction of these local land forms and primitive races by the introduction of improved varieties is analogous to taking stones from the foundation to repair the roof....The wheat stem rust which took 65% of the durum wheat crop in 1953, 75% of it in 1954, and 25% of the bread wheat in that same year, or the southern corn blight of 1970, are only early warnings of the value of genetic diversity and the liability of genetic uniformity...."

Plant Patenting.... The proposed amendments (HR 999 & S. 23) are designed to drop all legal barriers to patenting six vegetables (tomatoes, carrots, celery, peppers, cucumbers and okra), standardize patent coverage with European laws by extending coverage from 17 to 18 years and thereby facilitate U. S. entry in the International Union for the Protection of New Varieties of Plants (UPOV). UPOV is a Geneva-based organization active in promoting and coordinating plant patenting laws around the world....Patenting legislation is easier to pass than to enforce. It is almost impossible to prove in court that "your tomato is identical to my patented variety." Unlike new mousetraps, plants do not naturally lend themselves to patenting. No two are ever alike. Who should know this better than UPOV, which openly admitted in one of its official publications that "in reality one single plant is not exactly like any other" ....little wonder that enforcement of patents on new plant varieties has become a legal nightmare in the European nations that first instituted the laws.... Undaunted by the constraints of nature, European lawmakers have plunged ahead. Today, member nations of UPOV are busily phasing in a system of "legal" and "illegal" vegetables designed to facilitate enforcement of the plant patenting laws. A "Common Catalogue" has been established listing those vegetable varieties it is permissible to grow and whose seeds may be sold. Seeds of varieties not listed cannot be legally sold....Fines are stiff....

Sir Joseph Hutchinson, Oxford University, England says, "This is where Plant Breeders' Rights legislation is so damaging. For administrative reasons, diversity will not be tolerated. If a man has a right to a reward for his variety, those who have to implement his right must be able to identify what is rightly his. So they insist that his variety must be so uniform that there can be no doubt about it. If this were as far as they were to go, it would not matter greatly. But they also insist that nothing must be grown except varieties that are equally identifiable. Thus we are compelled by law to do our utmost to eliminate all variabilities from the most valuable stocks we have....We are in fact selling our birthright for a mess of pottage...."

Dr. Erna Bennett of the Crop Ecology and Genetic Resources Unit of the United Nations' Food and Agriculture Organization (FAO) in Rome estimates that by 1991, fully three-quarters of all the vegetable varieties now grown in Europe will be extinct due to the attempt to enforce plant patenting laws.... Some sources evidence little concern, contending that the deleted and soon-to-be-extinct varieties are inferior....They dispute the need to preserve these varieties. Others hotly contest such views. Lawrence Hills, Director of the United Kingdom's Henry Doubleday Research Association, goes as far as to assert that some of the varieties being lost are the "Goyas and Rembrandts of the kitchen garden." In his opinion the Common Catalogue scheme is nothing less than a bureaucratic blight on Europe's traditional vegetable varieties....

The proposed amendments to the Plant Varieties Protection Act would standardize our laws with Europe's....Is there a concrete reason to believe that our laws will not ultimately have the same disastrous results as similar laws in Europe are now having?....Will U. S. entry into UPOV encourage or facilitate further marketing of new varieties in the "gene centers"?....Are multinational corporations taking responsibility for the traditional varieties their new models displace? Are they seeing that these traditional varieties are collected and stored?....Is the U. S. government willing to take responsibility for seeing that this is done? (The present budget for plant collection is approximately \$35,000.) Is the U. S. government willing to take the responsibility for it not being done?....Who benefits from patenting? Very few. A study by Hope Shand of the Graham Center in North Carolina found that of the 73 patents issued for beans (as of March 1979), 79% are held by just four companies: Sandoz, Union Carbide, Upjohn and Purex. Sandoz and Upjohn hold 43% of the patents issued for peas. Two-thirds of the patents on lettuce are held by Upjohn, Union Carbide, FMC, Purex, ITT and Celanese. Nearly a third of the patents on wheat are held by five corporations. Eight corporations hold 42% of the patents issued for soybeans. Four corporations hold 44% of the patents issued for cotton.

Meanwhile, the lucrative market for patented seeds has created something of a black market in seeds. According to a consultant with the Mexican government, one major U. S. grain company obtained certified seed from the Mexican government and proceeded to patent it in Europe. And according to Dr. Glenn Anderson of CIMMYT (the International Corn and Wheat Improvement Center) in Mexico, companies are trying to patent varieties developed by CIMMYT from research that is currently free for all to use....

Proponents of the Plant Variety Protection Act and amendments HR 999 and S. 23 offer two major reasons for their position. First, they claim that plant breeders have a "right" to patent their new inventions. But let us not forget that literally hundreds of generations of our ancestors domesticated and developed our food crops over thousands of years. Does a plant breeder coming along in 1979, making a few changes here and there, really deserve the "rights" to the new variety as if it were his/her own?....Do the rights of plant breeders supersede the rights of the public to a competitive seed industry devoid of conflicts of interest with the agricultural chemical industry? Do the rights of plant breeders supersede the right of the public to have the genetic wealth of our agriculture safeguarded and the future of agriculture secured?.... The second major reason proponents of the amendments cite for support is the claim that the amendments will encourage the development of new varieties. We are unaware of any evidence that would support this claim....If such encouragement is needed, it can be accomplished by increased funding of government breeding programs at land grant institutions....

(Cary Fowler's testimony was 14 pages long and very complete. It poses questions and offers accurate statistics that you won't find elsewhere (example: which corporations are presently buying out independent seed companies here in the U. S.). I had hoped to bring you much more of it, but space will not allow. However, Hope Shand will send you a copy of Cary's testimony for $2 to cover duplicating and mailing. She said they had hoped to not have to ask anything for it, but their budget is very tight. The address is Rural Advancement Fund, PO Box 1029, Pittsboro, NC 27312 - Kent.)

****

<u>Testimony of Gary Paul Nabhan, Agricultural Botanist, for the Congressional Record, concerning HR 999, a Proposed Amendment to the Plant Variety Protection Act of 1970</u>--

I am a concerned scientist who has spent the last five years of my professional career involved with plant variety protection....I consider the Plant Variety Protection Act of 1970 to be inappropriately named--it protects big business, not the plant varieties which are crucial but threatened resources....I see such restrictive legislation as putting into private control resources which have developed and should continue to develop as part of the public domain. Such legislation gives credence to breeders who manipulate one or two genes of a traditional land race, then "lock up" the genetic resources of a crop which has evolved over thousands of years. Privately employed plant breeders can even cross two varieties derived from many years of genetic research done by the USDA or universities, then patent the hybrid as their "own creation"....to let a private company control for eighteen years genetic materials that were developed largely by others is to inequitably favor such opportunistic businesses ....

The 1970 Act and this proposed amendment are nightmares to enforce, since it is so difficult to technically define and verify what plant varieties are....One method utilized in empirical plant variety identification is electrophoresis or "plant fingerprinting"....problematic and costly....$5,000 estimate of costs to properly identify the material and to rule out alternatives....if a small seed company must pay $500 to file an application, $250 for certification of protection, then up to $5,000 if one of their cultivars is contested, plus legal and operational costs, the entire process becomes too expensive for small seed companies ....While Senator Frank Church and others suggest that the 1970 Act has given plant breeders incentives for producing new cultivars, there is no evidence that the Act has served to increase the number of new varieties available per year. To the contrary....the total number of newly recognized varieties attained a high point in 1968....the reason for the decrease in new varieties released since 1970 is that regional seed companies have been consolidated and bought up by multinational corporations that are not interested in diversity....the 1970 Act has encouraged a "blackout" in regard to scientific information on plant breeding projects. Since 1972, not a single private agribusiness breeder has published descriptions of breeding schemes or techniques for their new varieties in the "Cultivar and Germplasm Releases" section of HortScience, the most popular outlet for such information among university and government breeders. Once there was free exchange of data and crop materials between the public and private breeding programs....

I am also opposed to any attempt to alter the provisions of the 1970 Act to be in accord with UPOV. Other UPOV members have actually made the growing of certain valuable traditional varieties illegal, further restricting our genetic base. America does not need this. Rather than attempting to economically stimulate the proliferation of new and superficial "pseudovarieties," we should have as our priority the conservation of ancient but now endangered traditional crop varieties. These basic genetic resources are going extinct faster than we are collecting and preserving them....The U. S. government does not see the urgency of supporting collection programs, and the National Seed Storage Laboratory did not even receive a budget increase during its first 15 years of existence. The USDA has not even had one full-time plant explorer since 1970 ....Basic germplasm conservation is critically underfunded and understaffed in this country--it is an essential job that has taken second place to more superficial concerns. I suggest that Congress consider full-scale hearings on ways to upgrade our national commitment to the conservation of basic crop germplasm--the key resource that stands between us and hunger.

****

In a personal letter from Cary Fowler, August 10, 1979 -- "The hearings went well. They had originally planned one hour of hearings to be immediately followed by a vote. It turned into three hours of hearings with no vote taken, which is good because we would have lost it. The impact of the letters of protest coming in from around the country was terrific....there were only eight others testifying (six in favor, two opposed). The testimony of those in favor was quite weak I thought, though I suppose that when ITT and Union Carbide say anything, it is said loudly--those were the only two 'seed' companies that testified....The USDA essentially came out and confirmed that (1) there is a serious problem with genetic resources, (2) Europe's laws are a disaster, (3) we are patterning our laws after theirs, (4) but they insist that just because our laws will be like Europe's doesn't mean that they will have the same effects, and (5) they are not aware of any consolidation in the seed industry....We're not quite sure what will happen next...."

****

The following was reprinted with permission from "The Patented Seed" by Jennifer Bennett, Harrowsmith, September 1979):

"....Although the 1979 Act included all non-hybrid grains, field crops and flowers, a puzzling list of vegetables was omitted: okra, celery, peppers, tomatoes, carrots and cucumbers. It reads suspiciously like the ingredients on a vegetable soup label....Sure enough, lobbyists for H. J. Heinz Co. and the Campbell Soup Co....made it clear that they would oppose any legislation affecting their soup vegetables....The lopsided coverage of the 1970 Plant Variety Protection Act showed, if nothing else, that big business was concerned with seeds and that it had the clout to mold legislation as it wished....

"....In the U. S., where seed patents have now been legal for 10 years, research has neither increased nor become more dynamic. In fact, a 1978 report by the National Academy of Sciences states that 'Between 450 and 500 new cultivars are released each year. Most represent minor genetic advances and "fine-tuned" adjustments to changes in production, harvesting, processing and marketing procedures'....

"While it is possible, through an intelligent programme of seed breeding, to increase a plant's resistance to disease or even to pests—many primitive crop ancestors have 'built-in' protection--it is doubtful that a corporation dealing in fungicides or pesticides would devote a great deal of time and money to seed research in that direction....

"....Hybrids are profitable because the farmer or gardener must return yearly to the seed company for more seed, whereas he can save the seed of a standard variety....while hybrid corn is often markedly more productive, standard tomatoes and peppers may be little different from their hybrid counterparts....'We've tested a thousand or so different hybrids, but none of these are equal to the best standard varieties of tomatoes,' says Dr. Ernest Kerr of the Simcoe (Ontario) Research Station....

"....Plant material or 'germplasm' is vital to any breeding programme, with success often hinging on some highly unlikely parent stocks. These can include forages not far evolved from their 'weed' state or small, wild tomatoes that sit at the base of 'Ultra Boy's' family tree. Similarly, a bean variety that lacks certain desirable characteristics, but that possesses other strong genetic traits, may be used in crossbreeding to create a marketable variety. Many of these breeding varieties never make it to market themselves, and are often unnamed, identified by a CO number only. Those who oppose plant patents in the U. S. express fears that breeders, seeing potential profits and ways to cut competition, might register plant material on a wholesale basis...."

****

The following is from an extensive document entitled "Seeds," put out by the Saskatchewan Council for International Co-operation, Huston House, 2138 McIntyre St., Regina, Saskatchewan, Canada S4P 2R7:

Seeds - Genetic Wipe-Out and 'Plant Breeders' Rights' - With the isolated exceptions of Jerusalem artichokes and a few berries....all that we eat comes to us from one of the nine Vavilov (their discoverer) 'centres of extreme genetic diversity' in Asia, Africa or Latin America. A combination of ten millenia of agriculture and a diverse topography have given key food crops in these nine centres fantastic variety. Now in the 1970's, the Vavilov centres are being brought to the edge of extinction by global companies and the green revolution. 'Genetic wipe-out' puts the whole food system at risk and places First and Third World farmers in the hands of just a few multinational agrichemical companies....

"In the first week of Plant Breeders' Rights legislation in England, one company, RHM, bought out 84 seed companies, and when the smoke had cleared, over 100 companies had merged. Plant Breeders' Rights legislation has resulted in a massive buying spree by major agrichemical firms. Suddenly, Royal Dutch Shell has become the world's largest seed and agrichemical company....

"Dr. J. K. A. Bleasedale of England's Wellesbourne Station describes the Common Catalogue as a 'self-inflicted wound' and confides that the Catalogue is inspired by commercial interests and is intended to clear the European market for the patented varieties....

"These are issues too important to be taken from the People. Among these are the conservation of the world's plant genetic resources and the struggle to guarantee humanity a safe and nutritious food supply through an effective public breeding programme. These are issues which cannot be trusted to the profit motive. If you control the seed--the beginning of the food system--you can control the entire system. The government has no right or reason to turn over control of the Canadian food system to multinational agrichemical companies...."

****

Manchester Guardian Weekly (England) - "....When huge acreages are planted with a single variety or a few closely related varieties, entire harvests can be wiped out by one disease, by the temporary resurgence of an insect pest, or by changes in climate. At a time when the heavy use of insecticides is constantly upsetting predator-prey relationships and changing the mix of insect populations, and when increased fossil-fuel consumption is thought to be leading to world-wide climate change, the planting of fewer and fewer varieties is particularly ominous....As one U. N. official put it, 'When farmers clear a field of primitive grain varieties, they throw away the key to our future.'"

From "Seeds of Trouble," a November 11, 1979 editorial in the Washington Post - "....Access to and information about the contents of corporate seed banks is generally limited to employees of the particular company. Plant breeders in

Third World nations are thereby excluded from access to vital resources that came from their own countries to begin with....A recent international study recommends a U. N. emergency fund to collect and maintain genetic reserves in a number of regional banks (a single bank is too vulnerable to biologic accident). There are suggestions that no laws allowing exclusive rights to plants be allowed...."

Bob Bergland, Secretary of Agriculture - "The prediction that by 1991 three-fourths of all vegetable varieties grown now will be extinct may prove to be correct. Obsolete varieties are being replaced by improved varieties adapted to extended areas of agriculture."

In a letter from Sen. Frank Church, who introduced the Senate Amendment - "....There is absolutely nothing in this legislation which would make it against the law to grow garden vegetables. Neither does this legislation have anything to do with promoting uniform varieties of plants....This legislation bears no resemblance to what is done overseas....Plant variety protection has worked to protect the new varieties developed by the small seed companies against exploitation by the bigger seed companies....The results of the 1970 Act have been beneficial. The numbers and types of varieties of plants have expanded....Competition to breed new seeds has apparently increased....Be assured that I would not introduce legislation to outlaw certain seeds, banish backyard gardens, force genetic uniformity and hand over big profits to a few seed companies...."

In a Nov. 8, 1979 letter from Rep. George Brown to Rep. E. de la Garza who introduced the amendment in the House - "....I have been surprised at the outpouring of opinion directed at H. R. 999....I have received more mail on this issue than on any agriculture issue in recent memory. I was even approached on this issue at the U. N. Conference on Science and Technology in Geneva, which I attended in August....I have been concerned about the state of our germplasm research and preservation program for some time and am, quite frankly, distressed by the relatively low priority afforded to this area. This is the kind of program that offers no immediate, direct payback and thus suffers during periods of declining research outlays such as we are currently seeing in agricultural research. Yet it is an essential area if we are to continue to enjoy a position of leadership in agricultural production....The 1972 National Academy of Sciences study, 'Genetic Vulnerability of Major Crops,' raises a number of issues and specifically cites the Plant Variety Protection Act as a policy encouraging genetic uniformity....The European experience with plant variety protection also raises some questions, especially with the predictions, confirmed by Secretary Bergland, that three-fourths of the varieties now grown would become extinct under the program employed there....I feel, as a cosponsor of H. R. 999, uncomfortable moving this legislation until some of these issues have been resolved. Perhaps the time is right for our Subcommittee to take a look at the state of our germplasm research and preservation programs...."

****

I just called Hope Shand, Cary Fowler's assistant. She said....there will be more hearings, probably in March, but they could come at any time....The American Seed Trade Association is currently spending huge amounts on a massive advertising and public relations campaign designed to turn attention away from the issues....They are saying that Dr. Erna Bennett's quote (by 1991 fully three-quarters of all the vegetable varieties now grown in Europe will be extinct due to the attempt to enforce plant patenting laws) is not true....that she was only talking about "synonyms" (similar varieties having different names)....We called Dr. Bennett in Rome....she was dismayed at the twisting of her words and said she has never had any communications with the American Seed Trade Association....They also came out with statistics to prove that patents have been widely distributed with only a small percentage going to large corporations....I made the study for the statistics in Cary's testimony and at first I thought I had made mistakes....but after rechecking the patents and my figures carefully, it is quite apparent that they are literally fabricating statistics to prove their case.... we are very concerned because it is difficult to counter such tactics.

(The 1980 Seed Savers Exchange, pages 49-54)

****

Pat Mooney, quoted in The 1979 Seed Savers Exchange—....This fall, Canada may establish a "Patent" system for seeds similar to U. S. and European legislation. We are deeply disturbed by the Common Market's recent "Common Catalogue" Agreement which, plant breeders tell us, will wipe out three-quarters of Europe's vegetable varieties by making their sale illegal. Voluntary groups led by OXFAM are now struggling in the three-year grace period before 1981 to find and preserve doomed varieties.

Saskatchewan Council for International Cooperation is part of an international network of international voluntary agencies in most industrialized countries. With our counterparts overseas, we hope to raise our concern for "genetic wipe-out" with our own governments and with the United Nations Conference on Science and Technology to be held in Vienna next year.

One way to draw public attention to this issue might be a kind of modern "Johnny Appleseed" campaign, distributing thousands of varieties of seeds to concerned people and asking folks to grow them in their backyards and farms. We are also thinking of approaching City and Provincial (State) Governments to ask them to set aside a small plot of ground on ceremonial lawns and to hold well-publicized planting programs/events. A similar initiative may also be taken with schools....

(The 1979 Seed Savers Exchange, page 34)

****

Plant Patenting Legislation Passes

NOTICE: Shortly before Congress adjourned, the Plant Patenting Amendment passed the Senate and became law. It reportedly passed the lame duck session on a voice vote with few members present. Our work is now cut out for us. I will immediately start working full-time to inventory vegetable varieties, distribute endangered varieties through our new Growers Network and secure samples for freezing in our Heirloom Seed Bank. Your help with and support of all of these activities is desperately needed. These vegetables will either die or live on with our help.

(The 1981 Winter Yearbook, page 49)

PAT ROY MOONEY

(Speaking at the "Seeds of Yesterday and Tomorrow" Conference on February 12, 1984 in the Civic Garden Center in Toronto)

When Gene Whelan first suggested that he would back plant breeders rights legislation in a meeting in Winnipeg in 1977, he said that the reason for the legislation was to help feed the hungry. Pretty sweeping statement. And Wilf Bradnock, who is the Director of the Seed Section of Agriculture Canada and spoke at the same meeting, went even one better than Whelan. He said it was first to help feed the hungry and secondly to make a more beautiful Canada. Such hyperbole was often part of the government's rhetoric back in the 1970s when it first proposed plant breeders rights and before it was controversial. We've gone from that to now saying that if we don't pass the legislation, there may be some plant material from overseas that someone in Nova Scotia may be unable to obtain....The government's transition in thinking shows a major aspect of how government reaches policy decisions....In the real world of politics, governments reach political decisions first, and then justify them by all kinds of rational thinking and rhetoric. In fact, when the government first proposed plant breeders rights legislation, it assumed it was simple housekeeping legislation requiring little justification to anybody, and so hyperbole was in order. It was quite possible to make sweeping generalized statements about feeding the world's starving millions and making a more beautiful country and so on. That was all safe to say because they were only talking to the seed trade anyway. But when it became more debatable, we see the government constantly falling back into other possible positions. It's been an amazing progression over six or seven years now to see them boiling it down to the seed trade finally saying, "Well, really, it won't make much difference anyway. Why don't you just let it go through?" Not wanting to even say this is particularly good for the country, but not wanting to say it's going to be bad for the country either....

Why should we as Canadian taxpayers subsidize multinational corporations, which is exactly what is done when the state creates a monopoly for anybody. It's a form of government subsidy where we decide that we're getting something out of it and therefore we should be prepared to give somebody else an 18-year or 25-year or 30-year period of exclusive monopoly control over that variety, because we get something back. Let's look at what we are actually getting out of it....we should at least get something back for research. But that opportunity has been with us now for more than 50 years. Agriculture Canada, when it analyzes plant breeders rights, constantly refuses to look at the first major plant patenting law in the world, which was passed in the United States in 1930 (which covered asexually produced plants—ornamentals, fruits and so on). We have had over 50 years here to look at what that legislation did for the world and what impact it had. I can't find anybody who's looked at it in the government. But what's fascinating about that 1930 law passed in the United States....was that it didn't require reciprocity. Anybody could patent anything they wanted to in the United States, and I mean literally anything. You could just trip over it and get a patent on it. As long as it didn't move, you could patent it under the American law. And that was also available to us in Canada. There we were on the borders of the largest market in the world, on the borders of a country that didn't require reciprocity and where the patent approaches were extraordinarily simple. And what advantage did we take of that? Out of about 4,500 patents over the last 50 years, there are about a dozen or so that have come from Canada. It hasn't exactly stirred us up. It hasn't exactly been incentive for us, to export or to patent in another part of the world, at all. It had no impact on the nursery trade or plant breeding industry in Canada. And I don't think that the passage of plant breeders rights in this country will in any way be an incentive to get anything back in terms of patents in other countries ....there are always loopholes so that we could patent in another country if we wished to. There are many ways around that. Just establishing a post office box in another country often solves the problem of reciprocity....Several Canadian companies now have their own subsidiaries in the United States and are patenting there under the Plant Variety Protection Act of 1970. The United Cooperatives of Ontario are part of the FFR Cooperative network in the United States and they're patenting soybeans there like crazy. And they get the benefit from it here as well. So I think it's simply not accurate technically to say that the export markets are denied to us because of a lack of legislation in this country.

The government often argues that we will have trouble getting material into Canada if we don't pass plant breeders rights legislation....McKenzie Seed, which is based where I live in Brandon, Manitoba, tells me that they are importing about 1,200 vegetable and flower varieties every year, almost all of which are patented from other countries. I asked them what their problems were and they said none....a couple of years ago, the seed trade in Australia made similar impassioned pleas to the Australian government, which also does not have plant breeders rights. They said that Australia would be cut off and have no chance of being part of the world agricultural system if it didn't get on the bandwagon and pass plant breeders rights. They claimed that Australia was in desperate need of all kinds of varieties from overseas. It was being told that by the seed trade, but it was also being told that by the American government. Bernard Lesse, who is the head of the U. S. Plant Variety Protection Office in the United States, took three trips to Australia to make speeches about how poor old Australia was going to die on the vine because American varieties couldn't come into the country since there was no protection for the American varieties in Australia....Americans have a very nice law called the Freedom of Information Act. Forces within the U. S. government are trying to destroy the law right now, but it allowed us to track down Bernard Lesse's travel program, exactly what he did, and especially his <u>justification</u> for travel. (It's a great law, by the way. I really encourage you just for fun to occasionally write away for that kind of information. They sent me everything, including his laundry bills.) ....The justification for Bernard Lesse's travel to Australia to argue for plant patenting in Australia, as he reported it to senior officials in the USDA, was to get rid of the quarantine barriers in Australia which were preventing the import of American varieties. Not a single solitary reference to any barriers related to access because of plant breeders rights. The quarantine barriers were the problem. That was why Australia was having some difficulties bringing in some American varieties. If you look at it around the world, the same thing is true. These are quarantine problems, or regulatory problems....

The basic truth is that if you've developed a product and you've only got a few years to market that product, you're not going to turn down sales. That's dumb. Companies aren't usually stupid. Canada and Australia are a pretty marginal market for the world. The big market areas are the United States and Western Europe. If we need something that they've got or they think they can sell it to us, we'll get it. I've been to Germany where I've talked to seed companies who are selling seed to Canada. They'd prefer that we had plant breeders rights because their profits would be higher, but they're still bringing the seed into Canada through companies here....So I think that argument is a smoke screen. It's often a threat used, not so much by Agriculture Canada as by some of the major multinational companies, to try to pressure us with the fear that we'll be left out in the cold with the lights turned off if we don't pass this kind of legislation....

Companies in our society demand their right to a profit....seed companies have always been profitable, with or without plant breeders rights. They ranked second only to the pharmaceutical industry in the United States in the 1960s for their profitability....but there is a huge difference between their right to a return on their investment and exclusive monopoly control....we're not simply saying give that man a royalty for that variety. We're saying let that person set the conditions of sale for that variety. And that's a huge difference....With that ability, they can vertically integrate into other parts of the food system and into other parts of the seed system. They can manipulate the marketplace. They can withhold, play games, force retailers to take some of their lousy varieties or some that they have a backlog of in their warehouses, in order to obtain the good varieties. They can demand certain levels of advertising....You can have my one good one, my LG11 maize in France for example, as long as you take these other three major varieties which are losers since I've got a stockpile of them, and you must guarantee them this certain level of budget for advertising. And because the retailer needs to have that one popular variety, you get the other ones as well....And the Commission for the European Common Market concluded that the effect of plant breeders rights in Europe was that it led to vertical integration by multinational companies, first into the seed system and then right through it to control all the levels of seed multiplication and distribution. And that's not related to royalty. That's related to the exclusive ability to set the conditions of sale for the variety....They want exclusive monopoly control. They're not asking just for a profit. They're asking for a monopoly. And there's a world of difference.

Government officials also argue that the exchange of germ plasm is enhanced by plant breeders rights legislation. This is just simply incorrect....Let me just give a few examples....The Irish government passed plant breeders rights legislation in the summer of 1980. The Irish government has, for better or worse, a gene bank in Dublin where they have been collecting material for several decades. Now, when the United Nations Development Program (UNDP) and the

International Board of Plants Genetic Resources (IBPGR) and the FAO got together to do a survey of European gene banks, they asked each gene bank what their policy was on the exchange of germ plasm between countries....The Irish statement was particularly revealing because they're a little naive at times in their politics. The director of the gene bank wrote as his response, which was published by UNDP, that only countries that have plant breeders rights that are members of the Union for the Protection of New Varieties of Plants (alias UPOV) will have access to material in our gene banks....Canada has given material to Ireland for their gene bank. Nearly every country in the world has given genetic resources to Ireland. They only passed the law in 1980. It's not like they're only restricting the material they've obtained since 1980, it's everything....They were just naive enough politically to say that in public. Other countries do exactly that but won't put it in writing. Of the European gene banks that replied (and very few of them replied), most imposed some restrictions to the access to the material in their gene banks. It's also well documented that many of the gene banks in Europe, because they've had the legislation there for longer, offer an embargo clause to any company that gives any new material to them. They will embargo access to that material for five or ten or fifteen years, depending on the request of the company. So yes it's in the public sector, but they're simply acting as a holding station for the material the company has, and no it is not part of a free exchange program, and yes that's due to plant breeders rights.

The entire International Agricultural Research Center constituency have all said bluntly and clearly that they are opposed to plant breeders rights because it stops the free exchange of germ plasm between countries. These major research bodies, which are doing plant breeding for the Third World, include the International Maize & Wheat Improvement Center (CIMMYT) in Mexico, the International Rice Research Institute (IRRI) in the Philippines, the International Crops Research Institute for the Semi-Arid Tropics (ICRISAT) in India, the International Centre for Agricultural Research in Dry Areas (ICARDA) in Syria, the International Centre of Tropical Agriculture (CIAT) in Colombia and the International Potato Centre (CIP). In the summer of 1982, at a meeting of the directors of all these international research centers, they said that it was true that plant breeders rights increases the pace of genetic erosion, because of the marketing pressure. Agriculture Canada still says that's not so. But there is a whole crowd of scientists out there that think it is. And they're not politicians....

Then we get to the question of transnational corporations in all of this....I can list 839 seed companies that have been bought up by international companies since 1970 around the world, most of them in Europe and North America and a good number of them in Canada. And that concerns me. It concerns me just in the level of corporate size....I can look around at seed companies that were well established—Ferry-Morris, for example. First it was owned by Purex Corporation, the people who make bleach. Then Purex got rid of it because they couldn't make the profit on it that they wanted, so they sold to Limagrain which is a French company. But now I'm told that Limagrain is in trouble and may have to sell it as well. What happens to the breeding program when those kinds of international companies take over, play with it like they do almost everything else, get tired of it or can't make the money on it that they thought they could, and then jettison it? I don't think that helps plant breeding. I don't think it helps agriculture very much....The crowd of multinationals coming in and out of the seed business is pretty frightening. Some companies have been bought up to three times in the last decade. That just about destroys their breeding programs. It certainly disheartens any of the plant breeders that are involved....And that worries me very much.

It especially worries me in the context of who those guys really are that are doing the buying. The largest seed company in the world today by far is Shell Oil (Royal Dutch Shell). They only got involved in the early 1970s and at last count they owned 74 seed companies around the world. Their profits from sales of seeds are probably on the order of $700 million worldwide. According to the British Trade, they have 40% of the British cereal seed market and about a quarter to a third of the British garden seed market as well. But they're also heavily into Western Europe; through North American plant breeders they're very heavily into the United States and are looking for opportunities in Canada as well....two Swiss pharmaceutical houses located across the street from each other in Basel, Switzerland are also heavily involved. One of them is Ciba-Geigy, whom you know through throat lozenges and Green Cross pesticides, and the other one is Sandoz, whom you know through Ovaltine and a lot of other drugs. Ciba-Geigy now controls 27 or so seed companies and Sandoz is up into the high 30's....Other major companies include Stauffer Chemicals, Reichold Chemicals, Pfizer Genetics, British Petroleum, Atlantic-Richfield, Occidental Petroleum....All of these companies you either know from the area of petrochemicals or from what's in your bathroom cabinet--pharmaceutical companies....why is it that they're involved?....There are at least a half a dozen major reasons why these companies got involved in the early 1970s. Some of them are pretty straightforward, logical decisions such as having a surplus of cash and seeing agriculture as a long-term hedge against inflation. Whatever their reasons were initially, when pesticide companies become plant breeders, the realities are scary....Let's talk about that a little bit....

Back in the late 1970s, Gene Whelan and others said that it was absolutely absurd to suggest that a plant breeding institution would take the time and go through all the arduous work necessary to dovetail the breeding of a new variety with the development of a certain pesticide. That was the 1970s. In 1980 the president of Pioneer Hi-Bred, Thomas Urban, was asked by the New York Times, why it was that multinational chemical corporations were buying into plant breeding. His answer was, "The assumption behind the trend is that the new owners can improve the plants' resistance to the herbicides and pesticides that the parent company sells." And the same story came out last year from the Economic Commission for Europe when they were looking at the future of the pesticide industry in Europe. They pointed to the "great pleasure"—they really put it that way--of plant breeding companies being able to turn over concern about disease resistance to the pesticide companies. And went on to add that now that the two are together, it helps....It's more than simply a coincidence that Stauffer Chemicals had a dominant position in the maize pesticide market when it bought out three maize seed companies in the United States in the same market area where they sell their pesticides. It's also not just a coincidence that Reichold Chemicals, which had a major interest in lawn and garden pesticides, bought out lawn and garden seed companies. Their annual report even said that they were bringing the two institutions together in Florida to dovetail their research. And it's not simply a coincidence that KemaNobel, of Sweden, the largest manufacturer and seller of vegetable and fruit fumigants and fungicides, would decide to get into plant breeding and seed distribution and has bought out the largest network of garden seed distribution companies in Europe. KemaNobel is now Western Europe's largest marketer of garden seeds. It's simply not a coincidence that when Ciba-Geigy decided to get into plant breeding (next to Stauffer Chemicals they are the only company you could identify as having a dominant position in maize herbicides), they bought out Funk Seeds, one of the world's largest maize breeding companies. In fact, you have to work hard to find exceptions to that around the world.

We should be concerned that these companies will jointly advertise their crop chemicals and their seeds together to encourage farmers and gardeners to use more of the crop chemicals. When I first said this in 1978 and 1979, I was told that that was simply stupid and nobody would do something like that. The farmers wouldn't stand for that kind of thing. They're too smart for that. I don't at all question farmers' intelligence, but I know how the marketplace can advertise. In 1983 Shell Oil, one of the companies who had said most clearly that this would never happen, started to advertise rather unusually in German farm magazines. In the beginning of the year you had an advertisement for Shell's maize varieties on the front page and then a few pages later on you found an ad for one of its herbicides. By the end of the year they had changed the name of the company (before that they went by the name of Knickerson's for the chemicals and the name of Shell for the seeds) to Shell-Agra and the one ad was offering both chemicals and seed. Try this herbicide with this seed variety. So what was ridiculous four or five years ago, is now reality....

Companies can also make money by pelleting the seed. They can wrap the seed and the chemicals together, so as a farmer you're getting an option you can't refuse. You bought the two in a package. We were also told back in the 1970s that the farmers wouldn't stand for that. But in the 1980s Ciba-Geigy was marketing hybrid sorghum seed in the Sudan that comes wrapped in three chemicals. Two of the chemicals are there to protect the seed against things like Johnson's grass and so on. The third chemical is there in the packet solely for the purpose of protecting Ciba-Geigy's seed from Ciba-Geigy's leading herbicide Duo, which is normally toxic to the seed. In other words, the whole process allows them to increase the sales of the herbicide, besides getting a higher price on the product. Business Week magazine a couple of weeks ago had a big front cover on genetic engineering in the United States and they gave a litany of examples of research companies working to dovetail the research and make their herbicides nontoxic to the varieties, so they can sell more chemicals. What was thought to be impossible two or three years ago is happening now. It's still early in their programs and it's going to happen more in the years to come. Many of these companies have only been involved in the seed industry for seven or eight years. It takes longer than that to even develop a new pesticide or to develop a new cereal variety.

But the real dangers aren't in those areas. The real danger begins when a company has the dominant position in a market as a plant breeder, as a seed provider, and that same company also produces the chemicals for that crop. When a disease attacks the crop, you've got two options if you're that company. You can find an organic solution to the problem, a plant breeder's solution to the problem, and breed around the disease, or you can refer farmers to the chemical on the shelf. You can make money by doing nothing at all. You don't have to plot. You don't have to strategize. Al you have to do is wait! Now, has that happened anywhere? Yes. But the truth is, it's really too soon to see those implications for the reasons I mentioned a minute ago about the time it takes to develop all of this. We've got a long way to go before the real implications of this will come out. But we are already hearing from the U. S. government plant breeders about the problems of insect resistance in wheat in western United States. In a statement by the Office of Technology Assessment in the United States, they made it quite clear that they had assumed that with the passage of plant breeders rights legislation in the United States, the private companies would move in and handle all the problems of pest management for them. So the public simply withdrew its activity in that area. They are now realizing that the companies didn't do it. What that means is increased sales of seed because you've got to develop new varieties if your seed is being destroyed, and vastly increased sales of chemicals....

Most government or seed trade spokesmen will argue that one of the positive benefits of plant breeders rights legislation is an increase in research dollars. We need to have new investment in plant breeding to meet the new challenges of the future and the only way you can do that is by going to the multinationals. They'll put up the bucks to do the research, and as long as we have good governmental institutions to monitor them, there's really no problem, folks, you can all go to sleep. That's the argument. Well, I tried to find out where that increase was, where all this new money was, and I can't find it. I've got all the annual reports and I've got all the reports from the Securities and Exchange Commission and I can't find the dollar increase over the last decade....Sandoz, the Swiss pharmaceutical company, has been circulating reports that show....in ratio to dollars of sales, the research dollars invested by the seed companies was 2.9% in 1950. By 1965, 15 years later, the investment was up to 3.6% of sales. By 1969, four laters later (don't ask me why they picked these particular years) the research had gone up to 4.4% of sales. So from 1950 to 1969 research expenditures increased from 2.9% to 4.4%. And then by 1976 it had shot up to 5.3% of sales. They attributed this increase to the Plant Variety Protection Act of 1970 being passed in the United States. This sounds pretty impressive. A major increase in expenditures on research thanks to plant breeders rights. But when you actually look at the periods, you find that the research rate remained at about 0.14% per annum from 1950 to 1965. It went up from 1965 to 1969 to .2% per annum and dropped down to its lowest ever, .13% per annum, after the legislation was passed. In other words, the rate of investment declined following the passage of the legislation. Well, why did it go up in the 1960s so much? There is a very simple, clear reason. In the 1950s they found an opportunity to create hybrid wheat. A whole crowd of companies got involved in plant breeding, because they thought they could hook the farmer on hybrid wheat and make them go back to the market place every year. When they found they couldn't do that, investment actually started to drop off.

When there is new capital in the seed industry, it comes from buying out other companies, which you always have to do at a higher price than the company was previously assessed as being worth. So a lot of people were buyng out a lot of companies, and all that newly created capital goes into something that seed trade officials call "new investment in new development." But it's simply take-over and is of no benefit to anybody. Period....Let me give you two examples of double-think by the seed trade that I find particularly entertaining. One was a British argument to prove that plant breeders rights is really useful there. They stated that between 1973 and 1978 the amount of dollars going into investment in the seed industry in the United Kingdom went up by 500%. And all the parliamentarians just rolled over and said, "Wow, that's great!" Well, there are a couple of questions to be asked about a figure like that. For example, the bill was passed in 1964. What happened between 1964 and 1973? Why was it so special between 1973 and 1978? Why not give us the whole story? What else happened in 1973? Well, of course, petro-dollars came into the British market from overseas in 1973, that's true. But the real thing was that the British joined the Common Market in 1973. Suddenly the British seed trade was wide open to the whole of the Common Market. There were no barriers of any kind. And the British have twice the land size per farm of most European countries that are in the Common Market, so they're really an easy market for the seed trade. And the British have mildew problems, so they don't save their seed nearly so easily. They're back in the marketplace more often, which makes them especially juicy. So in 1973 a lot of foreign companies started to buy into the British market. And that's what happened in 1973. That's why there's a 500% increase. Nothing whatsoever to do with plant breeders rights of 1964. But it makes you kind of wonder, when the trade uses figures like that.

Similar claims about the legislation have come from the United States. Harold Lowden (one of the most interesting main characters I've met in the seed trade), Executive Director of the American Seed Trade Association, has said over and over again that the final proof that plant breeders rights works in the United States is because of soybeans. We now have a crowd of companies involved in soybean breeding in the United States that we never had before, and that's all happened since 1970. You can point to the year 1970 when the Plant Variety Protection Act was passed and say everything has changed since that year. And he's absolutely right. Everything has changed since that year. But it has nothing to do whatsoever with plant breeders rights....

In 1970, American exports of soybeans went up 800%. Land acreage shot up 400%. To argue the seed trade was instrumental in that in 1970 is absurd. The market situation changed dramatically for soybeans in 1970. If you go back to Business Week magazine and the Wall Street Journal in 1970, you find that soybeans were a fairly major part of their conversation during the year. But they never once mentioned plant breeders rights. The reason soybeans had suddenly become a Cinderella crop in the United States was—and you'll enjoy this, I think—because of earthquakes in fishing villages along the coastline of the Andes. These are the fishing villages that go out after anchovies which provide a protein supplement that competes with soybeans. The fishing villages couldn't get out there that year....That whole supplement market died away. Almost simultaneously the groundnut industry in West Africa collapsed. Why was that so special? Groundnuts were becoming competitive with soybeans. Because of their Common Market connection through France, West Africa had easy access to the European Common Market for their groundnut protein supplement. But in 1970 that period of trade protection ended flat-out. The market collapsed completely and soybeans swept in from the United States. Somehow everyone forgets all that. Somehow it's never mentioned. And all we hear is yes, there's been a major increase in plant breeding in the United States as a result of plant breeders rights....

There have been 42 companies, public and private enterprises, that have been involved in soybean breeding in the United States since 1970. Of those 42, 25 had programs prior to the passage of the legislation. That's clear because they had applied for their patents very quickly after the bill passed and everybody says that it takes at least seven or eight years to develop a new variety. So we assume that all of the applications prior to 1977, in fact, were based on research before the bill was passed....How many of those companies are still plant breeding out of the 42? Well, 15 of the 25 that were plant breeding in the beginning of the 1970s aren't there anymore. They've cancelled their programs. Out of the new ones that have come in--new only in the sense that they applied for their first patent in 1977--half of those new ones haven't done anything since 1980....In fact, if you look at it (and I have looked at each major crop in the United States) you see a drop-off in every major crop that has been patented since the beginning of the program. That's called monopoly. That's what multinationals do. That's what patents do. Patents are simply a way in which you carve up your turf, declare your territory and keep the little guys out of it. Patents are a barrier to entry for small countries and they're the way that multinationals divide up their market shares. And that's what's happening to us.

There are some points about germ plasm conservation that I think are going to have to be dealt with. We do not have a full and free exchange system of any kind....Private companies do have a major involvement in this. We don't know how major it is....The governments don't even want to ask....The Chicago Group, which is an investment analyst company involved in agriculture and biotechnology, analyzed private involvement in germ plasm conservation. It was a public study. They actually gave it to the State Department in the United States. The conclusion was that the private companies do a lousy job of germ plasm conservation, and I'm not exaggerating my statement. In fact, they went on to say that they are not only doing a lousy job, but are also hoarding the material, and not making it available to others. Now that's the industry talking about itself....

United Brands, the Chiquita people, has two-thirds of the world's germ plasm in bananas in storage in Honduras. Officially IPBGR says that material is freely available, and that's also officially the policy of United Brands....But Dr. N. B. Simmons, who is the big banana expert, was commissioned in a private study to go and try to find out what was actually there. Simmons replied back to FAO in a report which we have (it's never been published) and said that he can't get inside the Honduras collection and no one knows what's in there. That was until May of 1983. On May 11, 1983, United Brands wrote to IPBGR and to a number of other governments that they were cancelling their banana breeding program and were ready to dispose of their germ plasm collection. If you want it, you can have it--if you can finance it, which they can't. That's two-thirds of the world's conservation material for bananas. They will jettison it within the next few months if other countries can't find the money to save it. Now they're going to make it available because they're getting out of the banana business. And now they're saying how important it is....

Then there's the Firestone Rubber Company collection. We're not sure exactly what they have, because they won't let anyone in to see it. Firestone has a large living collection of rubber trees in Liberia, probably the best in the world and probably the largest in the world. In the 1982 FAO/IPBGR surveys, they simply said "Not available to be looked at." We wrote to them last year and asked them what they had and, because of the politics I think more than anything else, they wrote back and said, "Well, we don't know. We don't have that collection anymore. We got rid of it several years ago." I hope they didn't, but they might have.

There is a fundamental difference between plant patenting and the patenting of this tape recorder. When you patent a tape recorder, the patent holder doesn't give a damn about the raw materials that go into making this tape recorder other than their price....When you patent a plant variety, the raw materials are the genes and they are unique. They are special in the world. And yes, it's true that the final variety that's in the marketplace is available for anybody to use and any reputable scientist or disreputable scientist can take that variety and utilize it and use those genes. So the seed trade makes that into a great wonder and says, "Look how great we are. We actually facilitate the exchange of germ plasm between everybody because we let anybody have free access to this genetic material." But how could anyone stop that? It is simply impossible with the varieties in the marketplace to stop that. Since it absolutely cannot be prevented, they make it into a virtue. But the point is, these companies hoard materials in their gene banks which they have not found useful and will therefore never be in the marketplace. Or they monopolize material which they may someday use way down the road, which other people could use right now. But, because they like to peel off their innovations one after the other over the years so as not to push a lot into the market at one time, we are losing out because they are hoarding that germ plasm. That's the material we have to worry about—not the final variety. And many of the companies that I've talked to say that they're simply jettisoning the material they don't use. Why in the world would they want to give it to their competitors? If they can't use it, they just destroy it....

There's a lot more to be said about monopolization....in terms of who controls germ plasm in the world between the rich and the poor. The basic reality is that genetic resources come from Third World countries and from peasant farmers who created that diversity during the last 10,000 years. Plant genetic resources are the only raw materials in the world which are donated by the poor to the rich. And the rich are now saying that they can patent that material and sell it back to them....or the grandchild of that material. And that's simply indecent and unfair. And I don't believe that this highly movable object called government policy-making can stand up against the irresistible force of multinational corporations wanting their way in the marketplace, once we have opened up the door and accepted the concept of monopoly control of plant material. I don't believe that "Give us this day our daily bread" is a prayer we should be making to Shell Oil.

(For more information, subscribe to the "International Genetic Resources Program Report." A year's subscription is $10 (four to six newsletters). It is available from: IGRP Report, P. O. Box 1029, Pittsboro, NC 27312.)

(The 1984 Fall Harvest Edition, pages 42-51)

INTRODUCTION FROM THE GARDEN SEED INVENTORY

The Garden Seed Inventory represents your heritage as a vegetable gardener. The diversity and quality and number of garden varieties now being offered commercially is almost beyond belief. Gardeners in the United States and Canada are truly blessed. But it is quite possible that half of everything listed in this book could be extinct within the next few years! The major forces threatening this diversity include: plant patenting legislation; takeovers of seed companies by multinational corporations; plant breeding for machines instead of gardeners; the profit-motivated hybrid bias of most seed companies; and increasing bankruptcies of small businesses. These events and trends, which have brought such a massive amount of plant material to the very brink of destruction, must be examined in order to determine how to neutralize this threat.

Plant patenting legislation was first passed in England in 1975. At that time England became a member of the International Union for the Protection of New Varieties of Plants (UPOV), which is a Geneva-based organization that promotes and coordinates plant patenting laws around the world. Proponents of the legislation argue that it is necessary to keep unscrupulous competitors from pirating newly developed varieties and offering them under other names. Opponents say it stifles the free flow of plant material between breeders, which is essential if dynamic breeding programs are to be maintained, and allows breeders to make slight genetic manipulations to plant material, which they did not originally develop, and then "lock up" that material for 18 years.

Enforcement of the legislation quickly became a legal nightmare because it is almost impossible to prove in court that one plant is identical to another. But European lawmakers plunged ahead and phased in a system of "legal" and "illegal" vegetables designed to facilitate enforcement. They published a "Common Catalogue" listing all of the vegetables permissible to sell in Common Market countries and established stiff fines for anyone selling varieties not on the list. Dr. J. K. A. Bleasedale of England's Wellesbourne Station described the Common Catalogue as a "self-inflicted wound" and confided that the Catalogue was inspired by commercial interests and was intended to clear the European market for the patented varieties. In July of 1980 the legislation took full effect and 2,126 vegetable varieties became illegal to sell in England and the other Common Market countries which belong to UPOV. Dr. Erna Bennett, formerly with the Crop Ecology and Genetic Resources branch of the United Nations' Food and Agriculture Organization (FAO) in Rome, estimated that by 1991 fully three-fourths of all the vegetable varieties now grown in Europe will be extinct because of attempts to enforce plant patenting laws.

The situation, as it continued to develop in England during those years, was reported in the quarterly newsletters of the Henry Doubleday Research Association (HDRA). Lawrence D. Hills, Director of the HDRA, spearheaded the forces in England that opposed plant patenting and has tried, valiantly but unsuccessfully, to change the legislation. He was able to stir up enough national publicity that OXFAM made a grant of 300,000 Pounds Sterling to establish a Gene Bank at Wellesbourne. The Gene Bank, built as a government facility, was dedicated six years after the situation first became apparent. The dedication was a bittersweet event because much of the vegetable material, which Lawrence D. Hills and his colleagues had hoped to save, was already extinct. They are currently establishing Vegetable Sanctuaries in the kitchen gardens of several Stately Homes in England, where the varieties they have been able to salvage will be permanently maintained. They have also sponsored expeditions to collect peasant-varieties in various European countries which have recently joined UPOV, hoping to lessen the destruction.

In January of 1979, amendments to strengthen and extend the Plant Variety Protection Act of 1970 were introduced on behalf of the American Seed Trade Association (ASTA) and its 600 member companies. The amendments sought to standardize U. S. patent coverage with European laws so that the United States could join UPOV. The hearings for these amendments were unpublicized and the ASTA and the USDA expected no public awareness on the issue and no trouble running them through. They were wrong. Before the hearings even started and continuing through the entire next year, gardeners who were members of several grass-roots organizations working with genetic preservation generated more mail against the legislation than any agricultural issue in memory. Opponents used the various hearings on the amendments to try to educate legislators about the threats facing genetic diversity and to expose the bills as destructive, European-inspired, special-interest legislation. But the final battle was lost due to unbelievable political acrobatics on one of the last frantic days of the Carter Administration. The amendments were passed into law on a voice vote during a lame duck session of the United States Senate. The furor over the legislation died down all too quickly when there was no immediate move to install a system of legal and illegal varieties as had happened in Europe.

At that point opponents of the legislation hoped that the newly created public awareness about genetic vulnerability would at least result in increased support for the National Seed Storage Laboratory (NSSL) and for various collection programs. Unfortunately, that has not been the case. To say that germ plasm research and preservation programs in the U. S. have been given low priority would be a gross understatement. The NSSL did not receive a budget increase during its first 15 years of existence. The USDA has not had even one full-time plant explorer since 1970. The U. S. government only allocates a pathetic $40,000 per year to collect the rapidly vanishing genetic resources on which the future of U. S. agriculture will depend. And all this at a time when scientists around the world are voicing increased alarm as they watch the wholesale destruction of the centers of diversity for the world's food crops, caused by the introduction of the Green Revolution's hybrids and the newly patented varieties which are being marketed in Third World countries. The U. S. government is making a catastrophic mistake by not immediately initiating a crash program for plant collection worldwide.

Plant patenting legislation has definitely made seed companies attractive investments for multinational corporations. During the first week after patenting legislation was passed in England, one company, RHM, bought out 84 seed companies. When RHM's buying spree was finally complete, over 100 seed companies had merged. Shell Oil of Great Britain has bought out 74 seed companies since the passage of the legislation in England. Royal Dutch Shell became the world's largest seed and agrichemical company, almost overnight. Seed company takeovers in the United States have also reached epidemic proportions: Atlantic Richfield (ARCO) took over Desert Seed Co.; Celanese bought out Joseph Harris Co.; Ciba-Geigy (of Switzerland) purchased Funk's Seed; ITT now owns the W. Atlee Burpee Co.; Amfac took over Gurney's Seed & Nursery and Henry Field Seed & Nursery; Sandoz (of Switzerland) purchased Northrup King Co.; Upjohn bought out Asgrow Seed Co.; and Monsanto purchased DeKalb Hybrid Wheat. And these are just a few of the more than 60 recent North American seed company takeovers.

Multinational agrichemical conglomerates view seeds as a logical lateral extension of their financial interests. They are already manufacturing pesticides, fungicides and chemical fertilizers. With their newly purchased seed companies, they are now able to give commercial growers a package deal--seeds which will grow well with their chemicals. Some agrichemical firms have even started selling pelleted seeds, which wraps each individual seed in a small capsule of pesticides and fertilizers. Such tactics point out an obvious conflict of interest which occurs when major agrichemical corporations are allowed to buy out seed companies. In the past, excellent breeding programs have produced food crops which are resistant to a multitude of diseases and even to pests. But it is doubtful that such corporations, whose very existence depends on selling pesticides and chemical fertilizers, will spend any time or money to develop disease- or pest-resistant crops. And yet plant patenting legislation is allowing these same corporations to lock up valuable breeding materials for 18 years. Agrichemical and pharmaceutical corporations should be prohibited by law from owning seed companies.

The tremendous consolidation now going on within the U. S. seed industry also threatens to destroy most of the plant material available to gardeners. When a large corporation buys out a small, regional, family-owned seed company, it invariably drops the former owner's collection of standard varieties and replaces them with the more profitable hybrids and patented seeds. The new corporate owners are only concerned with profits and usually switch immediately to generalized varieties which will grow reasonably well anywhere in the country, thus assuring the greatest sales in the company's new nation-wide market. No thought is given to preservation or the fact that the collection of seeds being dropped is probably the reason that company was successful in the first place. Often these regionally-adapted collections represent the life's work of several generations of seedsmen within these families. Plants in these collections often are extremely well adapted to local weather, diseases and pests. We cannot allow this irreplaceable genetic wealth to be destroyed just for the short-term profits of a corporation which may not even own that company next year.

Such losses might be viewed with less alarm, if the varieties being dropped had been superseded by superior ones as was often the case during the first half of this century. But the garden seeds currently being dropped from the catalogs are the best home garden varieties we will ever see. Almost all of the vegetable breeding being done today is for commercial applications and such varieties seldom suit the needs of the home gardener. Most commercial breeding strives for an extremely concentrated harvest period, so that machines can do the picking all at one time. Fruits must have tough skins and solid flesh to withstand mechanical harvesting and then endure cross-country shipping. Varieties are bred for eye-appeal, so they will look good and sell good in grocery stores and in seed catalogs. Some are specifically developed to be stored for extended periods in refrigerated, controlled environments. And most are bred to produce well under optimum weather conditions and with inputs of chemical fertilizers and pesticides.

As long as our food crops are being bred for machines and large commercial growers, the varieties being introduced will continue to stray ever farther from the needs of the home gardener. Gardeners are most concerned with flavor. If that fantastic flavor is there, appearance means nothing. Fresh garden produce is only "shipped" from the backyard to the kitchen table, so skin and flesh can be as tender as possible. Gardeners want varieties that can be harvested all season long, so they can enjoy fresh produce right up until frost. Many of the older varieties have remained popular because their long keeping ability in fruit cellars allows gardeners to enjoy the fruits of their labor until the next spring. Gardeners need locally-adapted varieties which will grow well in their unique and increasingly changeable weather and withstand local diseases and pests.

Home garden varieties such as these are the ones which are currently being dropped from the seed catalogs. Far from being obsolete or inferior, these varieties are the cream of our vegetable crops. Each is the result of millions of years of natural selection, thousands of years of human selection and usually almost a decade of intensive and costly plant breeding and testing. Only the very best make it to the catalogs and each is unique and irreplaceable. But they are being allowed to die out due to the economics of the situation with no systematic effort being made by government agencies or lay organizations to store and maintain them. The survival of these home garden varieties represents the vegetable gardener's right to determine the quality of the food which that family consumes. If home gardeners allow their vegetable heritage to die out, they will be locking themselves forever into a position of dependence on the seed companies and the varieties that their owners choose to offer.

In the spring of 1981, the decision was made to attempt an inventory of the entire U. S./Canadian seed industry. Mail-order vegetable seed catalogs, no matter how small or obscure, were gathered from throughout the United States and Canada. Computer equipment was purchased and the work began late that summer. Early estimates mistakenly concluded that the inventory would be completed in one year and would include 120 seed companies and 3,000 non-hybrid varieties. The Garden Seed Inventory has taken over three years to finally complete and includes 239 companies and nearly 6,000 non-hybrid varieties. This gigantic project involved an exhaustive and nearly incomprehensible amount of work. But it had to reach completion, because the stakes are high. This preservation tool is capable of turning the present situation around. Revenues generated by this book are being used to buy up these endangered garden seeds and to develop networks of amateur growers who will permanently maintain them.

The Garden Seed Inventory's greatest value is that it shows which varieties are in the most danger before they are dropped. Many gardeners would gladly buy up a supply of seed, if they knew it was about to be dropped. But usually they have no warning that a favorite variety is in danger until it simply doesn't show up in a particular catalog one year and they are unable to find another source for it. This Inventory shows all of the alternative sources that are still available. Gardeners can now search through all of the varieties being offered to locate ones which are perfect for their local climate, diseases and pests. Gardeners in northern and high-altitude regions can use the Inventory to locate hardy and short-season varieties. Concerned individuals in other countries can use it as a model for similar inventories. Preservationists around the world can use it to buy up endangered commercial varieties, while sources still exist, and then permanently maintain them. And, because the Inventory focuses attention on the seeds that are the least available, many small, almost unknown seed companies will be rewarded and strengthened because they are offering unique or regionally-adapted varieties.

For the first time it is possible to accurately assess which varieties are being dropped and how quickly. The Garden Seed Inventory became increasingly more fascinating and more frightening as it grew towards completion. Fascinating because, when viewing the entire garden seed industry in detail, the amount of plant material available is incredible! Frightening because it is now apparent that over 48% of all non-hybrid garden seeds are available from only one source out of 239 companies! This study shows that: 2,792 varieties (48.3% of the total) are available from only one source; another 642 varieties (11.1%) are available from two sources; so a total of 3,434 varieties (59.4%) are available from only one or two sources.

	Companies Inventoried	Number of Varieties	Lost During That Year	Percentage of the Total
End of 1982	138	Incomplete	117	Incomplete
End of 1983	184	Incomplete	237	Incomplete
End of 1984	239	5,785	263	4.5%

These losses do not truly reflect the overall decrease in availability. Many of the varieties, which are now available from one or two sources, were available from three to six (or more) sources when this study began. And most of those companies have dropped the varieties in 1984. In other words, it appears that the sources of supply for these seeds have already disappeared. Although many of these varieties have not yet been dropped completely, they will be as soon as those few companies sell out their remaining supplies of the seeds. An almost unbelievable amount of loss is possible within the next few years. Hopefully an immediate and systematic effort can still rescue most of these endangered seeds. But there is no guarantee that any variety offered by a small number of sources will still be available in next season's. seed catalogs.

There has been only one other complete U. S. inventory of commercially available food plants. In 1903 the USDA published American Varieties of Vegetables for the years 1901 and 1902 by W. W. Tracy, Jr. It included variety names and sources, but no descriptions. This earlier inventory has been studied in depth and then compared to printouts of what is being kept in the National Seed Storage Laboratory. Only three percent of everything available commercially in 1901 and 1902 survives today in that government collection! It is depressing to see the huge lists of garden varieties available at the turn of the century and realize that almost all of them have been lost forever. Imagine everything that would still be alive today if the 1901/1902 USDA inventory had been updated annually and if endangered varieties had been systematically procured and maintained. But even though disastrous mistakes have been made in the past, there is no excuse for losing anything from now on. At this point we are just picking up the remaining pieces, but we must at least do that and do it quickly. Time is running out and we will never be given this chance again.

It is ironic that we presently have access to such a vast array of the best garden varieties ever developed, and yet so much of this invaluable and irreplaceable resource is in immediate danger of being lost forever. We are truly at a crossroads. It is still not too late to rescue from extinction what remains of our vanishing vegetable heritage. Just try to imagine what it would cost, in terms of time and energy and money, to develop this many excellent varieties. But they are already here. All we have to do is save them. We are the stewards of this sacred genetic wealth and we better start acting like it. If we don't, generations yet unborn will curse our stupidity and deplore the fact that we valued power and money more than we valued their survival.

(The 1984 Fall Harvest Edition, pages 4-9)

(Editor's Note: The Garden Seed Inventory, published by the Seed Savers Exchange early in 1985, is a computer inventory of 239 seed catalogs in the United States and Canada. It lists all of the non-hybrid vegetable and garden seeds still being offered and includes: variety name; range of days to maturity; a list of all its known sources; and the plant's description. This 7 1/2 x 10" 448-page book took over three years to compile, describes 5,785 varieties, and is being heralded by both gardeners and scientists as a landmark study. It was developed as a preservation tool, because most of our garden seed heritage is in danger of being lost forever due to garden seed industry consolidation and current economic conditions. If sufficient revenue is generated by The Garden Seed Inventory, a Preservation Farm will be purchased near Decorah, Iowa where endangered vegetables will be maintained and protected on a permanent basis. Softcover copies are still available for $12.50 postpaid from Seed Savers Exchange, PO Box 70, Decorah, IA 52101.)

FROM THE NEWSLETTER OF THE HENRY DOUBLEDAY RESEARCH ASSOCIATION

National Centre for Organic Gardening
Ryton-on-Dunsmore, Coventry, England CV8 3LG

A glance at a copy of a seed list issued at the turn of the century is an eye-opener. Page upon page of seed potatoes, tomatoes and other vegetables....Try and see how many of them have survived through to the present day and it's unlikely that you will find more than 20%....In today's economic conditions, it only pays to sell fewer and fewer varieties in ever-increasing numbers....

To this natural wastage can be added another force which started with the best of intentions, but has resulted in a wholly unnecessary destruction of vegetable varieties. The Seeds (National - List of Varieties) Regulation 1973, which came into force on 1 July 1973, made it an offence to offer for sale seed of a plant variety unless the name of that variety is given in the United Kingdom National List or the E. E. C. Common Catalogue of vegetable varieties....The chief purpose behind the legislation is to protect plant breeders from having new varieties poached by rivals and sold under another name. All new varieties have to be sent to the Ministry trials at Cambridge to ensure that they are sufficiently pure and distinct from other varieties to qualify for entry on to the lists. Unfortunately this has been applied retrospectively to all vegetables in cultivation which have been classified into varieties, and their so-called traditional and non-traditional synonyms and selections. The basis of this classification has been, in the main, the growth habits of the plants. Aspects such as flavour, pest and disease resistance and storage qualities have been largely ignored. Similar varieties which were once known under different names, are now said to be the same; or in official jargon the one is a synonym of the other....It is the intention of the Authorities to delete ALL non-traditional synonyms from the list on 30 June 1980, in total more than a thousand varieties....Furthermore, varieties are being struck off the list every month as a "maintainer" who is "responsible" for each variety decides that he no longer wishes to keep it....

The varieties which disappear most quickly from the list are the older types as growers join the mad scramble to produce more and more F1 hybrids. This works to the detriment of the gardener, for plant breeders concentrate almost exclusively on producing varieties for food processors and aim for qualities such as colour, ability to hold water and thereby increase yield, simultaneous ripening for machine harvesting, etc. If you are looking for varieties with real taste, resistance to bolting, long harvesting periods, tender skins in the case of tomatoes for example, the choice is ever receding. The position would not be so serious if all deleted varieties were automatically consigned to a "living museum of vegetable varieties" for possible use in the future. No one knows what sorts of qualities plant breeders will be looking for in 10 years' time, let alone 100 years....

From Lawrence D. Hills, Director, Henry Doubleday Research Association -- ....We can only distribute seeds as experimental material, because of the Common Market Regulations. Therefore, we will want to be able to buy seeds from you....We would very much like to be able to reach out to your members in various parts of the United States. We are looking in particular for vegetable varieties which crossed to America before the 1914 War at least from temperate states. We are particularly interested in things from Oregon which will grow very well for us....

(*The 1979 Seed Savers Exchange*, pages 34-35)

Time is running out for Lawrence D. Hills, Director of the Henry Doubleday Research Association. In July 1980, plant patenting legislation will take full effect and 2,126 vegetable varieties will become illegal to grow in England and the Common Market countries. He writes, "We particularly need seed stocks to pass on to the International Vegetable Gene Bank which is being started at our National Vegetable Research Station at Wellesbourne in Warwickshire. This is still not in operation, but it is owing to the agitation I began in the spring of 1975 that it is getting off the mark and will be paid for by OXFAM and other bodies apart from our Agricultural Research Council which is a government body....

Lawrence sent me a list of all the pea and bean varieties that were known to be in England in 1905. The lists of these soon-to-be-extinct varieties of peas and beans filled 15 legal-sized pages. (I also know that enough onion varieties to fill two and a half pages of single-spaced typing will be excluded from the Common Catalogue and will therefore be outlawed.) Any member who has access to a large number of these bean and pea varieties, write to Lawrence Hills for these lists. He wishes to obtain as many of these varieties as he can by spring 1980, so they can multiply them for the Gene Bank.

(The 1980 Seed Savers Exchange, page 44)

HYBRIDS: THE ULTIMATE ADDICTION

(The following was reprinted with permission from "Remnant Review," which is edited by Dr. Gary North....)

I was speaking with my brother-in-law earlier this year about the problems of organic farming. He has a tiny garden that produces a lot of food. He told me about the problems one of his suppliers has been having with the government....I contacted the supplier. He has produced non-hybrid seeds for decades in California. He is about to leave the country to head for Australia. In a recent form letter to his customers, he explained what is happening:

"We now face a $2 annual registration fee, per variety of seed, in each of several states and there is a possibility that all 50 states will soon require this fee. The 165 to 200 varieties of seeds we sell would cost $400 per state to register each year -- 50 X $400 = $20,000 per year cost.

"Besides this registration fee, all seed packages must now be marked with new generic names. For example, it is now illegal to sell one of our packages of "Black Eye Peas" unless it is also marked "Cow Peas," even though the package manufacturers for 40 years have issued them as Black Eye Peas, and we have thus been selling "Black Eye Peas." This new law makes obsolete thousands of seed packets we have in storage. Possibly 200,000 already-printed packets will have to be scrapped....and replaced at a cost of $5,000 to $6,000. This would be a net loss.

"Everyone should be made aware that there is a world-wide movement afoot that is a threat to all of us....Most of our old-time seed companies that started 30 to 75 years ago are now being financially gobbled up by big multinational business giant companies, such as Anderson Clayton, Cargill, Celanese Central Soya, C. Geigy, Food Machinery Co., I.T.T., Pfizer, Purex, etc., etc. They are now the monopoly 'OPEC's' of the seed industry. Their backing of the Plant Protection Act passed in 1970, plus Senate Bill S. 23 will aid the "Bigs" in gobbling up the small older companies. The strategy seems to be to reduce competition and survival chances for small seed companies which are and were producing 'open pollinated' vegetable and cereal seeds....

"Since 1970's Act, the 25 principal backers (the big corporate giants) have purchased at least 50 of the largest old family-founded seed companies of the USA, plus many other companies, world-wide....

"(Our company) will soon be ordered not to use the mail for interstate transport of our seeds that we've been selling for 40 years nation-wide on a very small family farm basis. Our seeds have received high praise and enthusiastic claims of superiority for all these 40 years. We've religiously never sold hybrids. We have tested some but usually found them very unsatisfactory flavor-wise, and 5 to 100 times as expensive seed-wise to repurchase each year as compared to our own farm cost of reproducing our open pollinated....We've had it, and can only sell inside our state of California....

"P. S. As of this printing, Atlantic Richfield bought out Dessert Seed Co."

The man who sent out this report on hybrid seeds is being attacked by the Department of Agriculture. It seems that in 1978, he sent out eleven packets of seeds — total value: $7 — to someone in Massachusetts. That person has moved and cannot be contacted by the seller. But the seeds supposedly didn't germinate at the rate established by law by the Department of Agriculture. He is now being told to sign a compliance statement and pay the Department of Agriculture $4,350. No wonder these small firms are selling out to the huge multinationals....The noose is being tightened. And it's not just a financial noose; it's a genetic noose....

(The 1981 Fall Harvest Edition, page 63)

American Seed Trade Association (Yearbook and Proceedings of the 96th Annual Convention, 1979) - "Vegetable Statistics Committee....This committee will study lists of garden varieties and endeavor to eliminate varieties that have become obsolete or have been superseded by improved varieties. Obsolete varieties will exist as long as someone will produce them. Because of the limited volume and high cost of production, it is desirable that they be eliminated...."

(The 1981 Winter Yearbook, page 62)

Seed Banks Serving People Workshop

(Editor's note: The Seed Banks Serving People Workshop was held at the Tucson Botanical Garden on October 13 & 14, 1981. It was organized and sponsored by the Meals for Millions Foundation and funded by the Wallace Genetic Foundation.)

Cary Fowler

Gary Nabhan

Bruce Bugbee

Skip Kauffman

SEED BANKS SERVING PEOPLE WORKSHOP
(Tucson Botanical Garden, October 13-14, 1981)

I recently had the pleasure of being part of a unique workshop that was held in Tucson on October 13 and 14. I would be surprised if there is ever another that even closely resembles it. I came away riding a warm wave of friendship and enthusiasm like I have never felt before. It was so incredible that I was even a little down just because it was over.

I met people that I have been wanting to meet for a long time, some with whom I'd been corresponding for up to five years. These included: John Withee whose Wanigan Associates built up a collection of over 1,200 heirloom bean varieties; Dr. George Larke who directs the Family Gardening Council which is a group of tomato experimenters; Cary Fowler of the National Sharecroppers Fund who has campaigned vigorously in behalf of genetic diversity; Karen Reichhardt whom Diane and I really enjoyed finally meeting; Forest Shomer who directs the Abundant Life Seed Foundation which is working to preserve the plants native to the North Pacific Rim; Rob Johnston Jr., a fine young seedsman who runs Johnny's Selected Seeds in Maine; Dr. Bruce Bugbee who has written articles for the SSE on home storage of seeds; Carl and Karen Barnes--Carl is an SSE member who has started an organization called CORNS to preserve as many open-pollinated corns as possible; Craig and Sue Dremann who run The Redwood City Seed Company; Jan Blum--SSE member from Idaho; Phil and Polly Germaine--SSE members from New Mexico; Steve Spangler who runs Exotica Seed Co. and has done a lot of plant exploration in Ecuador, Mexico and Hawaii; Steve Facciola--SSE member who is working with Steve Spangler on a sub-tropical rare fruit nursery; Skip Kauffman and Peggy Hass from the Organic Gardening & Farming Research Center; Jack Doyle from the Environmental Policy Center; Carolyn Jabs--a garden writer whose excellent articles have done much to spread the message of the SSE; Kit Anderson--garden writer with Gardens For All; Dr. Eric Roos--Plant Physiologist with the National Seed Storage Laboratory; Dr. Howard Scott Gentry--retired USDA plant explorer presently working with NEWCAST at Arizona State University; Dr. Richard Felger--Ethnobotanist and Senior Research Scientist at the Arizona-Sonora Desert Museum; and Dr. Robert Bye--Ethnobotanist currently working in Mexico. I met so many people so quickly that I know I'm leaving some out and I hope they will forgive me.

I'm saving the best for the last--the staff of the Meals for Millions/Freedom From Hunger Foundation/Southwest Program: Gary Nabhan--Project Manager; Cynthia Anson--SW Program Director; Mahina Drees--the Conservancy Garden Coordinator; Susie Terrence--Urban Community Gardens Project Manager; Jane Nyhuis-MFM Horticulturalist; and Mary Wolken--Coordinator. Their organization should be very proud of the incredible job they did organizing and putting on the Workshop. They are some of the most dedicated, hard-working, caring people I have ever met. I'm holding them personally responsible for all of the good that came out of the Workshop. The effects of what happened there in those two days will be felt for years to come. I just want to take this opportunity to thank all of them. What a crew!

I transcribed the following from over seven hours of tapes which I recorded at the Workshop. The panelists and people giving demonstrations only had time to give a tantalizingly short taste of what they were doing. I have concentrated on the speakers I thought you would like to hear and the ones we can learn the most from. This is only about a third of the speakers and less than half of what each of them said. If there are mistakes in the following, just assume that the mistake was in my transcribing and typing, not in what was said....The first morning started with Cynthia Anson, SW Program Director of Meals for Millions, welcoming all of us—

"We are excited and optimistic at having so many people here who are interested in preserving plant diversity, in the whole matter of genetic conservation and in learning more about the practical skills that will allow that to really happen. We have people here from as far away as Maine, Washington State, Idaho and Mexico City. Some of the people here have wanted to meet each other in person for a long time and it seems like this may be a unique opportunity for people to get together for the first time and really talk about something that many people have been working with, some for their whole lives....

"Cary Fowler will be the first speaker. Many of you know Cary already. He has been active for a number of years as an advocate for the cause of genetic conservation. He is presently Program Director of the National Sharecroppers Fund. He operates a small farm himself in North Carolina. He is a co-author of the book Food First which many of you are familiar with, and he's presently working on a popular book on genetic resources. He has also in the past several years served as a consultant to the U. N. Centre on Transnational Corporations. Cary will attempt to set the content of the Workshop in the context of the whole problem of genetic conservation and the different perspectives which many people have to offer...."

Cary Fowler

A couple of weeks ago when Cynthia wrote me and asked me to give this introductory lecture, I admit I felt really pleased and honored. It wasn't until I was flying out here on the plane that I realized how clever Cynthia is. Because what could anyone say to an audience like this that you don't already know and know better than I do? I am very aware that some of you have spent the better part of your lives working to preserve genetic diversity. So in all frankness, I want to say that it is with a fair amount of humility that I welcome you to this conference.

I expect that everyone here has vivid memories of the first time that they really became aware of the crisis of genetic vulnerability....The first time when it really took hold with me was some years back and I was reading an article entitled "Genetics of Disaster" by Dr. Jack Harlan of the University of Illinois. It was a very powerful article which talked about the loss of genetic diversity and the consequences this would have. And Dr. Harlan, who had long had a history of trying to talk to the public about this problem, concluded that article with a question—"Will people wake up only after it's too late?"

I think it's probably too early to answer that question now....this conference is evidence that indeed people are waking up. I find it quite gratifying to see the human diversity at this workshop. We have scientists here, geneticists, plant breeders, environmentalists, politicos, gardeners, networkers, collectors, farmers, seed company owners, many different kinds of people. I don't think that this diversity is any accident. Genetic resources are threatened by a number of factors. If they are going to be preserved, they will be preserved by many different means....(Cary went on to describe many examples of the losses that are currently occurring world-wide --Kent)....Finally, and perhaps most tragically, we are even losing some of the genetic diversity in our gene banks. It's probably a rare gene bank in this world that cannot give a few examples of some major and important losses. Perhaps the best sorghum collection that had ever been assembled was lost at the University of Illinois. A valuable pea collection at the University of Saskatchewan is gone.

Just as the diversity is being lost in many ways, it is going to take many different actions to preserve what diversity we have now....Many people at this workshop have come to recognize the importance of encouraging farmers and gardeners to actually preserve some of the older varieties. In West Germany the national gene bank has begun to put ads in the newspapers requesting that farmers and gardeners send in their old varieties for storage. In Czechoslovakia they're using a system of expert gardeners and retired agriculture professors to do growing out of the varieties at their seed bank. And of course here in this country we have folks like Kent Whealy who are trying to encourage and to make it possible for gardeners to grow some of the heirloom varieties of vegetables....

It is a shame that with all of the might and power this country can assemble on occasion, we spend only $40,000 a year of government funds on collection. This compares rather unfavorably, I think, with the $80,000 cut glass bowl that the Reagans gave Prince Charles and Lady Diana a few months ago. And it compares unfavorably with the $608,000 that we spend each minute on the military. It makes me want to call up the Army and ask them for a couple of minutes of their time....

(From a speech the second day) -....Shortly after I came to work for the Sharecroppers Fund, we published a small directory called "The Graham Center Seed Directory, (subtitled) Gardeners' and Farmers' Guide to Traditional Varieties of Fruits, Nuts and Vegetables." It was a guide to sources of those varieties.... We had several purposes in producing this booklet. One was to get gardeners and farmers in touch with sources of some of the old varieties. We knew that a lot of people out there wanted to use the types of varieties that Kent and others have, but they just didn't know where to locate them....also we felt that the companies who had these varieties were mainly small companies, they weren't the Burpee's of the world, and they deserved that kind of support....

In the Directory....we also had an essay which talked about the need to conserve our genetic resources and the problems that were occurring around the world....We wanted to somehow educate people, who were already somewhat sensitive about the need to preserve old varieties, so that they would become even more sensitive and somewhat sophisticated in knowing and understanding the broader picture of genetic vulnerability. Our hope was that people who were interested in growing an heirloom variety of tomato might also....act in the public arena in behalf of genetic conservation work. The ultimate goal of all of this was....to create public consciousness about genetic vulnerability that we hoped would translate in very definite terms into greater support for collecting missions, for the National Seed Storage Laboratory and for others around the world that were doing this kind of work.

I had been reading a lot of scientific reports from what I guessed were very frustrated scientists who had been talking for decades about the problem of genetic vulnerability and had somehow never been able to reach the public.... we knew that to break out of this bind, something had to happen...."The Graham Center Seed Directory," tiny as it is, was the only way we could come up with at a certain point to make that step. We found that people were very interested. In short order we went through four printings without doing any advertising....

Shortly after "The Directory" was published, we accidentally came across the information that the Plant Variety Protection Act was about to be amended. The PVPA is an Act which establishes essentially patent-like conditions for new varieties of crops. We had done some research already on its effect in Europe. We thought it was a harmful effect....We thought it contributed to consolidation in the seed industry. We didn't know what effect it was having on seed prices, but we knew that seed prices had risen faster than any other farm input since 1970 when the Act was first passed....But we were most concerned with the potential impact of the Act promoting further increases in seed exports which were displacing traditional varieties primarily in Third World countries without the benefit of having an adequate system to ensure that those old varieties were collected and preserved....

We created some waves over the Plant Variety Protection Act....but by and large I think that some good things happened because of that Act. We made some enemies, but we also made some friends....No matter how you feel about the Plant Variety Protection Act, some of you were in favor of it and others of you were opposed, but the net result at this point, since it's passed and is water under the bridge, is that we did focus some attention on the problem of genetic vulnerability. And I would be very happy if that attention got translated into increased funding for Fort Collins or into more funding for collection work....

(Back to the conclusion of the first speech) -....This conference is coming about because of the labor and work and love of the people here locally on the staff of Meals for Millions Foundation....This is the first time that anything like this has been done and I think that we should all be proud of them and proud of ourselves.

Gary Nabhan

(The next speaker was Gary Nabhan whom I've had the good fortune of getting to know really well the last couple of years. Gary is an Ethnobotanist and Project Manager for Meals for Millions. He was instrumental in starting their Southwest Traditional Crop Conservancy Garden & Seed Bank. He also does a lot of private plant exploring, going down into Sonora and tracking down really rare traditional food crops and then reintroducing them on reservations in the U. S. to native peoples who have lost them.)

Like Cary said, we really only have time to give a little taste or sampler of what we are all doing. We have some time constraints and I hope all of us can stay within them. After each panel is an audience participation session when we will be taking questions from the group in general and we hope you will take this opportunity to bring additional points out. Also there will be a lunchtime open microphone, so that other people can express additional opinions and points of view. I have the task of attempting to keep within my eight minutes first, so if I can do it perhaps the rest of us can try too. If not, drag me away from the microphone, quick....

Around 1904 Robert Forbes, who was head of the Arizona Experiment Station, decided to visit some Indian agricultural fields on the Colorado River to get an idea of what could grow in this desert area that was frustrating so many of the Anglo-Americans who were new to the area. He wrote about his experiences traveling down the Colorado River looking at the same kind of agriculture that had gone on for centuries there. He wrote, "The Indians, especially the Chemehuevi who are at present the most successful farmers on the river, grow beans, cowpeas, watermelons, Turkish winter muskmelons, devil's claw, a soft maize that matures in about 7 weeks, a black sweet corn, winter squashes, pumpkins, a little wheat, barnyard millet, sorghum and a seedy grass, Panicum sonorum, useful for both grain and horse forage. These plants are nearly all planted as soon as possible after the overflow floods from the Colorado River, the seeds being placed in the bottoms of open holes 6 ft. to 1.5 ft. deep, out of which the stalks and vines issue up while the deeply rooted plants are moist in the soil long after the surface soil is parched and dry...."

Going to the Chemehuevi Reservation about a year and a half ago, I asked Nina Murdock if any of these crops were still around. Only one on the list was still available to the Chemehuevi. All of the others had sort of gone down the drain after the dams were put on the Colorado River, and the Chemehuevi no longer had these seed stocks that they had had since prehistoric times. Fortunately, however, some of these seed stocks are still available in the region in tiny refuges of cultivation both south of the border in Sonora, in the Sierra Madre in Chihuahua and on other reservations in Arizona.

The Meals for Millions program here in Tucson is after that diversity that once was all over this region and is now only found in parts. We find that dealing with plants on a regional basis gives us a foundation not only for helping the Chemehuevi find something that they once had years ago, but it keeps us in touch with what we can suggest to other agriculturalists in this area....We believe that seed banks are a backup system, that the health of the agriculture in a region depends on these things being grown where they are naturally adapted and in fields where they have evolved over the centuries....

There is a wealth of information that can help us start a process of seeing what crops were grown in an area and which localities these things were found in: historic surveys of traditional crops, things that anthropologists have put together on a particular tribe's crop repertoire--this kind of material is where we start. We go from there to talking to farmers themselves, particularly elderly farmers, because sometimes a listing in a book may say that Hopi sunflowers are rare. But when we get up to the Hopis they say, "Aw heck, I know someone who has a whole field of that. What I'm really interested in is this." And I think we have to keep that in mind. That the best way to really identify the material that we need to spend time collecting right now, is to start with the literature but very quickly get into the areas where people still have knowledge of these things.

We also use other sources like correspondence with international guardians of particular crops, like sunflowers or squashes or tobacco, to really help sort out what they have collected vs. what we can collect. For instance, Dr. Whitaker, a squash expert, recently sent us a whole collection of things from Sonora and Chihuahua which is the southern half of our region. And we decided that if we were going to go for squash, that area's pretty well covered, so we should really take a look at what is left north of the border. Then we discovered that none of this material had gotten into the National Seed Storage Lab or other collections before, even though there was still quite a bit of it around.

Finally we set priorities in terms of our modest collecting activities, on what growers and gardeners really want in our area. For instance we grew a lot of Hopi sweet corn this year, because we felt that other people are dealing with blue corn on a larger scale and the Hopi sweet corn was kind of being neglected. And we also really go for the things that are particularly rare.

You were all given a handout on endangered wild relatives of crops and this is a way that we have begun to identify what materials we should really go after. For instance, there are a lot of wild relatives of crops in the southwest and in the U. S. in general, but this list shows that over 100 are on the U. S. endangered species list right now and only six of those are in seed collections that we know of. So by putting together lists like this, immediately you can see that there is a lot of work to do—enough to keep all of us busy for a long time.

I'd like to turn things over to Kent and have him talk about a survey of vegetable varieties that he's been doing....

Kent Whealy

I'm Kent Whealy and I direct an organization of vegetable gardeners known as the Seed Savers Exchange that maintains and also exchanges heirloom vegetable varieties. I'll be telling you more about the organization tomorrow, but what I really want to talk to you about today is some work that I have become very interested in recently.

During this last year I have turned my attention to the preservation of vegetable varieties that are currently being dropped from commercial availability. Each year hundreds of vegetable varieties are dropped from seed catalogs not because they aren't delicious and unique, but because it is only profitable for large seed companies to stock the varieties which sell the most. Also the tremendous amount of consolidation going on within the seed industry right now has escalated these losses tremendously. Multinational agrichemical conglomerates are buying out family-owned seed companies at a tremendous rate and proceeding to drop their collections of standard varieties in favor of more profitable hybrids and patented varieties. In many cases the deleted varieties represent the life's work of several generations within these families and are extremely well adapted to local weather and pests and diseases.

During the first half of this century, most deleted varieties had been superseded by superior ones. But most often that hasn't been the case during the last 20 years. Far from being obsolete or inferior, the varieties being dropped today are literally the cream of our vegetable crops. Each is the result of millions of years of natural selection, thousands of years of human selection and usually almost a decade of intensive and costly plant breeding and testing. Only the very best make it to the catalogs and each is unique and irreplaceable. But they are being allowed to die out, due to the economics of the situation, with no systematic effort being made by government agencies or lay organizations to keep them alive or to store them. We must stop this shortsighted destruction.

I am presently working on a computerized inventory of all non-hybrid vegetable varieties that are currently available in the United States and Canada. The Inventory includes variety name, the range of the days to maturity that are represented, a complete description of the variety and a list of all its known sources. I will update this Inventory each year.

The Inventory will be valuable in many ways. This will be the first time that we will be able to clearly assess what is being dropped and how quickly. It will be valuable to gardeners in two very specific ways. Many gardeners will buy up varieties if they know that they are in danger of being dropped. Most times, however, they don't know that a favorite variety is in danger until it simply doesn't show up in their catalog one year and they are unable to find another source for it. But most gardeners only deal with three or four companies and the Inventory will cover somewhere around 130 companies. The Inventory will also let gardeners search through everything that is available to find varieties that are specifically suited to their climates and problems.

The Inventory's most important value is that it will clearly show which varieties are the rarest and in the most danger before they are dropped. I believe that we should place in frozen storage all varieties that are available from only one or two sources. I estimate this would be less than 1/3 of the approximately 3,500 non-hybrid vegetable varieties that are still available commercially. I would like to meet with anyone who wants to work with me in accomplishing this.

I am a vegetable gardener and I believe that the standard varieties that are available today are the best home garden varieties that we will see. The vast majority of the vegetable breeding being done today is for commercial application and such varieties are seldom suited to the needs of the home gardener. This....(holding up the *Garden Seed Inventory*)....is the vegetable gardener's heritage. We are the stewards of this irreplaceable and sacred wealth with which we have been blessed. This Inventory will either be the vehicle for the preservation of these varieties or it will be a very clear window through which we will watch them disappear.

(From a speech the second day) - Yesterday was only the second time that I've ever spoken before any group, much less a group that's half scientists, and to tell you the truth I was terrified. But all that has just faded away. It is really heartwarming to see the cooperation and understanding that is being displayed here....

In the early 70's my wife and I were living in the extreme northeast corner of Iowa. Diane's elderly grandfather was teaching us to garden and he gave me three heirloom varieties that his family had brought over from Bavaria four generations earlier. Well, the old man didn't make it through that winter and I realized that if his seeds were to survive, it was up to me. I began to wonder just how many varieties die out that way each year....About that same time I was lucky enough to come across the writings of Dr. Garrison Wilkes and Dr. Jack Harlan, which have had a great influence on me....

At that point I had no idea how prevalent heirloom vegetable varieties might be. I knew the ones I had were excellent....So I started writing letters to gardening magazines and by the end of the first year I had contacted and traded seeds with six other heirloom seed savers. One of the six--Lina Sisco--died the next spring, but by then three of us were growing the Bird Egg Bean that her grandmother had brought to Missouri in the 1880s....so that's how the Seed Savers Exchange started, and the more energy I poured into it the more it grew ....during the last six years approximately 800 different members have offered an estimated 2,000 heirloom and unusual vegetable varieties to over 15,000 interested gardeners who have made an estimated 150,000 plantings of vegetable varieties that aren't in any seed catalog and in many cases were on the edge of extinction. Even I can't imagine the impact that this kind of an exchange is having....

If I had to pick out the one thing that I enjoy most about my work, it would be helping folks find lost varieties....It really makes me feel good when I receive warm letters from gardeners thanking me for helping them locate vegetable varieties for which they've been searching for up to forty years....

Searching out heirloom vegetable varieties in isolated areas and ethnic communities across our country is an area in which laymen can make a very valuable contribution....it is often essential to approach such folks on an equal basis, as a fellow gardener who is interested in seeing their treasures live on....Yesterday I told you that there are currently 3,500 vegetable varieties commercially available in the U. S. and Canada. From what I have seen there could easily also be that many excellent heirloom vegetable varieties being kept by backyard gardeners. So if we can keep the commercial varieties alive through these rough times, and if we can locate and make the heirloom varieties available to gardeners, we may be able to double the amount of diversity available to backyard gardeners....just try to imagine what it would cost to develop that many varieties....But it's already there....and all we have to do is save it....

Bruce Bugbee, Ph.D.

One of the things that concerns me is that when people take seeds for long-term preservation, they make a lot of physiological errors. They make good genetic choices on the material to save, but during the process of growing the mother plant and then harvesting seed from that plant and then storing it for a long period of time and then consequently taking it out of storage and then replanting it, there are a lot of physiological things that are not done just exactly right. They end up with seeds that don't store just as long as they should or during the process of storage they undergo genetic mutations and change in storage. So I'm going to talk mostly about the role of the mother plant in seed vigor and also mention a few things about container types during storage....

Taking care of a mother plant is a lot like taking care of a person, not until they are 15 or 16 years old which is what we do when we grow vegetables for eating, but it's like taking care of the plant until it's 100 years old. And you have to be careful to continue to take care of the plant so that the seeds are going to be the highest quality possible. I could summarize everything that I am going to say by saying that you have to have as optimum an environment as possible and be especially careful that the plant has good nutrition....

Some of the specific reasons that nutrition is important are that there have been some studies recently that have shown that high protein in certain kinds of seeds, specifically wheat and oats, is related to really high vigor of the seeds. Whether this is related to vegetable seeds isn't known yet. But at any rate, nitrogen is related to high protein in seeds and you want to be sure and have high nitrogen in the plants in addition to keeping the plants vigorous during the time that the seeds are maturing on the plant.

It turns out that potassium is much less important because it's readily translocated in plants and goes right into the seeds, so it is not so crucial in terms of seed vigor. However, phosphorus is more important because it is stored in seeds as a compound called phytin which is very useful in absorbing free radicals and preventing some of the problems of seed deterioration in storage. It's not clear whether you can make a seed store longer by really loading it up with phosphorus, but it may be the case....

Calcium can be real problematic in seeds because it's not readily translocated into the seeds and one of the big problems with leguminous species, large-seeded legumes, is that after they germinate, they get a problem called collar rot. It looks like damping off....and this is because of lack of calcium in the seeds. So if calcium nutrition is inadequate, the seed vigor is going to be really poor, especially in the large-seeded legumes. If seeds have adequate micronutrients, they'll get a good start even in soils where those micronutrients can be limited or where other factors in the soil cause micronutrients to be unavailable to the seeds....

I also want to talk about types of containers during storage, because I see a lot of errors being made by people in storing seeds....It is really fairly easy to get the seeds to low moisture and fairly low temperature. But in order to keep the seeds at a low moisture content, they need to be in containers that are absolutely moisture proof....The error most often made is in assuming that something like polyethylene is moisture proof, and it is not! Thin polyethylene, 2 mil plastic, allows moisture to permeate through the bag and into the seeds, and if they are in a high humidity environment outside, the seeds are going to pick up moisture and their storage life will be correspondingly reduced.

The other way that moisture gets in is through the seal, through the lid. Even if the walls of the can or container are completely moisture proof, if the lid doesn't have a good seal then moisture gets in and the seeds pick up that moisture and consequently storage life is reduced....

The type of container that I really like the best is just a big glass jar with a wide mouth that has a rubber gasket underneath the lid, not a cardboard gasket but a rubber one. You can see the seeds inside the container....and because it has a screw-on lid, you can turn the lid down tighter periodically and keep that really tight for long periods of time....

(From a talk the second day) -....I think we should always encourage nonprofessionals to do their own research and to get the answers themselves. Rodale Press just published a book called Improve Your Gardening With Backyard Research. I thought that was great....When we confront certain segments of the population that are very resistant to change, they are resistant to you coming in and saying, "THIS IS THE WAY TO DO IT!....Studies show that THIS IS THE WAY TO DO IT!"

A much better way is to say, "Here's how to set up a little project to determine which way is best" and let them find out for themselves....If you teach people to do research, even if it's simple research, they are going to learn the answers themselves and really make solid conclusions that will last a long time....That's one area I think gardening magazines could help out with more, is encouraging people to be more experimental. I think Rodale Press has done that a lot lately and I've really been pleased to see that encouragement of trying to get people to be their own researchers and come up with the answers themselves....

Vegetable Seed Storage Facility Proposal
(Presented by Skip Kauffman and Peggy Haas)

(One of the most exciting things that happened at the Workshop was the presentation of a proposal for a Vegetable Seed Storage Facility by the Soil and Health Society to be located at the Organic Gardening and Farming Research Center. They are proposing that a $200,000 cool storage facility be built to store endangered vegetable varieties that are currently being collected by individuals and private programs. Skip Kauffman and Peggy Haas from the Organic Gardening and Farming Research Center were there to present the proposal informally and to gather feedback. Please realize that this question and answer period was a lively exchange that went on for well over half an hour, so this is just a small fraction of what was said.)

PEGGY: I'm Peggy Haas and I'd just briefly like to go over some of the seed storage proposal with you. I think you all received copies of the proposal—A Vegetable Seed Storage Facility....basically, we want to build a seed storage facility in which we would store domestic vegetable varieties. We would try to concentrate on things that aren't easily available from the seed companies. We would also try to get some things that are available from seed companies, perhaps doing like Kent Whealy has suggested, concentrating on those things that are the least available and most endangered. We would like to develop some sort of growers organization so that we could have people in certain regions of the country growing things that are adapted to those areas....We could store those seeds for them and then they could determine how the distribution might work. For you to get a better understanding of what the whole thing is all about and so that we can also get some feedback from all of you, probably the best thing is just to open it up for questions....

QUESTION: Do you see yourself becoming a seed company?

SKIP: No! Absolutely not!

QUESTION: Why are you leaning toward cool storage as opposed to frozen storage?

SKIP: First of all, we see this as a working collection rather than long term storage. We would like to be able to get seed out to people who need it. And this is the type of thing that we would have to go in and access fairly often. There is also the expense involved with long term frozen storage. We thought we would try to get this thing rolling and then see what sort of support we get from individuals who support the Soil and Health Society and within groups such as you people present. Then, if things look good, maybe make the investment a few years down the road for the type of long term storage that would be needed.

COMMENT (Gary Nabhan): One of the suggestions that is brought up in the proposal is the creation of an advisory committee that would include people from the different seed exchange and regional growers networks that are starting to crop up. Certainly Kent's is the largest, but now we see it diversifying with The Gourd Society having a bunch of people grow squashes and pumpkins. They're already four or five years ahead of us in knowing what people really like and knowing what seeds are around. And I think having a group that would include representatives like people from those organizations would really provide that link that would keep us in touch with the grassroots....The difficulty that I see, in getting totally dependent on the networks in helping guide the growing out periodically, is that there are still a lot of open variables. We don't know if these organizations are going to have a lifetime of 3 years or 50 years because none of them have been around that long....I think that is one of the real problems of linking up with the networks, but also one of the graces of the Rodale Proposal is that even if one network went under, perhaps because of organizational difficulties, the others might be able to take up the slack, so that we don't lose the work of 5 or 10 years of 50 or 100 individuals. So in terms of its effect on these organizations, it will lend a lot of stability to what is being done by individuals.

QUESTION: What do you estimate the cost of running the facility each year will be?

SKIP: In the proposal we talked about $100,000 a year to run it after the building is built. Mr. Rodale has said that we could have company support to build the building, to put up a cool storage laboratory to begin with, if that's the way it goes. And then to start to generate funding, but that's what we were talking about on a yearly basis.

QUESTION (Rob Johnston): Maybe I missed something, but it seems to me that you are talking about all the maintenance of the inventory being done by the amateurs. Is that your intention?

SKIP: It reads like that, doesn't it? I think that realistically we would look at our Organic Gardening and Farming Research Center as being a place to do seed increases of certain things that we can. But we've talked about amateur grower groups taking some responsibility to see that it be done. Do you have suggestions on that?

ROB: I think this whole thing is fascinating! I could talk about it for a long time. But that was just one question?

QUESTION: Could Botanical Gardens do increases and demonstrations that would also serve as education?

SKIP: I don't think that's in the proposal, but it's sort of been batted around as we talked about it--Could Botanical Gardens do increases and demonstrations that would also serve as education?

COMMENT (Kent Whealy): There are supposedly over 350 "period gardens" around the country that represent different time periods and sometimes also certain ethnic groups. I have about a half dozen Directors of such gardens contact me each year because they have a very hard time locating appropriate varieties. Sometimes they will restore whole villages complete with people in costume, but will be unable to locate garden varieties typical of the period they are trying to represent....We could supply starts of such seeds to them and in return they could either multiply them for us or at least maintain some varieties on a permanent basis.

SKIP: You see we don't have any sort of network like that. And like I keep on saying, this is really the early stage of the thought process and this meeting is helping us get suggestions.

QUESTION (Cary Fowler): I like to think of all this great diversity of vegetable material as a public resource. I know that all organizations sooner or later insulate themselves in some way or another. My question is whether you have considered setting up either an advisory group of non-Rodale, non-company people to give you technical advice and furthermore perhaps even something like a Board of Directors of people who actually have some power or responsibility to give your organization input concerning how these materials will be used.

PEGGY: Actually we have thought about that after Gary Nabhan's suggestion. He has given us a list of names of various people who might be good to ask if they would like to be on this board of advisors, provided the seed storage facility does go through....We think it is a very important aspect of the whole thing to have people from the outside giving advice and guidance....

John Withee

When I got down into Massachusetts and got a piece of land that was big enough to have a fire in the backyard, I decided to revive my family's tradition of cooking beans in the ground....When I was a youngster, it was my job on Fridays to clean out the hole and start the fire and get the rocks hot. Then my mother would have the pot of beans ready and we'd drop them into the ground on Friday night and the whole family would eat them on Saturday. I decided to revive that....

I went to the local stores to find my supply of beans and I didn't find certain ones that I had remembered. So I went back home to Maine and I wrote back and I visited various places and began this collecting. The first thing I knew, I had as many as 50 varieties and I'd never heard of anyone that had as many as 50 varieties of beans....My way of collecting was simply in traveling a lot, to drop into every food store I could find, especially in ethnic areas, and talk with people. I'd put little notices in free publications and got replies. And by swinging a lot, I got a few hits....

It kept going like that until the number was getting up pretty large and then I had to worry about how I was going to keep them alive. They only keep a few years and they've got to be regrown. So then I dreamed up this retirement scheme....I retired five years ago and I said this is going to be great for me in retirement, and I figured it all out, you see....

I decided I had to print up something, you have to send something to people, so I really took on an ambitious task. I decided to describe the beans so that people would come nearer to finding what they wanted. And that took a lot of time. Then I began to seek help in growing them out. And this was done by offering people some packets at a low price, which proved later to be <u>very</u> low.... At the time they would order a couple from my little catalog, I would also send them a couple more that I wanted them to increase for me, tying them into the whole idea. Well, not everybody got infected. There were about 10% that caught the bug and became extremely good supporters of the whole idea. Some of them were generous to the tune of sending me back 10 or 15 pounds of a variety and others would tend to send me back six or eight beans and say they had a poor crop. Or they were candid about it and said they'd liked them well enough to eat them and promised to do better next year....

I think the real reason why this thing could expand and did expand....was that I felt that the way to continue this was to be personal about it. I detested form letters....so everyone got a handwritten reply. I didn't even have a typist. If they were sending me a bean to be identified, the best way to get that bean was to send them a look-alike. I'd tell them frankly that I didn't know what the devil it was that they had sent, but it looked like this one and here's some for you to try and if, by chance, you have enough of yours, would you send me a few. And then we would talk about our recent illnesses or how bad the weather is here and it was very personal....the network was a person-to-person thing....

One of the things that has come to my mind here a number of times during this meeting is this business about variety names in beans. They mean absolutely nothing to me. I don't care whether a bean has a scientifically accurate variety name. If they send me a bean and it's Uncle Quimby's Baking Bean, so be it....I'm not about to decide the difference. I'll leave that to the experts, if they ever get interested, to do that. Collecting like this, I am sure there are a high percentage of these that are different names for the same variety of bean. However if a bean is grown in New Mexico, even though it looks exactly like one that we have in Massachusetts, I like to think that at least it's a different strain. But I'm not going to get into that, because I'm not the scientist. But I have always collected these as given to me. When a name isn't suggested, I carry it under a No Name label with my accession number and then try to prevail upon the donor to give it a name. And though I might suspect that it is exactly the same as one I already have, treat them different and you will get more cooperation back, I have found. There is no real purpose to it anyway. Isn't the whole enjoyment of it in just moving them about and having everybody interested in collecting and holding onto them? Rather than focus on storage, which is necessary, I think we as growers should focus on growing and getting them distributed as much as we can....

Back in the 40's at the University of New Hampshire, there was a horticulture professor named Hepler, who became interested in the local varieties that people brought to him at the University. He decided to grow them out and even developed some new varieties of his own. He carried this on to the point where he even started a small seed company. And his son when he was growing up, took it over from him and called it the Billy Hepler Seed Co. They had a number of varieties in there which were of New Hampshire and Maine origin and their given names were kept. Prof. Hepler also bred new characteristics into some of them, offering them under such lost names as Brilliant; we haven't found that one yet. His son carried on the seed business until he left for college, couldn't carry it on and sold the business lock, stock, and barrel to Farmer Seed and Nursery out in Minnesota. And you know what happened. It wasn't more than two or three years when the whole Billy Hepler collection was gone. They either changed the names of the seeds they offered or they just discarded them. Anyway, in that process of working with these beans, Prof. Hepler called them "heirlooms" and that's where that term was first used....

Many of the commercial beans....even Kentucky Wonder which has been in catalogs for many years was originally a home-saved seed, as many have been. When I was a high school kid running a little market garden, we had a bean that was very popular in the farmer's market downtown that was called Bountiful. It was a flat green snap bean. But it went off the books and there was a long time when I couldn't find it anywhere in the catalogs. But it came back again and is now a popular bean, very widely grown. So I like to think that even commercial beans should be treated like we do the heirlooms or other lost beans, because they sure as heck are going to be in a short while, if the seed companies decide that the sales aren't great enough. I don't for one minute ever criticize seed companies for dropping off lines because of lower sales, but I do think they ought to have the decency to at least keep their own little gardens to keep some of the old varieties going that they used to have. They might come back....

If you have a bean that you don't know the name of, don't be too worried about finding its correct name. That takes time and it takes growout and expert opinion and decision. Give it a name. It's yours. Who's going to criticize you? Who's going to know? So you find a bean and I can say that looks like the regular Colorado Pinto bean, but you say you grow that up in Kansas? All right, it's Kansas Pinto, just as simple as that. Or it's Your Name Pinto, you know—Uncle Quimby and Aunt Sadie and all the rest....So I just think that what we ought to do is keep spreading the word....keep passing them around.... cooperate....

Forest Shomer
(Director of Abundant Life Seed Foundation)

It is interesting the variety of names that people have given to their organizations. We have exchanges, banks, companies, foundations and even libraries. I think these are all viable ways of getting seeds in and then getting seeds around. The most important thing to me is to keep the seeds moving.... because then they become part of the fabric of our lives and they mean something. And if they mean something, they'll get grown. If there is a crop failure, then it will be very important that they get grown the next year....

The role of non-profit organizations in genetic conservation, as I know it, is to educate and inspire....it's as simple as saying, "Let's see, I've already collected some seeds that other people haven't. I'm going to get up and tell others what I've already learned. I won't portray myself as an expert, but I do know something about it." And almost each one of us here can do that. If you've collected anything at all, if you are just willing to go out and speak about it....the only thing that it takes is enthusiasm. Because I'm motivated and you're all motivated too. And that's contagious. That's the spark. Like the spark that's in the seed, we are all seeds and we can pass that spark on. Just like with vegetables—you grow your vegetables and you're so proud of them that you turn your neighbor on to some of your vegetables. Certainly they'll eat them and maybe they'll catch that spark and want to grow some vegetables. That's something that we've all caught from somebody that we can communicate to somebody else. That's the inspirational aspect....

I now have three or four cohorts who are skilled enough to go out and give a seed talk. They've grown enough seeds, they've handled enough seeds, they recognize the subtleties and they have the enthusiasm. And they can go talk to their neighbors as they all do, or go talk to a group and explain—This is what we are about, this is what we are trying to do, do you want to become involved

John Withee

Forest Shomer

Rob Johnston, Jr. Giving His Workshop

in it? If they do, then we have some more things for them. We can tell them where they can study if they want to read. We have some literature or we can refer them to other people scattered around the region. We tell them, "Next time you're up around that town, why don't you go visit these people? They're really into it. They've collected a lot of seeds. You should see their gardens." Then they go see their gardens and think--I've got some of those flowers or those vegetables too, but I've never thought to collect them. Then they go home and start collecting seeds. They're seed people. That's initiation....

A friend of mine had some land, some very special land, and I wanted to keep that land open for the community of people around that person. He's a teacher and had a lot of friends and he took them all up to this land. Then someone came along, much to my surprise, and said, "I can give a donation, a large donation, to a non-profit organization. Do you have tax-exempt status?" Of course, I didn't. But I thought about it for a few weeks and then went for it. It took about nine months. Just about that time the land got sold. Never did get the land, but we wound up with a non-profit tax-exempt foundation. I'm really glad for that because one of the things that happened along the way was that the IRS asked a lot of questions. Before you can get a tax exemption, you have to be clear about what you're doing. They won't just give you an exemption so you can go out and do whatever you want. They got me so clear that I haven't had any questions about what to do ever since....

A non-profit foundation has no dividends, no stockholders, there are no investors. It belongs to itself, you might say. And if it should ever disband, its assets must be transferred to another non-profit organization. That's the way they all work....that creates a need for members, members pay dues and can make donations, and it sets us up for receiving grants, if we should ever be so fortunate....I'm the Director, but I'm not the owner. I passed that up five years ago and I don't want to go back to that. But I can draw the wages that I need....

One of the projects that I have going right now is a calendar that is a record of close to 350 species we've collected over the last seven years and when we first collect them and when we last collect them, and it will be footnoted. It will make it clear and will remind you that it's the third of August, so it's time to go out and get the hyssop seed....

When you do any reading about seed production you invariably find reference to the J. L. Webster Seed Production Services, British Columbia Department of Agriculture about 40 years ago. It's a collection of 15 little pamphlets ranging from about 8 to 40 pages. Webster covered the seed industry in British Columbia thoroughly. He described in detail how to grow parsnip seeds, how to grow corn seed, how to grow pea seed, how to grow flower seeds, what the yields you might expect would be, how to fertilize them, what to treat them with (he wasn't too heavy into chemicals either; this was before DDT even) and some overviews. Basically what you need to know if you want to farm seeds. But they've been unavailable for years and years and several times I've gone to ask the people in Victoria to reprint it again and they just can't even hear it. So I'm going to print it again. It's going to take a couple thousand dollars and will be forthcoming sooner or later....

In any case, if you start doing research on seeds, not seed storage but seed growing, you'll find Webster's name in every chapter of all these books. That's because he got it down and he had hands-on experience. I'll just leave you with his opening words in the first of the 15 documents. He said, "Home grown seeds are best."

John Kimmey

(While enjoying a delicious buffet of Mexican food the second day, we were all shown just how good an idea the lunchtime open microphone was. For more on John Kimmey's activities, see the article on the "Taos Pueblo Native Seed Co.")

I have one grandfather that I have been traveling around with and have spent a lot of time with who is over 100 years old, a Hopi man. Whenever he gives a presentation of any kind to a group, he always starts out the same way. He has a cloth and on that cloth is a symbol. It is called the emergence symbol. The emergence symbol is a spiral. It starts in the center, works its way out in a clockwise direction, and then it comes to some conclusion at the outer rim. What Grandfather told me about that and what he says to people when he talks to them is that this represents the grain of life. He says everything in life moves in this way, in this manner. And he says whenever you undertake anything, whether it's a project that's going to take five minutes or that's going to take your lifetime, the only way that you can possibly have any success with it is to start at the heart of it. Start at the center of it and work your way out and then you will succeed.

I think of that very much in relation to seed, that we are dealing with essences. We are dealing with the essence of life. When we hold a seed in our hand, we hold history. We can't even comprehend how much history is in that seed. Both what has come before and potentially what will come after us. And we are only a moment in that transition. That because it's in our hand, there seems to be something that the Great Spirit has in mind for us....We are the Great Spirit's agents in this world. That the perpetuation of life is exactly what the Great Spirit created all living things to do. And I feel that we are adolescents as a race of people, just beginning to reach the point of initiation into further degrees of knowledge. And my great prayer is that each year that we come back together, we can share deeper aspects of that knowledge and help guide one another through this time....

Rob Johnston Jr.

I'm Rob Johnston and my company is Johnny's Selected Seeds (Albion, ME 04910) which is in its ninth year....One of the reasons that I got involved in peddling seeds was that I didn't see anyone doing first-hand development of varieties for short season areas. There is no guesswork involved when items are being researched in that climate....

Seed growing, seed saving, is not generally technically difficult. A conference like this tends to flood people with information. It's difficult to grasp all of it.

People leave thinking that there is a lot of stuff that they are going to have to study and learn. I really believe that it is more a case of just doing it than learning too much about it. There are three processes that people go through when they do anything: they decide what they want to do, then they figure how they are going to do it, then they do it. Most people get hung up on one of the first two, generally the second one--how are you going to do it? I say just eliminate #2 altogether. Just--what?--do it! Because it's really not that difficult. When you're saving seeds from something, you're just mimicking what the plant naturally does anyway. There is no engineering that has to take

We have a number of older varieties in our catalog that we have commercialized which have a uniqueness or usefulness in their own right that makes them worthy of growing in this modern day, if you will. They don't <u>only</u> have sentimental or old-fashioned appeal, or academic appeal when looking to the long range; they're good right now. And if we list an old thing like that and tell people that it's good, you ought to grow it, people generally will and it will sell....

Just a small percentage of our customers we expect will ever get involved in saving their own seed, in seed growing. And yet I really don't think that the numbers are that important. A small number of people getting involved isn't necessarily insignificant, as long as the practice of doing it continues to exist. In 1976 I wrote a small booklet called "Growing Garden Seeds" which tells people in plain language the basics of saving seed. And the thing has sold an incredible number of copies....If one percent of the people that bought that book really get involved in saving seeds, I would be happy. I think that it would be a big step....

I think it's important that an individual or a single project should specialize to the degree that's necessary to do an adequate job on the materials that are being handled. I think that it would be a mistake, for example, to say that you are going to grow all of the native vegetables that were grown in your area starting in the year 1900. You might be able to get all that planted, but whether you would get it all successfully harvested and properly recorded and so forth, even if your time was worth nothing to you, would be very doubtful. The opposite of that would be just taking the woods strawberries that grew naturally in your region, becoming familiar with them, collecting the seeds and maintaining a viable supply of the seeds. Now that sounds like a very specific project, but it would certainly be of much more value....than to make a real mishmash of a big project and then just burn out and come out with nothing in the end. So I would suggest that people stick with the crops that they enjoy. If you have certain crops that really intrigue you, work with those and forget the other things because somebody else is going to have those other crops as the ones that they are most interested in....

We currently have a sort of a hip pocket project on rice....I would feel very pleased if at the end of my career, we could have developed a variety of rice that could be grown at our latitude....

Generally speaking, the plant pathology departments at universities are quite willing to be of help on questions of seed pathology. It's very rare when a private organization can afford the salary of a qualified plant pathologist. It's really nice to be able to go to the university with all your questions, that in order to have answered otherwise, you'd have to have a salaried person on full-time....

QUESTION (Gary Nabhan): Are constraints within the commercial seed industry difficult for a small company carrying a lot of varieties to maintain in terms of purity and germination testing and all of that? In other words, if you're a small seed company with a lot of varieties, are you fighting upstream because of that?

The smaller the operation, generally speaking, your quality control will be a much bigger percentage of your budget than it will be for a larger company. If you have 100 pounds of a certain kind of seed, it's going to cost you the same amount of money for quality control as somebody that's selling a ton. So this is a real factor and it's a difficulty....

It's just really fantastic in this conference that we have a mix of people from, well I don't like to use the cliche, but from all walks. I'm starting to get the impression really clearly that people by nature are willing to cooperate....

****

If I were to pick out the moment during the Workshop that made me feel the best, it would have to be a compliment given me by Dr. Howard Scott Gentry. Dr. Gentry was a USDA Plant Explorer from 1950 to 1970 and is currently an Economic Botanist with NEWCAST at the University of Arizona. Gary Nabhan warmly referred to Dr. Gentry as his academic grandfather. To tell you the truth, I was awed by the elderly man's presence.

The panel I had just been on with John Withee and Forest Shomer and Mahina Drees had been rather lively and now we were breaking for lunch. Dr. Gentry came up to me and said, "I want to thank you for your appearance here and what you are doing....We need somebody besides bureaucracy, we need grassroots people doing work of this kind. Your work has exciting potential....This is a great moment....I just wanted to thank you...."

****

All too soon, it was almost over. People were getting up and making resolutions that we should do it again next year. Some suggested that we start immediately to line up funding to hold it again. Forest Shomer got up and said he thought we ought to set it up like any other conference. That he would gladly pay the 60 bucks for the weekend and his travel expenses to get there. Then he made what was, for me at least, one of the greatest statements of the whole conference. Forest looked out at all of us and grinned and said, "I think we ought to take this traveling medicine show on the road!"

(<u>The 1981 Fall Harvest Edition</u>, pages 26-41)

The Growers Network

(Editor's note: The Growers Network was formed when the SSE was asked to take over John Withee's Wanigan Associates bean collection in 1981. John had spent the previous 14 years assembling a bean collection of 1,186 accessions. Today the SSE's Central Seed Collection contains all types of garden plants and totals nearly 5,000 strains. Nearly a third of this huge unique collection is grown out each summer in a Preservation Garden near Decorah, Iowa. This chapter contains the series of articles about our Growers Network during the time that it was being formed.)

**Kent Whealy with Wooden Seed Drawers
Which Hold Air-Tight Bottles
Containing The SSE's Collection
Of Over 2,200 Bean Accessions**

THE SSE'S GROWERS NETWORK

Winter, 1981—The Beginnings....

Wanigan Associates

Wanigan Associates is a group of about 400 bean collectors who are working to save heirloom bean varieties from extinction. John Withee is the guiding force behind this unique organization. John's "hobby," as he calls it, began with his search for one long-remembered bean. The beans being traded and grown within his organization now total about 1,100 varieties. Quite a hobby!

John retired three years ago and hasn't had a spare minute since. Wanigan Associates has snowballed to the point that it's more than he can handle. John has requested that the Seed Savers Exchange take over his organization. I fully realize what a trust this is and the responsibility that goes with it. I feel this is a tremendous opportunity and hope to combine our two networks into an even stronger organization.

I intend to drive to Massachusetts at the end of February to meet John. Along the way I will visit Miles T. Roberts in Villisca, IA; Paul Dennis in Des Moines, IA; Russell Crow in Capron, IL; and Ralph Stevenson in Tekonsha, MI. All of these men have very large bean collections. I will try to collect three or four packets of each variety along the way. I also intend to make up some flat wooden carrying cases with glass fronts and many small compartments. I hope to take a small sample of every colored bean in each collection (John has said that about half of all his bean varieties are distinguishable by their seed). For the first time we will be able to see and examine the entire collection.

I will take four packets of every variety of which John has sufficient seed. Two of these packets will be entered in the two frozen collections of the Heirloom Seed Bank. The other two packets will be distributed to two different gardeners participating in our new Growers Network. There will be no choice of varieties in our Growers Network, which is simply a Preservation (multiplication) Program. However, our members offer approximately 400-600 varieties each year to anyone looking for a specific heirloom bean.

John uses Prof. Meader's method of classification, which catalogs a bean under the name it had when acquired. Although the seed of two beans may look exactly alike, if they have different names they are considered to be different varieties until they are proven by growing to be the same. This is as it should be, but I think it is time for us to start growing these look-alikes to weed out the duplicates. (John estimates that 1/4 of his collection are possible duplicates.) I request assistance with this problem from USDA growers, university horticulturists or anyone with the knowledge to grow these possible duplicates and determine which are unique. This is a job for experts....

I've never been more excited about anything in my life. I've heard that 180 varieties of Burt Berrier's beans survive in this collection. I welcome all Wanigan Associates to our ranks and trust you will be pleased with our organization. Let's all work to multiply this fantastic collection every year through our Growers Network.

Growers Network

Many gardeners write me saying that they do not have any heirloom seeds to offer, but that they would gladly work to increase such endangered seed stocks. Until now the SSE has never systematically distributed seeds, but has merely let its members trade among themselves. Such haphazard trading does not ensure that the varieties will not be lost. With this Growers Network we can: maintain large collections that are endangered; offer aid to any gardener needing help multiplying a variety because of infirmity or whatever; multiply varieties for the Heirloom Seed Bank; and distribute seeds to non-listed members who wish to join our ranks.

This year we will mainly be multiplying the collections of Wanigan Associates and hopefully Gleckler's Seedsmen (Merlin Gleckler had a stroke and has gone out of business and, although I am negotiating to obtain samples of his seeds, I would like to hear from any gardener who is presently keeping any of his varieties). We may also be multiplying Edward Lowden's collection (mainly cold-hardy and sub-Arctic tomatoes) and I would like to hear from any northern or high-altitude gardeners who would grow these tomatoes.

I have written our Seed Saving Guide to teach you how to keep varieties pure. With some cross-pollinated crops, unless you are adept at making hand pollinations you will have to be aware of any garden within 1/4 mile. But with tomatoes and beans, there is virtually no problem with crossing, and they're mainly what we will be multiplying.

I would rather see you multiply two varieties successfully than to take on 20 and fail. You will have to be the judge of how much you can handle. I have no way of knowing: how good a gardener you are; how good your soil is; if you have watering facilities; if you are in a drought area; if you are planning a three-week vacation in the heat of the summer; if you are skilled at growing tomato and pepper plants from seed as part of your yearly gardening routine; or if you live in town where cross-pollination from neighbors' gardens will ruin corn, watermelon, squash, etc. In many cases I will be entrusting you with rare and endangered varieties. You must be aware of the responsibility involved.

Your letters of request must reach me before March 15, 1981, when I will be sending all seed samples. I realize this will be too late for southern gardeners to start tomatoes and peppers from seed, but there is nothing I can do about it this year. Tell me how many varieties of each vegetable you can handle. Example: two pole beans, four bush beans, two tomatoes and one watermelon. Depending on which collections I can secure, such an order might be filled entirely with beans in a one-to-two ratio of pole beans to bush beans.

The packets of seeds I'll distribute will contain about 1/4 cup of bean-sized seeds, down to about a level tablespoon of tomato-sized seeds. Along with each packet of seeds I will send three empty packets (all marked with the variety name) which you will be expected to return after harvest with the same-sized samples. Even if you have a complete failure, return these three packets. Please realize how disastrous mix-ups will be and how well you will have to keep track of the varieties.

There will be a charge of $1 for each variety requested to cover the costs of buying and printing packets, postage on samples, printing of forms, etc. Send all requests for seeds to: Growers Network, c/o Kent Whealy, Rural Route 2, Princeton, MO 64673.

(The 1981 Winter Yearbook, pages 50-51)

Fall, 1981....

Last spring I distributed 404 packets of seeds to 74 gardeners who participated in our first Growers Network. That multiplication is now being returned, but until it's all in I can't really tell you how successful the program was. It's looking really good, though. As with any new program, it takes awhile to get the feel of it and to get your methods down. John Withee had given me advice, especially about the correct timing for mailings to different parts of the country, so I had some excellent guidance. I am really glad this started last year and I had a chance to get most of the bugs out then, because if we are to accomplish what we must this next year, our Growers Network will have to grow to 4-5 times its present size! Hopefully a lot of the Wanigan growers will join us. And Rodale's Soil and Health Society has made our Growers Network a project for their Backyard Researchers, which should also add 50 or more growers to our network.

Multiplying the Wanigan Associates Collection

As you know, the SSE has taken on responsibility for multiplying the Wanigan Associates heirloom bean collection. This is the largest collection of heirloom beans that has ever been assembled. It now numbers over 1,200 accessions. Many of the people who were the original sources for these beans have passed away. The remnants of Burt Berrier's collection are here. There will never be another opportunity like this and there will never be another collection like this. It is irreplaceable and its survival depends on how well our Growers Network functions this next year.

John was unable to make any distribution this last year because of health problems. The accession list for the collection shows that the last time the beans were grown out was in 1978, 1979 and 1980, with about a third being multiplied each year. In most cases, beans must be renewed within four years if you are to keep them going. So we are already within a year of losing the third of the collection that was last multiplied in 1978. And if we are not able to also multiply the ones that were last grown in 1979, then next year we will be in exactly the same shape with them.

I am awed and humbled by the work that John Withee has done with this collection. The best way that we can really show him how appreciative we all are is by working together to assure that the Wanigan Associates collection lives on through the years. We all have so much to learn from John. I really hope that things go well for him and that soon he will have the time to share his knowledge with us.

I am proud that John has trusted us with the collection. I also feel the weight of the responsibility very clearly. From the moment the UPS truck delivered those three big boxes of beans, they've been--well, it's hard to explain, but--like a presence. At first it was so strong that I would wake up in the night worrying that I would fail and they would die. The collection is so huge. If you took out a packet and looked in it and put it back and that took you one minute, it would take you 20 hours to go through them all. But I was forgetting that I am not alone, and also the kind of people that all of you are. I'm now confident that we can succeed, but I do desperately need your help.

If you want to help grow out some of the Wanigan Associates collection, send me your name and address and tell me approximately how many bush varieties and how many pole varieties you will be able to handle. This will give me a rough idea of how many we are going to be able to multiply and how I should proceed. (If you live in an area where you may be troubled by damp weather at harvest time, it may be a good idea for you to concentrate more on pole varieties. As their ripening proceeds upwards, it will give you a better chance to pick dry pods over an extended period than with the bush varieties which only give you one or two pickings of dry pods at the end of their season.)

The number of varieties that we will be multiplying is so large that there is no way I will be able to send each of them to two different growers as insurance against their loss. So you will be the only one growing those particular beans. I would rather see you multiply four varieties successfully than to take on 20 and not do as well with them. But if you have confidence in your skills as a grower, don't hold back, because it's now or never.

I will tell you right now that I do not intend to charge for the seeds that I distribute through the Growers Network. They're not mine. The seeds were given to me; I'm just spreading them. I just don't feel that's right, since you will be going to the trouble of growing them and also to the expense of sending back the samples of them....

This is the greatest challenge our organization has ever faced. To those of you who decide to participate in this project, I really appreciate your help and I'm very grateful for it.

By this time some of you probably think that the only things that the Growers Network will be multiplying are beans. Not so! We are slowly building up a pretty good inventory of other vegetables as well. If you are keeping seed of a vegetable that you can no longer maintain or if you are keeping a rare or unique vegetable variety that you would like to see added to the inventory of the Growers Network, send me a sample of new seed (if at all possible) and a complete description of both the plant's characteristics and its history. Please also tell me if the plant will only do well in certain climates or under certain conditions.

(The 1981 Fall Harvest Edition, pages 22-23)

Winter, 1982....

I want to welcome all of you to the second annual Growers Network. We are very fortunate in having about 50 gardeners from Rodale's Backyard Researchers participating, along with a good number of Wanigan Associates who have also joined this combined effort. Add to them the best growers from the SSE membership and this can't help but be an exciting and successful Growers Network. This multiplication is at the very heart of our efforts, and I am extremely grateful to all of you....

A potential problem, that was brought up repeatedly at the Seed Banks Serving People Workshop in Tucson, concerns purity. When we send out Aunt Mary's sweet corn for multiplication, how can we be sure we are getting Aunt Mary's sweet corn back? This question of purity (or lack of it) is something we must all concentrate on, whether it's the multiplication for this Growers Network, seeds that you intend to offer through the SSE, or any variety that you don't want to be lost due to contamination.

I want everyone to carefully study our "Seed Saving Guide." It will be expanded in this fall's Harvest Edition, but it is complete enough now to let you know what you must do to keep varieties pure. In many cases I will be entrusting you with rare and endangered varieties, and you must be aware of the responsibility involved. Please don't think that I'm trying to lecture you. Many of you know more about keeping plants pure than I ever will. But others of you are coming into our programs for the first time and may not realize the precautions you must take, and they are essential.

We are very fortunate to have obtained: Aunt Mary's sweet corn that W. W. Williams of Ohio has kept pure for 40 years; Earliest Catawba sweet corn that the late Phil Hewitt kept pure for 40 years; tomatoes from the collections of Ted Telsch, the Rev. C. Frank Morrow, Ben Quisenberry, Edward Lowden and Gleckler's Seedsmen; squashes from the collections of Tom Knoche and Glenn Drowns; beans from the collections of Russell Crow, Ralph Stevenson and John Withee's Wanigan Associates; and many, many more. My thanks to all of you who have contributed your seeds so generously....

(The 1982 Winter Yearbook, page 4)

Fall, 1982....

The multiplication of the Growers Network is streaming back in right now. I won't actually know what percentage of return we're getting until everything is logged in, but the whole project seems to have been very successful. I got a tremendous distribution on the Wanigan Associates bean collection. This year the Growers Network had 351 gardeners participating. I was able to send out over 2,400 packets of seeds. It's too bad the weather wasn't better this summer, but we certainly can't control that. I just hope most of you weren't plagued by strange weather the way we were. It rained here for six weeks straight when it normally would have been dry.

The more I work with the Wanigan Associates collection, the more admiration I have for John Withee and what he has accomplished. He spent 14 years putting together this fantastic bean collection and then simply gave it to us. There's no way that we can ever really thank him sufficiently for that, except by working together to keep the entire collection alive so that gardeners in the future will be able to enjoy growing them, too.

And not only did John do it, but he did it right. John always stressed to me the importance of keeping a good record of original sources. He did a fine job of that considering the incredible rate at which he was receiving material during the last few years. I want to stress how very important it is to keep a list of not only the name of every seed you receive, but also the person's name and address who sent it to you. Then if that seed is ever lost, you will be able to possibly get another sample. We can also get back to these sources for the seed's history, which is not only fascinating but will help establish how long the variety has been kept in a certain area. And that may determine whether a specific Historic Garden will be willing to maintain it permanently. Also, as people send seeds to the Growers Network, we can tell immediately which are duplicates of ones we already have without actually growing them out, but only if we know your source. It simply isn't that hard to make up a form and log in the seed and its source when you receive it. It literally makes the difference in having a collection of living, documented heirlooms as opposed to just a collection of seeds.

I'm setting up the Growers Network in this Fall Harvest Edition in an attempt to get the Growers Network applications back much quicker. Last year I set it up in the Winter Yearbook and by the time everyone had sent in their applications it was well into March. That's much too late for many of you to start tomatoes and peppers indoors (yes, there are seeds in the Growers Network other than beans). This year I will start sending out seeds just as soon as the Winter Yearbook is mailed (early February).

This year I have decided to only distribute beans, tomatoes, peppers and lettuce through the Growers Network. The cross-pollinated crops (corns, squash, watermelons, cantaloupes, cucumbers, etc.) will only be sent out to selected growers who are already experts at hand-pollination. This does not mean that I am questioning anyone's ability as a grower. It's just that this thing has gotten so big that I don't know all of you anymore and when I'm distributing extremely rare materials which may be lost forever by crossing, or trying to get back seed for long term storage, we simply can't afford any mistakes. If you know that you are proficient enough to work with cross-pollinated crops, write me and describe your qualifications and what you wish to work with. Later this spring I will send you lists of what's available....

I intend to make up and distribute only two packets of each of these seeds next spring. Even that will be over 3,000 samples, which is more than everything I sent out last year combined. You can immediately see the problem I'll have, since one out of every 10 people will want Jacob's Cattle Gasless. I really can't promise that you will receive anything you request, especially if you are just requesting names. (It is much better to go after beans which the Source Code shows originated or were last grown in your area. Those should do the best for you. Ones that you have read about in an article about John, or ones that you simply like the names of, may not do all that well in your location.) I will try my best to send you beans that are similar to what you request and that are from your area. It has to be this way these first couple of years while we are building up our stock of seeds. The real purpose of this project is to save this endangered collection. If you aren't here to help with that, if you are going to be really disappointed when you don't get everything you request, you might want to question whether or not you should be participating.

Seed Data Forms — I've had forms printed which ask a series of questions about seeds: origins, source, history and a complete description of the plant and its habits. I would like a sample of any bean that does not appear on this list (except for commercial varieties). But first write and tell me how many Seed Data Forms you will need....I am especially interested in locating any collectors who are keeping beans that you received from Burt Berrier (but which did not come from the Wanigan Collection). Also, if you are, or know someone who was, the "original source" for any bean listed here that has no Source Code, send me a couple of beans to compare with what I've got. If they match, I'll send you a Seed Data Form to fill out.

I also want to start building up the following collections within the Growers Network: tomatoes (especially), peppers, lettuce, potatoes, peas, eggplant and okra. I will also distribute to selected growers: sweet corns, popcorns, squash/pumpkins, watermelon, cantaloupe and cucumbers. But please send for the Seed Data Forms before you send me any seeds. I am especially interested in heirloom varieties. Please do not send commercial varieties unless you're sure they have been dropped from all catalogs.

Wanigan Bean Project at the Rodale Research Center

When John Withee asked me to use the Growers Network of the SSE to distribute the Wanigan Associates bean collection, it was a tremendous job just to turn the collection over. The Rodale Research Center sent four young women up to Massachusetts in the summer of 1981. They threshed all of John's beans, answered correspondence and attempted to get the huge collection into a manageable order. At that point the entire collection was split into three parts: one part stayed with John, one part was sent to the Seed Savers Exchange and one part went to the Rodale Research Center.

Peggy Haas, who was with the team that went to Massachusetts, sent me a report this spring about their work with the Research Center's collection. They are increasing every accession which was in really short supply (contained only one to twenty seeds). They are also increasing the oldest seeds in the collection (those last grown in 1977). They hope to increase their own seed supplies, and also provide the Seed Savers Exchange with increased seed which will be distributed through the Growers Network. During this process they will be observing a total of 370 Wanigan beans growing in the field and taking notes on their characteristics.

I couldn't be more pleased with their project. It complements the activities of our Growers Network perfectly. My distribution was a massive one. I distributed all of the beans they are growing to "selected" growers. In addition, I distributed all of the other beans in the collection at least twice and in some cases up to a dozen times. Between the two programs and the two approaches, I think we will be able to save everything possible from this tremendous collection. I really appreciate the assistance and support that the staff of the Rodale Research Center has accorded me. I don't think I have ever met a more professional group of young people.

I also want to thank Louise Schaefer who is with Rodale's Soil and Health Society. Within the Soil and Health Society there is a group of gardeners known as the Backyard Researchers who are working on gardening projects. Last year Louise ran a newsletter article in the "Soil and Health News" describing our Growers Network and encouraging interested Backyard Researchers to participate. Almost 100 of our 351 growers this last year were the result of that article. And she ran a similar article this fall, too. I really appreciate the way the Rodale organization has supported our efforts. Their help has been invaluable.

Thank You

And now I want to thank the people who have really made this project a success--the gardeners who participated in the Growers Network during the summer of 1982. You are the ones who made it happen. You planted, protected, nurtured, harvested, dried and even paid the postage to return seeds to me. This collection is yours, as far as I'm concerned. It was right on the edge of being lost, but you wouldn't let it happen. Because of your efforts, gardeners in years to come will be able to enjoy this fantastic collection. I think we all have every right to be very proud of what's being done here.

(The 1982 Fall Harvest Edition, pages 67-71)

Winter, 1984....

Early last spring, John Withee sent me the remainder of the Wanigan Associates bean collection (actually, I think he sent me the complete contents of a couple of sheds). The Rodale people picked it all up, repacked it into 33 large, sturdy boxes and shipped it to me. In addition, I also had over 300 smaller boxes of seeds from our 1982 Growers Network.

A few weeks before that, James Eagle, one of our members from Florida, called to ask if I'd be interested in some half-pint hospital formula bottles with excellent airtight lids. To make a long story short, James washed and packed and shipped me over 3,300 of these jars and only charged me for the shipping! What a tremendous amount of work he did.

Then I called Russ Crow, a SSE bean collector and friend. He'd been in a car wreck a week earlier, had torn up some ribs but was all right, was off work for a month and was sitting around the house bored. For room and board and a bus ticket, Russ agreed to come down and help me with the beans.

I took the computer data on the beans that was in The 1982 Fall Harvest Edition, added the new varieties to it and printed out all of the information on

Kent Whealy and Russ Crow
Finished Sorting John Withee's Last
Shipment and the Rodale Increase

three sets of gummed labels (one set for the jars and the others for two packets of seeds). I rented the basement of a large community building in Princeton and hauled all of the seeds into town in a farm truck. That same day a semi delivered 140 cases of bottles from Florida.

We really went at it--shelling, cleaning, sorting, labeling jars (and color coding them for last year grown) and making up nearly 2,500 packets of the seeds that were old or in low supply. At this same time Jim Schooler, a woodworker friend who made the beautiful bean display cases which most of you have seen in various pictures I've published, was building five units of seed drawers. Each unit is 10 drawers high and each drawer holds 80 bottles.

At the end of two weeks, the mess we had started with had been transformed into 2,200 labeled jars of seed and over 2,400 labeled packets of seed ready to be sent to the 1983 Growers Network. At last the entire bean collection was in perfect order and literally at my fingertips! There is no way that I can ever thank James Eagle or Russ Crow enough.

I also want to thank our Growers Network for the tremendous job they have done these last couple of years. When John Withee sent me his collection in the fall of 1981, I was really apprehensive. About half of the 900 samples were either tiny (a dozen seeds or less) or 3-4 years old. Two years later, we now have ample supplies of new seed (1-2 years old) of over 800 of those beans. Your efforts have been really incredible.

New additions to the Growers Network collection have continued to arrive in a steady stream. We are now keeping over 2,000 beans, 500 tomatoes, 200 squash, etc. With all of the disruption involved when my family moves to Iowa this spring, I know that I won't have time to distribute the 2,000+ samples of seeds like I have the last couple of years. So this spring I will distribute only beans and there will be no choice of varieties. We will concentrate on renewing those beans in the collection that are still in short supply or older. If you are willing to participate under these conditions, send me your name and address and tell me how many varieties you want.

(The 1984 Winter Yearbook, page 253)

Fall, 1984....

The Growers Network Seed Collection has grown very rapidly these last few years. The SSE is currently maintaining over 3,500 endangered vegetable and garden seeds (2,000 beans, 500 tomatoes, 200 peppers, 200 squash, 140 corns, 100 muskmelons, 100 potatoes, 100 lettuces, 100 peas, etc.). This invaluable and irreplaceable collection has become so large that the time involved (for distribution, processing after return, storing and redistribution) makes the project unworkable in its present form. It is absolutely necessary to change the way our Growers Network is structured. Instead of multiplying and returning the seeds each year, we want you to select certain seeds that interest you and grow them for at least five years before turning them back in. I realize that these new guidelines may not be acceptable to some of our growers. But the project can simply not proceed as it is now structured, and the changes are necessary to add long-term stability to our Growers Network and to ensure the survival of the Growers Network Seed Collection.

We are looking for growers who would be willing to participate under the following conditions: 1) adopt specific seeds and agree to maintain them for a period of at least five years; 2) follow our guidelines for purity while these varieties are being grown; and 3) offer a supply of these seeds through your Yearbook listing so that they will be available to others. If these conditions are acceptable to you, send me your name and address, and during the first week of January you will be sent a list of approximately 1500 varieties to choose from.

(The 1984 Fall Harvest Edition, page 156)

Winter, 1985....

The SSE's Growers Network is switching over from a program of grow and return to one of long-term maintenance....We have developed a strong, solid core of growers who know that they will be participating in the SSE over the long run. These dedicated and committed members are the ones we hope will participate in our new Growers Network.

Things have been so incredibly hectic around here that we still haven't had time to send out the list of Growers Network varieties from which you can choose. It's getting late enough now that perhaps we had better start off slowly this first year and begin by distributing only the beans from the collection. Glenn Drowns has all of our seeds in perfect order and already has the packets made up. We have a printout of over 850 bean varieties of which we have a large enough sample of new seed to distribute. If you are interested in "adopting" some of these beans, please write and ask for a copy of the bean list.

(The 1985 Winter Yearbook, page 247)

THE ORIGINAL AUNT MARY'S SWEET CORN

(From a letter from W. W. Williams of Ohio)--"Dear Kent....Back in the early 1930s, my father and I both worked for the gentleman that discovered and developed (Great) Aunt Mary's sweet corn....To this day, I have kept this corn true and alive. Personally, I don't think there is another sweet corn that can compare with it....if you are interested in seed from this true Aunt Mary's sweet corn, please let me know....I keep a gallon or so of seed each year and I never throw the old seed away untl I'm sure my current seed is fertile...."

You better believe I was interested! With all cross-pollinated crops, there is no way to tell how much they have changed over the years due to outside contamination. Our Growers Network is very lucky to have been given both W. W. Williams' Aunt Mary's sweet corn and Phil Hewitt's Earliest Catawba which these two dedicated gardeners have kept pure for over 40 years. I wrote back to Mr. Williams and he sent me over half a gallon of Aunt Mary's sweet corn and the following history:

"....This started when Mr. Lee Bonnewitz went to visit an aunt in Virginia. She and her family lived back in some hilly farm country and this was during the Depression when everybody was having a rough time trying to exist. During his visit, his Aunt Mary served sweet corn, which he told all of his friends was the most delicious sweet corn he had ever tasted. Mr. Bonnewitz made a deal with his aunt for the seed of this corn, which had been in her family for years but didn't have a name. He brought back about a half bushel of seed on the cob which really looked terrible, but he was determined to salvage and develop it.

"His first planting was a small patch in the rear of his house and he paid my father $1 per hour to sit and guard this corn against any theft. After this first crop he started selective picking and finally ended up with an 80-acre farm totally planted with Aunt Mary's. This is about a 90-day corn and is a little difficult to get the seed from, since it is very sweet and will spoil in the husk. So you need to watch it carefully when it nears the seed stage....What I really like about this corn is that the early ears run up to about 10" long with very deep sweet kernels. You can pick from the same planting for up to two weeks. The later ears will be smaller, but every bit as tasty...."

(The 1981 Fall Harvest Edition, pages 23-34)

PHIL HEWITT PASSES AWAY

Phil Hewitt was a member of the SSE in 1978, 1979 and 1980. During that time many of us got to know Phil quite well. He always listed a purple-seeded, eight-row sweet corn which he believed was the Catawba variety which was developed by a New York minister in 1903. He had seen one other ear on exhibit in a museum at the Experiment Station in Geneva, New York. He had kept the strain pure for over 40 years. During that time he continually tried to find another grower who was keeping the same corn.

Mark Kane wrote an excellent article about the SSE in the June issue of Organic Gardening in which he mentioned Phil and his purple corn. Mark also wrote that government seed banks in Colorado and Iowa had no sample and that the purple ear of corn which was on display in New York labeled Earliest Catawba, was made of wax. Because of that article, I had two people write me who were keeping eight-row Catawba. I sent their names to Phil because I knew he would be delighted to finally have the chance to verify that his purple corn was indeed Catawba.

But I received back a letter from Diana Sammataro, a neighbor of Phil's in Connecticut, which said "....I am sorry to tell you that Phil Hewitt passed away....I am sending under separate cover a box of his purple seeded corn that Pricella Hewitt (Phil's sister) gave to me....Phil was a good friend and I understood his love of plants. He was very proud of the red corn...."

(In a letter from Phil's sister)—"I think a lot of people are missing Phil as he was so knowledgeable about so many things—bees, plants, flowers, animals, literature....as far as I know, the corn that Diana sent you was from 1980. Phil had dried a lot of ears and I gave them all to her with the condition that she supply anyone from Seed Savers that asked for it....Phil was very proud of that corn and I am sure he would have been happy if he knew you were going to keep it alive....I do not have the 'green thumb' that he had. He could make almost anything grow. My other brother and I have a small garden growing in the same place where he had his garden, but it is not at all as luxuriant as his gardens were...."

If it hadn't been for Phil's sister and his neighbor Diana knowing about the SSE, we might have lost Phil's corn. It might be wise to tell trusted family members about possibly sending your seeds to the SSE, if anything should ever happen. I don't like to talk about such things, but it might be a good idea.

I have two gallons of Phil's corn which will be offered through the Growers Network. I hope to get the two gardeners who are keeping Earliest Catawba to grow a little of Phil's corn and compare it. I just wish Phil could know how that turns out.

(The 1981 Fall Harvest Edition, page 24)

MOON AND STARS WATERMELON FINALLY LOCATED

For the last four years, members of the SSE have been trying to find Moon and Stars watermelon (some folks call it Sun, Moon and Stars). Last spring a crew from the television station in Kirksville, Missouri came over and filmed a ten-minute report about the SSE. The interviewer, Ron Heller, and I just sat in front of the fireplace and talked. I had time enough to make a pretty good statement about what we are doing. As usually happens when written articles are published, there was a flurry of calls and letters from people keeping old varieties just after the film was aired. Merle Van Doren who lives about 100 miles from me wrote that he was keeping Moon and Stars watermelon and said that he would save me some seed.

I went down to Merle's place when his melons were ready. Jan Malmgren and Larry Hollar were here from The Mother Earth News doing an interview and taking some photos which will be in the January issue. We tried to get Merle to have his picture taken with the melons, but he shied away. Merle told me that some folks would give a pretty penny to get their hands on some of that melon seed. I told him I was afraid I was just going to have to give it away to as many people as I could. He just grinned at me.

Moon and Stars is a round dark green watermelon that looks just like Black Diamond, except for its bright yellow spots which range in size from a pea to a silver dollar (although some are bigger). Merle says it's about a 90-day melon and that it's much more disease-resistant than Black Diamond.

Merle also gave me the crown fruit from a pure white cushaw pumpkin. It hasn't been available commercially since Shumway changed hands a couple of years ago. He also gave me seed of a four-foot-long, slim, orange, odd-shaped squash that he calls Kentucky Squash. I've never seen it before. Its name has been lost because it changed hands a half dozen times as it traveled from Kentucky to Missouri. All of these seeds (plus those of a hundred or more other unique vegetables) will be available through the 1982 Growers Network.

(The 1981 Fall Harvest Edition, pages 24-25)

Stories

Ralph Mostoller Displaying Beans

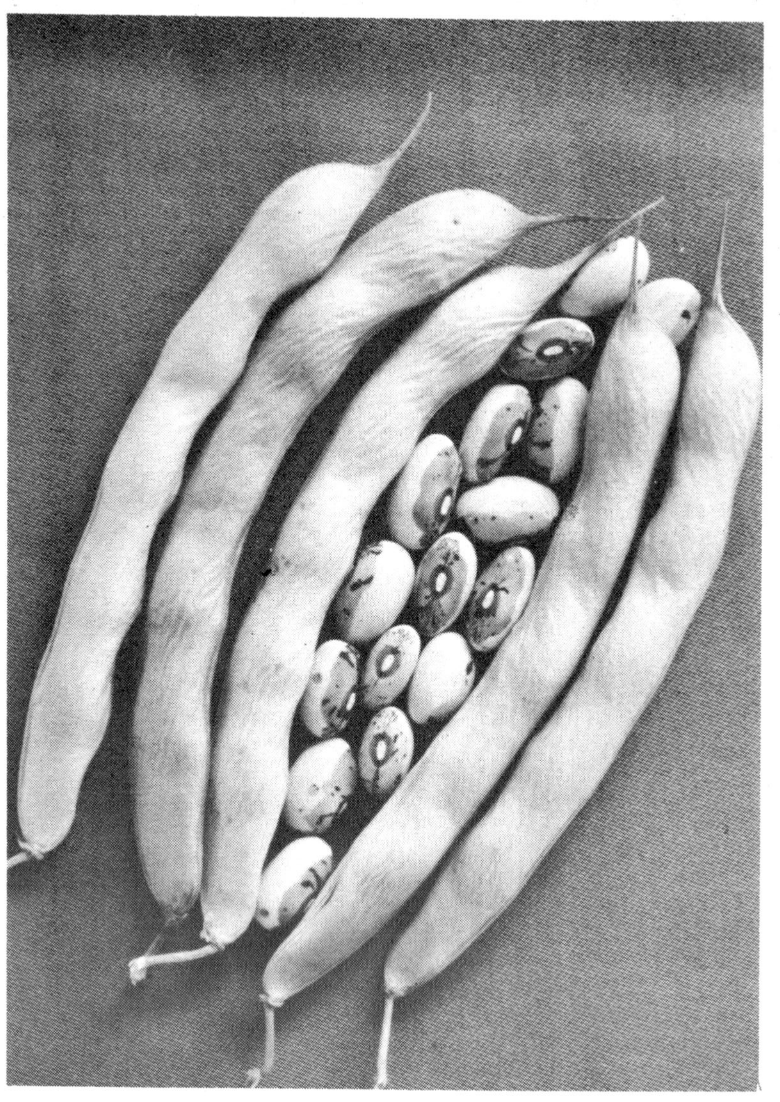

The Mostoller Wild Goose Bean

MOSTOLLER WILD GOOSE BEAN
by Ralph V. Mostoller

"This is the forest primeval.
The murmuring pines and the hemlocks,"
--Longfellow

Those opening lines from Longfellow's Evangeline are reminiscent of the valley of the upper Stoneycreek when, in 1776, Frederick Mostoller, his wife, two sons and two daughters, came from eastern Pennsylvania to settle in Somerset County, then a part of Bedford County. The oldest of the children, John, was only six years old at the time. While virgin pines, hemlocks and rhododendrons thrived in the valley, various hardwood trees grew on the slopes and chestnut trees topped the ridges. Not until after 1881, when the Somerset and Cambria Branch of the Baltimore and Ohio Railroad was extended from Somerset to Johnstown, did extensive lumbering of these virgin forests begin.

However, during the Civil War period, one of Frederick's grandsons, Joseph, built a water mill to saw timber a few hundred yards downstream from where Wells Creek empties into Stoneycreek. Then a race was built to divert water from Stoneycreek to power the sawmill. Though the sawmill has long since disappeared in decay, the course of the old race can still be traced. A bridge was built over the race where it intersected the road leading from Lambertsville to Mostoller Station. A few older persons still remember the so-called "race bridge."

In September of 1864 one of Joseph Mostoller's sons, John W., was mustered out of the Union Army, just in time to help his father in the construction of the millrace. The following June, in 1865, another son, David, after serving two enlistments, was honorably discharged from the Union forces. Having a wife and several children to support, David quickly joined his father and brother in lumbering and in the operation of the sawmill.

That fall when the waterfowl were migrating, severe weather caused a number of ducks and geese to land on Stoneycreek, and one of the geese ventured into the millrace. In modern times the ducks and geese would probably have landed on one of the many artificial impoundments. Since this part of Pennsylvania was unglaciated, it lacked natural ponds and lakes, and landing places for waterfowl were necessarily streams and rivers.

David was the first to discover the migrants and told his brother John, noting particularly that one of the geese was on the millrace. Both owned guns and were proficient in their use, having been hunters and having served in the War Between the States. John had especially distinguished himself, even though only a private in the Union Army. The following is from an article in the December 9, 1925 issue of the Somerset Democrat following his death at age 82:

"Thus both Fate and Death have held in seeming abeyance natural laws to permit long life to the young soldier of twenty-one, who on the battle front of Lynchburg, Virginia, on June 18, 1864, by his own initiative gave the command, led the charge and captured a Rebel battery on the hillside, and thereby saved an entire brigade of the Union army from otherwise certain destruction, or capture."

For his meritorious service he was awarded a Distinguished Service Medal and was thereafter addressed so many times by his friends as "General" that many people never knew his given name. The medal is on display at the Union Street entrance to the Somerset County Courthouse.

In my possession is a powder horn which belonged to my grandfather, David, that may have been used to load one of the muzzle-loaders owned by the brothers. There is still some uncertainty as to which of the two brothers stalked the goose on the millrace. The lone goose, without the protection of the sentinels of the flock on the stream, was soon bagged by either David or John. Whichever, it must have been an exciting experience for him to bag so large a bird and he probably felt a glow of pride as he carried the dead goose to his mother's home nearby. As often happened in the early history of our country, the task of preparing fowl for the kitchen table fell to the women folk--in this case, Joseph's wife Sarah. After the goose had been plucked, the carcass was placed in a dry sink for further preparation. It was then that Sarah noticed a much distended crop or craw. Upon closer examination, she found that it contained a large number of beans, which she removed and placed on the ample log window sill to dry. They were different from any beans that the Mostoller family had ever seen.

In the spring of 1866 the preserved beans were planted to learn whether the variety was worth perpetuating. They grew and proved to be a shell bean variety that required poles rather than snap beans or bush beans. While not as prolific as some varieties, they were later multiplied and distributed among relatives and friends and came to be known as "Mostoller beans." Some of the younger generation know them as "wild goose beans." The beans are mealy and quite tasty when served as baked beans or in bean soup. Now, five generations later, housewives with cold freezers, freeze the shelled beans while green and tender and later prepare them for the table as they would baby limas.

The beans, when dried and shelled, average .35 x .8 x .7 mm in size. They are basically white but about one-third of the area around the "eye" or hilum is pinkish buff splotched with reddish brown. An occasional bean will have the hilum surrounded by a solid reddish brown area. In good soil the plants grow quite tall for pole beans. When I first planted them they grew taller than my poles which prompted a neighbor to facetiously remark that I must have planted Jack-and-the-Beanstalk beans by mistake. Gardeners who have grown the beans for many years relate that they mature in a shorter time than most beans, which leads them to believe that the original stock grew much farther north where the growing season is shorter. Indians are known to have planted corn and beans in the same cultivated area. It is possible that the unfortunate goose may have pilfered the beans from the Cornplanter Indians along the upper Allegheny River.

In 1974 some of the dried beans were sent by a resident of Washington, D. C., to the Beltsville (Maryland) Experimental Station of the United States Department of Agriculture. Dr. J. P. Meiners, Chief, Applied Plant Pathology, wrote to the donor as follows:

"....the 'Wild Goose' beans are very interesting and I plan to increase the seed and send it to Dr. Dolan at Geneva, New York, who is very much interested in heirloom beans....these beans are particularly interesting since they do not fit exactly the published description of the Wild Goose variety."

Of note, when writing about beans, is that one of John W. Mostoller's best friends was William Atlee Slick, also a Civil War veteran. Mr. Slick was a distant relative of the founders of the W. Atlee Burpee Seed Company who named a variety of beans the "Fordhook Lima Bean." Fordhook House is located in the parish of Acton, Middlesex County, a few miles from London, and was the ancestral home of the Atlees for many generations.

Much of the information about the time, locale and persons involved in the original acquisition of the beans was related to me by William J. Mostoller, son of John W. Mostoller, Jr. Details about the cleaning of the goose were told to William when he was a boy by his great-grandmother, Sarah Mostoller. Additional information was supplied by Mr. and Mrs. Willard G. Constable when they lived in Riverside (near Johnstown). Mr. Constable at that time was growing beans and other vegetables in the rich flood plain along Stoneycreek in Riverside. He gave me some seed beans in 1977 and I have been growing and sharing them with others since then. Mr. Constable received his seed beans from a friend of one of the Mostoller families.

The beans have now been planted by Mostollers, their relatives and friends for 117 years and are definitely "heirloom beans." I now have the satisfaction of knowing that they will continue to be grown long after I am able to cultivate them.

(Submitted by Ralph V. Mostoller, One Cleveland Street, Johnstown, PA 15902.)

(<u>The 1984 Fall Harvest Edition</u>, pages 148-150)

HOW KOKOPELLI GOT ME

by John Kimmey

(Among the Hopis, Kokopelli is the mythological character associated with fertility and germination. He is often known by non-Hopis as "the hunchback flute player." Kokopelli's unique image has been inscribed in stone or painted on pots throughout the Americas over many centuries....The hump on his back is, to many, a pack of seeds from which he plants as he moves around. The flute is said to be the source of spirit which is breathed into every seed.)

****

It was July, several years ago, when I was a guest for four weeks on Third Mesa at Hopiland. It hadn't rained for three weeks and the 100-degree-plus days hung heavy on the land. It was midday and my host snored peacefully in his cool stone house.

I was restless. I quietly closed the screen door behind me and stepped into the heat of the kisnovi, the plaza. All was quiet as midnight as I looked around for some movement. Only one dog stirred, just to change his prone position within a thin slice of noonday shade. Everyone else seemed to be practicing the ritual of brief hibernation which Tawa-Grandfather Sun imposed like a strong drug.

"Mad dogs and Englishmen," I mused, "out in the midday sun." I wasn't even aware of where my feet were carrying me as I descended the mesa's edge on a path long ago pecked into the soft rock during cooler days.

When I reached the base of the cliff a lizard scurried down to the bottom of a dusty path at my feet. I followed it, as if this creature was my guide. After about a fifteen-minute walk in one direction, the path suddenly turned north around a pile of rubble from the cliff. Before I could see around the other side of the rocks, I faintly heard a voice, singing. I slowed my pace and peeked past the last outcropping.

There before me was the widest expanse of cornfield I had yet seen here. It didn't even seem Hopi, it was so big. The singing became clearer though I saw no one.

It was the soft powerful voice of an elder, I realized. But where was he? I waited several minutes, listening to this singing cornfield. Then suddenly a white head appeared among the green clumps and moved through the rows slowly, the singing never stopping. I suddenly realized what my eyes had been showing me. This mid-summer crop was lush and beautiful. Approximately a dozen ears ripened amongst each small clump. A quick estimate indicated that I was looking at perhaps 1200 clumps of corn stalks.

The ground was dry and cracked from the long drought, yet the corn hardly seemed wilted as were most of the other fields I'd seen around the village.

Complaints I'd heard from farmers near the house where I was staying led me to believe that all the corn was dying of thirst, yet this field looked as if it had received rain just a few days earlier!

I quietly went back up the trail to the village, the old one not seeing me. My host was awake and asked where I'd been. When I told him what I had seen and heard, I watched his concern for my whereabouts turn into an amused smile.

"I see you found Titus' field," he said, chuckling.

"But why is his field still so healthy? Does he have a secret source of water?"

Grandpa just laughed. "Of course not. What he has is Navoti."

"What's that"? I asked, thinking perhaps some secret fertilizer existed which was only available to a certain clan.

"He has the Hopi Way," Grandpa explained after a pensive pause. "He knows the old songs to sing which refresh his corn children. He makes his prayers correctly at planting time. Most important of all, he knows not to worry, as this damages the plants as much as the drought. Instead of worrying, which makes his children nervous, he goes to them in the heat of the day and sings the old songs which give his children courage."

"But Grandpa, surely the other men see the difference in his corn. Why don't they learn his songs and sing to their corn?"

My elder Hopi teacher sighed. "That would do no good. Navoti is not alive in the others' seed."

Upon my return from that important month on the mesa, I drove north up the length of the Rio Grande Valley to my home at Taos. As I passed each of the nineteen Pueblos, I felt as if something was calling me. I noticed, perhaps for the first time, how little the farmlands of the native cultures were under cultivation, even for alfalfa.

I felt it was as if the seeds were calling. I realized that the source of power I felt was trapped in sheds, pots, coffee cans and lard buckets tucked away in dark corners, and in long braids of corn lashed together into "ristras."

The seed which was calling me was the old seed, gathered in the years before supermarkets and commercial seed packets which appear on store shelves early each spring.

It was the seed which Grandpa had told me about; that which still held the Navoti of ages past. After perhaps fifty years its vitality still remained within. The high, dry climate had helped preserve an ancient power from an era when men sang to their plants. Now the song of the seeds sang out to me, desperate to be heard before finally perishing into oblivion.

When I arrived home I went immediately to my adopted dad's place in Taos Pueblo. I told him of my unusual experiences. After a thoughtful pause in which he gave my vision time to bond with his own experience, he finally looked up and into my eyes. His face reflected all the wisdom of his eighty-two years, his mind seeming to swim through time to happier days.

"Just last week I was over at my sister-in-law's place looking for something Momma sent me to find. I opened an old trunk which I recognized as that of Momma's father. There, amongst old tools, pieces of rawhide and mementoes, were three lard cans. Inside were kernels of dark blue corn. I brought it over here."

Dad rose slowly and disappeared into a side room. He returned with the buckets of what looked like blue jewels. I looked at them and heard singing such as I had heard that first time in the lush Hopi cornfield just weeks earlier.

That following spring we were ready. My brother had returned home for the occasion and by May 15 we were in the field dropping the seed into the holes we had made with our planting sticks, all the while singing as we went. We prayed that our vision of the old seeds' vitality was correct.

By the middle of June our plants covered the field, their broad green leaves singing in the afternoon. By late summer, the nine-foot-high plants called attention to themselves and received a great deal of interest from the eldest members of the pueblo, who hadn't seen plants of such stature since their childhood.

My interest in the importance of our family project had grown to the extent that I was reading everything I could find on the subject. I discovered that agronomists and biologists the world over were deeply concerned with the rapidly declining rate of genetic diversity of domestic crops.

It seemed that the "green revolution" of the sixties and seventies had replaced the perpetuation of native seed with newly developed hybrids which were invented in laboratories for maximum yield and homogeneity of product. These seeds are sterile and therefore need to be produced each year through scientific duplication, as had the original hybrids.

Meanwhile the world is undergoing drastic environmental changes due in part to the pollution and rampant development of lands to accommodate the unchecked population increase in the temperate zones, where most of the world's food is grown.

The hybrids, designed years earlier for a stable climate, are not responding to the changes in weather patterns, now oscillating wildly, causing moisture in heretofore arid lands and droughts where moisture used to be predictable.

I also learned of the race to buy seed companies by the petro-chemical conglomerates such as Shell Oil, AT&T, Union Carbide, etc.:

> "More significant than the corporatization of seeds, though, has been their patenting. The Plant Variety Protection Act (PVPA) passed in 1970, allows seed companies to obtain patent protection for their own seed varieties.
>
> "The act represents a severe departure from American patent traditions, allowing for the first time, a claim of ownership over living and sexually reproducing organisms. Protecting plant varieties with patents has made the seed business enormously profitable, as seed companies may now charge royalties on the sale of all patented varieties." ("Seeds of Disaster" - Mother Jones Magazine, December, 1982)

I continued uncovering more and more alarming evidence that my lesson at Hopi was correct. I sent letters of inquiry to many small seed companies. This led to the exciting discovery that I wasn't just a lone coyote, baying at the moon. There was, in fact, a whole pack of us who had come to the same conclusions.

Kent Whealy of the Seed Savers Exchange in Princeton, Missouri (now in

Decorah, Iowa - ed.); Gary Nabhan in Tucson; Forest Shomer and others were appealing publicly for a national movement to take place in which small farmers and backyard gardeners survey their own regions for heirloom varieties of crops. They have started organizations and publications to inform and instruct us in the proper procedures for planting gardens in which the seed is saved and distributed through exchanges.

They expressed to me a great deal of interest in the blue corn which I described growing at Taos Pueblo, and invited me to a conference which was to take place that fall in Tucson. My wife, my brother and I took some ears from his harvest and attended the two-day event at the Tucson Botanical Gardens.

The theme was "Preservation of Genetic Diversity." Such a wide range of people were attracted to the event that even the organizers were amazed. The group included noted ethnobotanists, backyard heirloom crop growers and an Indian with Taos blue corn, as well as the director of the U. S. Federal Seed Bank in Ft. Collins, Colorado.

Among the guests was Wilfredo Saguano, a Peruvian expert on corn for the Pioneer International Seed Company. P. I. is the only international dealer in seed that remains independent of the conglomerates. They co-sponsored the conference, presumably looking for allies. Saguano was very excited about the Taos blue corn. He told me of a potential market for it in South America, among native tribes who have been forced by geographic restrictions imposed by governments to interbreed their races of corn in order to maintain sufficient yields. Many tribes, he said, have no living members who can remember the races, red, blue, white and yellow, being grown in their pure form. He also said he felt that geneticists may actively seek out the most nutritious and tenacious of the ancient varieties of our food crops to interbreed them through cross-pollination. This would naturally reintroduce qualities which were sacrificed in the initial breeding process.

I began to envision our Rio Grande Valley producing an abundance of food, using the tried and true ancient methods to produce the optimum genetic stock possible, along with a revival of traditional religious practice to support the agriculture.

Whether the destruction of the future food supply is influenced by fascist conspiracies or whether it is the product of greed and ignorance is not the most significant factor. The most important consideration is that what is happening is an entropic process. Entropy is a very commonly employed principle of science. Essentially it is a state of disorder in the life process in which a condition of decay of matter takes effect. Once this state is reached, there is no turning back. The life form decays.

If mankind depends upon invented, patented varieties to feed its already impossible population, while allowing multinational profit-mongers to destroy future food supplies and varieties, then we have surely reached entropy.

I have always been a fighter and this time is no exception. We must take some time now, not later, to go to our communities and "survey" mainly the older generations. Search out old varieties of seed and get their histories (origin, name, characteristics) and then find folks with land and put two and two together and voila....you have a "Conservancy Garden." The next step is to locate a convenient cellar with no leaks, get together a few volunteers and establish a seed bank and library so any member of the community can come get free seed, as long as they return some of their harvest.

You can take it a step further and create a directory of seed-savers and have a Seed Savers Exchange. I'm sure Kent Whealy wouldn't mind a little help. He is the originator of the only national Seed Savers Exchange. Of course there's the Federal Seed Bank in Ft. Collins, Colorado. According to its director, they need all the help they can get, as they are miserably underfunded to do any kind of significant collection.

The mystical and practical worlds agree. Start saving (and growing) the seeds of the real food. You build your future, your body, mind and spirit. Just remember, if you don't sing to 'em and love 'em as your children, they'll never hear your song in time of drought.

(John Kimmey, in respect for Kokopelli's spiritual guidance, has this season planted a field of blue corn in the shape of a spiral, or symbol of emergence. The seed was recently grown from fifty-year-old seed found by his adopted father at Taos Pueblo....John and others have recently founded Talavaya Center which is a non-profit organization dedicated to preserving the genetic diversity of the native crops of the northern Rio Grande Valley and to providing assistance to agricultural researchers and small farmers worldwide. They are currently multiplying over 200 varieties of Native American crops....John teaches and resides near Santa Fe, where he offers his knowlege of nearly twenty years of experience with Southwestern native elders. For further information, write: John Kimmey, Talavaya Center, PO Box 9289, Santa Fe, NM 87504, or call 505/984-0373.

(The 1984 Fall Harvest Edition, pages 141-145)

GRANDMOTHER'S SEED BOX
(From a Letter from Charles E. Harkrider)

I grew up on a farm near Stanberry, Missouri that had been homesteaded by my grandparents. My grandmother came from Kentucky in a covered wagon. Her parents had immigrated from Germany....My grandmother saved seeds from year to year. I'll never forget the pole beans she grew. She carried seeds in an apron pocket and would walk through the cornfield near the barn. Where the corn was thin she'd dig at the bottom of a larger stalk with her hands and lovingly plant a few seeds. What beans she grew! They had strings, large long tough ones, but when you pulled them off you had a large long plump bean. My grandmother also used these strings stripped back to hang beans together on the barn rafters to dry and make what she called "leather britches." She would cook some of them with open-pollinated white field corn to make succotash. Those seeds had been passed down from generation to generation originating in Germany....

This white corn was the eating patch and had long white ears with a large velvety cob. These cobs were the wiping cobs and were used for that purpose. My grandfather also made pipes from these cobs and smoked tobacco he grew. He'd hang this tobacco in bunches he called "hands" in the corn crib to dry. He'd fell a sugar maple tree, and with a bit he called a "ship bit" he'd drill holes in this maple log and tamp the holes full of tobacco. He'd let the tobacco cure in this log and then split the log for firewood and recover the plugs of tobacco. This was his chawing tobacco. The stems of this tobacco were boiled into a strong tea. This was the only insecticide I remember my grandmother using.... The white corn was also used to make hominy using lye she made from wood ashes....The seeds from her garden were dried and always stored in a homemade seed box....

I went to the Army in 1941 and shortly after that my grandparents died. My father wrote me he was selling the farm and moving to Kansas City to work in a defense plant. I wrote my father to take a portable grainery that was on skids, store the seed box packed in dry hay along with my shotgun and rifle wrapped in oil-soaked gunny sacks, board this grainery up tight, and move it to an uncle's farm nearby....When I came home my uncle had died and his draft-dodging son lived on the farm. He professed to have no knowledge of any grainery being stored there and said Paw must have gotten rid of it....

Years later I was on a hunting trip to that area. The farm had been sold and the buildings abandoned. I found my grandmother's seed box in the barn loft, less any seeds. My name had been burned into the bottom of the box, possibly by my grandfather. The box was covered with copper and had SEEDS hammered with a nail. The copper had taken on a beautiful patina....Naturally I took the box....I went to where my grandparents' home had been. I found the patch of what my grandmother called potato hill onions. The ancestors of these onions had come from Germany (where before then?) and came to Missouri in a covered wagon. I filled my hunting jacket with these small bulbs and put them under mulch in the corner of my garden. I have given starts of these onions to many gardening friends....How I would like to have seed of my grandmother's large oxheart thin-skinned tomatoes. They were actually as large as an ox heart....

(The 1984 Fall Harvest Edition, page 153)

ORIGIN OF THE LEE TOMATO
by W. Roy Mason, Major, C.S.A.

One day in June 1862, General Lee rode over to General Charles W. Field's headquarters at Meadow Bridge and asked for me. I would say here that on leaving home to enter the army I carried a family letter of introduction to General Lee; and on account of that, and also my relationship to Colonel Charles Marshall, an aide on his staff, my visits at army headquarters were exceptionally pleasant. When General Lee approached me on this occasion, he said, "Captain, can General Field spare you a little while?" I replied, "Certainly, General. What can I do for you?" "I have some property," he answered, "in the hands of the enemy, and General McClellan has informed me that he would deliver it to me at any time I asked for it." Then, putting aside his jesting manner, he told me that his wife and Miss Mary Lee, his daughter, had been caught within the Federal lines at the White House, the residence of General W. H. F. Lee, his son, and he desired me to take a courier and proceed with a flag of truce to Meadow Bridge and carry a sealed dispatch to General McClellan. At the Federal headquarters I would meet the ladies and escort them to Mrs. Gooch's farm, inside our lines.

I passed beyond the pickets to the second bridge, where I waved my flag of truce and was asked by the Union officer of the guard to enter. When I reached the picket, the officer said he had been ordered not to permit any flag of truce to pass through his lines until he had communicated with the headquarters of General McClellan. I waited on the bridge, and when the courier returned he had orders to bring me before the general. The officer insisted on blindfolding me, and positively forbade my courier accompanying me. I was then led through the camps, where I could hear the voices of thousands laughing, talking or hallooing. After riding an hour, a distance, as I supposed, of three or four miles, I reached headquarters and was relieved of my bandage. The general came out and gave me a hearty welcome; and when he heard that I had been blindfolded, he was so indignant that he placed the officer, my guide, under arrest. I had never seen him so excited. He asked me into the house, produced his liquors, and gave me a dinner of the best, after which we discussed the situation at length. He asked me no questions which it would compromise our cause to answer, but we calmly reviewed the position of things from our separate points of view, and he inquired anxiously after all his old friends. (General McClellan and my brother-in-law, General Dabney Maury, C.S.A., formerly captain, U.S.A., had been classmates and devoted friends, and the general had visited my father's house and my own at Fredericksburg.)

About 3 o'clock in the afternoon, looking down the road, we saw a carriage approaching. The curtains were cut off, and it was drawn by a mule and a dilapidated old horse, driven by a negro of about ten or twelve years, and followed by a cavalry escort. General McClellan, jumping up hastily, said, "There are Mrs. Lee and Miss Mary, now." As the carriage stopped before the door, General McClellan, greeting the ladies with marked cordiality, at once introduced me, and remarked to Mrs. Lee that the general (her husband) had chosen me as her escort through the lines, and that by a strange coincidence he (McClellan) had found in me a personal friend. He offered to accompany us in person to the river, but this was declined by Mrs. Lee as entirely unnecessary.

When we reached Mrs. Gooch's farm and our own pickets, cheer after cheer went down the long line of soldiers. Near the house we were met by General Lee and a large number of officers assembled to honor the wife and daughter of their chief.

Before leaving for Richmond, Mrs. Lee handed me from a basket under the carriage-seat, two fine tomatoes, the finest I had ever seen, remarking that she supposed such things were scarce in the Confederacy. The seeds of these tomatoes I preserved, and some years after the war, General Lee ate some tomatoes at my table, and praised them; whereupon we told him, to his astonishment, that those were the Lee tomatoes, and that they had been distributed all over the State under that name, from the seed of those given me by his wife.

(Battles and Leaders of the Civil War - Part No. 11, copyright 1884 by the Century Co., Union Square, New York. Material submitted by Alex Genz of La Crosse, Wisconsin.)

(The 1984 Fall Harvest Edition, pages 151-152)

LOST HERITAGE
by Lorraine Lilja

They sat patiently waiting on the hard wood benches of Ellis Island. Their faces reflected the nervousness they felt, but their eyes--their eyes were shining with hope.

Maria DeSalla stood near a window, her arms stiffly pushing fists into the stretched pockets of her black coat. Within one of those clenched palms was the only bit of Italy she could take along: seeds of the plum tomatoes that would guarantee decades of manicotti, lasagna, spaghetti marinara, braziola.

Katie Murphy sat in the front row, her arms encircling a blue-eyed toddler. At the bottom of the canvas sack on the floor were four seed potatoes--ones that had survived the famine--traced back as far back as family lore and human memory stretched.

There were cabbage seeds in the lining of Karl Schmidt's suitcase - chili pepper seeds in the band of Jose Sanchez' hat - rye grains filled the toe of Ivan Ivanovich's wool sock.

These brand new Americans brought along these grains of life that gave them confidence to start life anew in a strange land. They planted their seeds and harvested their crops, preserving some seeds for the next growing season.

But succeeding generations lost touch with the earth, and failed to learn the lesson that their ancestors knew so well. The simple acts of self-sufficiency--a small premium to pay for security. The decades spent depending on the commercial distribution system have left few of us nurturing the seeds of our heritage. And this loss is not insignificant.

If your inheritance includes heirloom seeds, care for them well. Share them with old friends and neighbors. Ask every master gardener you know if there are old varieties in his or her garden. Plant and preserve them--for tomorrow.

(Reprinted with permission from the "Gardens for All News" (now National Gardening Association), published by The National Gardening Association, 180

Flynn Ave., Burlington, VT 05401.)

(The 1982 Fall Harvest Edition, page 119)

THE MAYFLOWER BEAN

A letter from Carl Davis from Missouri: Dear Kent....About 20 years ago, I went into the St. Louis Seed Co. and asked for a brown and white spotted pole bean. The clerk said, "If you know where we can find them, we'll pay five hundred dollars for the seed." We made two trips to the Carolinas and haven't found them. They are supposed to have come over on the Mayflower....

A letter from Maynard Philbeck from North Carolina: Dear Kent....In about 1973, long before I became interested in saving seeds, I happened by a local country store just as an old acquaintance drove up. We began talking about our gardens. He said, "You may not believe this, but I have a variety of beans that came over here on the Mayflower." He offered to furnish me with as many seeds as I needed, but our garden was already chock-full, so I declined. I put the incident out of my mind, not realizing at the time its extreme rarity.

A few years later, the man died of a heart attack. When I read in the SSE that this bean was on your Lost Vegetables List, I got in touch with some members of his family, but by that time his possessions were either dispersed or discarded. No one knew of any seeds that he might have kept. I keep asking old-time gardeners around here if they have or have ever known of a bean called the Mayflower bean. No luck so far. I have a man who travels extensively in the mountain regions of this state on the lookout for the bean too, but so far nothing.

I have a feeling that one of these days that bean will show up, maybe somewhere close around. I thought you might be interested in how close I came at one time to getting my hands on the Mayflower bean. Times change and ignorance marches on. But one bright, memorable day I am going to ask a backcountry gardener if he has ever heard of the Mayflower bean, and he's going to say, "Sure. I have a whole field of them growing just over there...."

(Note from Kent - Maynard, you should get your local newspaper to do an article about the Mayflower bean. We live in an isolated area, too, and everyone around here reads the weekly newspaper. I sometimes think that the real key to locating heirlooms is with stories in weekly rural newspapers. I bet you'll find the Mayflower bean not 20 miles from your late friend's home. Good luck.)

(The 1982 Fall Harvest Edition, page 122)

SHIITAKE MUSHROOMS

Did you ever go mushroom hunting? We start hunting the morels (two different kinds) when the nights get really warm and damp in the spring. Everybody around here really gets the fever. They might as well let school out! Aaron and I really hunted them this year, searching the little backwaters all up and down the river. We found one place that was just thick with them. There is really no feeling like crawling through underbrush to get one and then looking around and seeing a dozen more poking their heads up through the leaves. Aaron started kidding me that he could smell them. One time he sniffed the air, told me that he smelled some, jumped over a big rotten log and fell completely out of sight. After a few seconds he raised back up with a huge grin on his face and held up a cluster of three morels. We both laughed so hard we fell down.

Next to the morel, the shiitake is regarded as the finest-flavored of all mushrooms. Dr. Yoo Farm (P. O. Box 290, College Park, MD 20740) sells two sizes of kits. The regular kit sells for $29.50 and contains a drill bit, an instruction booklet and 500 shiitake spawn chips, which is enough to inoculate 1/2 to 3/4 of a cord of oak logs. His small kit is $16.50 and contains a drill bit, instruction booklet and 200 shiitake spawn chips. From his catalog - "Highly rated Japanese Tree Mushroom is called the King or Ginseng of Mushrooms. It is grown on fresh cut oak logs. Inoculate once and harvest for five years after 1 1/2 or 2 years waiting. Easily 500 lbs. can be harvested from a cord of oak logs....fresh Shiitake mushrooms are selling for $7.00 a pound at Balducci's in New York City....they are delicious whole or sliced, sauteed in butter with some mild minced shallots and parsley, salt and pepper and a light splash of lemon juice and served on toast or with scrambled eggs or steak...."

Another source is Kurtzman's Mushroom Specialties (573 Harbour Way, Richmond, CA 94801). This catalog ($1) carries a wide range of spawns for commercial mushroom growers and serious hobbyists. They also carry a large selection of books and booklets on mushroom culture and cultivation as well as an excellent selection of field guides.

I was sent an article about an experiment in which shiitake mushrooms were grown on tree logs under forest conditions at the Agricultural Research Center in Beltsville, Maryland. It is described in Hortscience, Vol. 16 (2), April 1981, pages 151-156. The tests show that the highest yields were on black oak and white oak logs. Total yields were about 90 g. fresh weight mushrooms/kg. fresh weight of log during the five-year crop cycle. If you are interested, request the magazine or a copy of the article through your local library. Also, I still want to hear from any individuals who have grown them on logs in forest conditions.

(The 1981 Fall Harvest Edition, page 58)

HOWARD JONES

(Many of you remember Howard Jones, who was one of our Foreign Members from Guatemala in the 1977, 1978 and 1979 SSE's. He was with a missionary training center which teaches agricultural principles and supporting oneself by gardening. I was always fascinated by his stories of life there: how they had just finished cutting nine acres of barley and triticale by hand; his friend who was managing a 200-acre farm and had just spent a month and a half with 16 men weeding 155 acres of wheat by hand; the controversy which was raging among the brothers over whether they should buy a team of oxen or a tractor.... During 1979 I lost track of Howard and his wife Mary and didn't hear from them until just recently when I received this long, newsy letter - Kent.)

Dear Kent....Last December my father had a serious operation and asked Mary and me to help with the farm....We are in Palouse County in northern Idaho. The May issue of National Geographic has pictures and an article on this area. We are in a dry land farming area that grows wheat, dry peas and barley. Agribusiness is the word to explain the operation. The chemical companies must sell untold millions of dollars worth of products. We are farming about 900 acres. The spring work chemical bill was over $18,000....The work is divided into spring and fall work. After harvest in August, one-half of the acreage, which was the pea crop, is prepared and planted to winter wheat. The wheat ground is worked rough and left fallow all winter and planted to peas the next spring. There are no row crops, so all weed control is done chemically. Right now there is a big push for no-till methods....The soil is a heavy loess, consequently we have big machinery. We have two H-D 11 Allis Chalmers crawlers and a D-5 caterpillar crawler. The harrows and spring-tooth cultivators are 50' and 35' wide. Everything is large. The hills in this area are a problem, so we use hillside combines....The biggest problem of this area is soil erosion--up to 50 tons/acre per year. A national disaster, actually....

We use salt-based fertilizers and lots of herbicides and insecticides. Each year we need more and more potent poisons and the problems just get worse. Soil erosion, cheat grass, plant disease and insects are the worst problems, in that order....The soil erosion is from reducing the organic matter in the soil with nitrogen fertilizers. Soil compaction comes right after. Cheat grass and pepper grass, etc. are salt-loving plants—lots of fertilizers, lots of salt, lots of salt-loving plants. The salts form a layer in the soil due to their movement with the soil moisture. These salts form a barrier that excludes oxygen from the soil below that level. Consequently the anerobic pathogens migrate up to the root zone of the plants and cause disease in the crops. With insufficient oxygen in the soil, the plants are not able to assimilate nutrients that are oftentimes chemically bound in the soil. A nutrient-stressed plant is open to insect damage.

We are looking now to balance the soil conditions, then foliar feed the plants as they require feeding, rather than give massive dosages once or twice a year and hope the plants can get what they need. The book that we're reading now is An Acres USA Primer. Other books that have excited us are Louis Bromfield's books, Weeds, Guardians of the Soil, Joseph A. Cocannouer, and Sir Albert Howard's Soil and Health....

(The 1984 Fall Harvest Edition, page 154)

Historical Gardens

"If you tell children about plowing with oxen, they will forget it tomorrow. But if they plow behind a team of oxen, they will remember it for a lifetime."

— Tom Woods

A PROPOSAL:
Heirloom Seeds in Exchange for Permanent Maintenance

(This letter was sent to some 400 historic gardens during the summer and fall of 1982.)

There are several hundred Historic Gardens in the U. S. representing specific periods of time and/or ethnic origins. Most of these gardens are being maintained as projects of various Living History Farms, Outdoor Museums, Botanical Gardens, State Historical Societies, etc. The directors of such gardens are contacting me with increasing frequency because they often have a very difficult time locating appropriate varieties. The Seed Savers Exchange has been able to locate a substantial amount of heirloom plant material, but has difficulty finding dedicated individuals or organizations who are willing to maintain these materials on a permanent basis. I see a great opportunity for the Seed Savers Exchange and Historic Gardens to work together to the advantage of both.

What I am proposing is an exchange of heirloom seeds in return for permanent maintenance. Some of the varieties in the SSE have been kept by the same family in the same area for one hundred fifty years or longer. Certainly some of these local heirlooms will be more appropriate than older commercial varieties which may now be only distant cousins of the materials you are seeking. Our publications will give you immediate access to all of the material in both our Growers Network and the listings in our Winter Yearbook which comes out each February. We also have a well-established Plant Finder Service which has been very successful. Many gardeners have used it to locate varieties they have been trying to find for up to 40 years. You could easily use the Plant Finder Service as an "open forum" to describe the period you are representing and the plant materials you are trying to find.

The only thing I ask in return for all of this access is that you write and tell me: 1) what you have been successful in locating, 2) its source, 3) that you do intend to maintain it permanently and 4) that you will turn it back in to the SSE if you decide to stop growing it. With the help of a small computer, I am now able to keep track of varieties that are being permanently maintained. Of course we would welcome any of you as active members in the SSE, if you have sufficient time to allow for the active distribution of seeds in addition to maintenance.

I don't feel that your participation will put any real strain on the SSE, because most of you are only looking for a very few specific varieties that are appropriate to your situation. I feel that the SSE has much to gain by having your maintenance as a living back-up to our preservation efforts. You are also in the unique position of being showcases for historic varieties. Your gardens can easily be used as educational tools to teach the many gardeners among your visitors the importance and need for diversity and preservation. You can generate a tremendous amount of local publicity through planting ceremonies and other seasonal events. You can also grow crops entirely for seed and make possible a second wave of distribution to local gardeners who have become interested in historic varieties because of your efforts. What unique opportunities you have!

I'm really enthusiastic abaout a cooperative effort between the Seed Savers Exchange and Historic Gardens which will benefit all concerned. What I am proposing here is not so much a formal program, as just an invitation to join us in our efforts to locate, distribute and preserve what is left of our vanishing vegetable heritage.

(The 1982 Fall Harvest Edition, pages 45-46)

THE OLIVER KELLEY FARM
Tom Woods

The Minnesota Historical Society is currently developing the Oliver Kelley Farm as an historic farm of the 1850-1870 era. During this period Oliver Kelley, the founder of the Grange (the first national agricultural organization in the U. S.), lived here....Besides our work with historic vegetable and fruit varieties, we are restoring about 70 acres of native prairie. Our program also involves cultivating about forty acres with oxen and horses and equipment typical of that period. In the field we will be growing crops which we know Kelley grew, and in many cases the exact same varieties Kelley and other Minnesota farmers grew during that period. The source for our documentation of varieties has been Kelley's own articles for papers, farmer diaries and the 1860/61 issues of the Minnesota Farmer and Gardener....Kelley was very active in seed exchange programs, both regionally and nationally. He was always searching for a better variety for his difficult, sandy soil and northern climate. We feel that it would be very fitting for us to also become actively involved in seed exchange programs.

(From earlier correspondence with Tom Woods, Site Manager, Oliver H. Kelley Farm, 15788 Kelley Farm Road, Elk River, MN 55330)

****

KENT: Tell me more about Oliver Kelley's seed exchange work.

TOM: Oliver Kelley founded the first county agricultural society in Minnesota, the Benton County Society, in 1852. One of the things that he did as corresponding secretary was to keep the Society's seed cabinet at his farm. He encouraged farmers to bring a sample of every seed they grew, just a half-pint or whatever, to put in this cabinet. He served as the repository for those seeds and then distributed them to other farmers....This was an area in 1850 that was just about as far north as anybody had farmed and they didn't know what would grow. So Kelley, particularly, and other farmers like him were trying new varieties constantly. They were trying everything they could think of.

KENT: Were most of the new materials they were trying from areas back East, or were they actually from Europe?

TOM: Some of the seed varieties and apple varieties were northern European, because they were actually looking at similar climates. And some of the apple trees were Siberian varieties. He tried a Siberian wheat....He also got part of his material from the Patent Office. The Patent Office was the only national branch of government at that time doing anything to improve agricultural practices. They were responsible for introducing new crop varieties. For example, they would import Turkish flint wheat and send out thousands of packets to farmers, ask them to grow it and report on their success. Kelley was one of the people who did this reporting. In fact his experiment is included in some of the Patent Office Reports. Kelley also wrote a column for a paper called The Frontiersman. In it he continually talked about the crop experiments he was trying and about the seeds that he was getting from the Patent Office. He was also sending things back to the Patent Office and Eastern farmers. Wild rice was one of the things he was interested in....

KENT: Most of the plant material that comes through the Seed Savers Exchange is not very well documented. And I realize that documentation is such a necessary and strict part of what you are doing. But I also know that we have a lot of material that's not available elsewhere, which I believe could be of great value to certain historic gardens.

TOM: A modern seed catalog may carry an item with the same name as one we read about in 1850 journals. Is it the same? How can you document that? I don't know myself. We feel that many heirloom varieties are from different ethnic groups and much of that material would be valuable for certain ethnic sites. Also many sites may not be as rigorous about documentation as we are, and heirloom varieties that were being kept during a certain period by families native to their areas may be much more appropriate than what they are presently growing.

KENT: It's always sort of amazed me that many sites will restore whole villages and have people in costume and yet the things they are growing really aren't appropriate at all.

TOM: I recently wrote an article about that very thing for our Midwest Open Air Museums Coordinating Council Newsletter which I edit. Outdoor museums and living history farms are frequently developed in order to show actual processes, so that visitors can become immersed in a lifestyle of a different era. When you talk about living history farms, major ingredients are animals and crops. Traditionally, museums have always concentrated on buildings and artifacts. I believe that living history farms should also be emphasizing other things and must definitely consider plants as part of their collections. Even though there is definitely a growing interest in historic varieties right now, that's not happening as much as it should be....Most museums still concentrate on buildings, artifacts and crafts. Studies have shown that visitors most like to see craft demonstrations and least like to see crops. How do you impress upon a visitor the importance of this being a long red beet from the 1860s as opposed to a Ruby Queen that you can buy out of any seed rack? Most people don't have the expertise to know the difference....However, some things have a very obvious visual difference. For instance, we grow one corn called Improved King Phillip. It's a brown-seeded corn that took the country by storm in the mid-1850s. Many visitors really recognize its strikingly different appearance....

With our work here we try to interpret the shift in agriculture from subsistence to market. The introduction of market agriculture allowed our whole society to become more modern. The Improved King Phillip corn was really important, because before it and other corns like it, farmers were growing flints exclusively. That was all that could survive here. Flints are northern corns and dents are native southern corns. They started experimenting with dents. They grew a corn called Illinois Dent which was a 120-day corn. But the growing season here is usually about 100 days, so it rarely made it. They were growing Saskatchewan White Flint, which we are also growing. They started trying to cross with the dents to get the increased production of the dents and still have the shorter growing season of the flints. The Improved King Phillip was one of the first crosses that was really successful in this area....

I believe that to really interpret the significance of agriculture's role in society, you've got to talk about crops. You've got to talk about the way crops developed over time. And when you've got a living history farm that avoids that issue, you've got nothing. Really! You might as well just have a building with a bunch of furniture in it....Many of the sites around the country that are maintaining historic gardens are funded through public funds, through tax-exempt donations. One of their major responsibilities is education, and participating in an exchange of plant materials would be fulfilling that responsibility.... It's part of their professional responsibility, I think, to become involved in what's going on in the field. Not only to be aware of it, but to also be an active part of it....I think that your attempt to include historic gardens in the activities of the Seed Savers Exchange will be very well received....

(The 1982 Fall Harvest Edition, pages 46-48)

THOMAS JEFFERSON'S GARDEN AT MONTICELLO

by Peter Hatch
Superintendent of Grounds

When Thomas Jefferson referred to his "garden" during the fifty years he regarded Monticello as his home, like most early Americans he was referring to his vegetable or kitchen garden which, along with his orchard and vineyards, composed eight acres on the sunny southern slope of his "little mountain." Probably no early American gardener kept as extensive a notebook on his garden as did Jefferson. The daily activities of sowing, manuring and harvesting between 1809 and 1825 are recorded quite precisely in his "Garden Kalender" which, with his "Garden Book" and most of the letters he wrote or received having to do with horticulture, have been edited and published by the American Philosophical Society in Philadelphia in 1944. The Garden Book, edited by a University of Virginia botanist, Edwin Betts, is an extraordinary work detailing Jefferson's fascination with the natural world. It includes dreamy visions of romantic grottoes as well as concise notes on the culture of as many as 250 vegetable and 150 fruit varieties. Jefferson once said that "The greatest service which can be rendered any country is to add a useful plant to its culture." His experimental efforts at raising sesame, olives, sugar maples, leguminous cover crops, upland rice, eggplants, tomatoes and other food crops offer a traditional foundation upon which to clarify present-day efforts from seed saving to tree crops.

The garden and orchard at Monticello are remarkable for a number of reasons, but the most striking is the scale and scope of the area. The 1,000 foot long vegetable garden terrace was literally hewed from the mountainside, so that one nineteenth century visitor described it as a "hanging garden." The garden plateau itself was supported by a massive stone wall that averaged six feet high. Perched atop the wall was a brick garden pavilion so precariously situated that it was blown down in a violent storm in 1824. Ranging below the "great wall" was the 400-tree orchard, terraced vineyards and "berry squares." Surrounding the entire complex was a 10' high solid board or "paling" fence that ran for 4,000 feet.

Such a dominating visual landscape can be re-created quite precisely due to the wealth of documentation in The Garden Book. The Thomas Jefferson Memorial Foundation, which owns and operates Monticello, is committed to an ongoing restoration of the area. This commitment began in 1979 with two years of archaeological excavations that complemented the documentation, confirmed garden details and enabled the restoration to begin. Archaeologists uncovered the stone wall, robbed in the twentieth century and covered with eroding soil, traced the fence-line through an uncovering of post-hole stains, and searched for the nature of garden walkways. Tree hole stains from 70 of the original orchard trees were also uncovered, revealing a planting pattern identical to one drawn by Jefferson in 1778.

The experimental nature of the garden and orchard is especially noteworthy. Not only did Jefferson import the finest pears from France and cabbages from Holland, but he also cultivated beans, corn and gooseberries brought from the west by Lewis and Clark as well as apple and peach varieties newly discovered in the New World. When he cultivated as many as 37 varieties of beans and 15 of English peas, he was not only providing for the family table, but primarily sorting out superior types so that, for example, the finest Ricara snap beans could be distributed to his neighbors.

The garden faces southeast and is carved from the hillside so that the warming sun is captured and entrapped. Jefferson organized the garden with a border against the northern bank, and so had a clear advantage in his pea contests. Every spring these contests with neighbors determined who could bring the first English pea to table. Peas were often sowed as early as January, tomatoes were occasionally planted as early as February, and greens were commonly gathered throughout the milder winter months. The garden was divided into twenty-four "squares" adjacent to the northwest border. These "squares," at least in 1812, were arranged according to which part of the plant was being harvested—whether "fruits" (tomatoes, beans), "roots" (onions, beets) or "leaves" (lettuce, cabbage).

Jefferson was ever the scientist in his Garden Kalendar as he cooly recorded that his "Ledman's peas" failed in 1813 "from drought" or that in 1809 his Hotspur peas were "killed by frost Oct. 23." He planted his squashes and melons together for the interesting variations that might occur with cross-fertilization. Occasionally Jefferson would be overcome with ornamental fancies for his garden and would plant an arbor of different shades of scarlet runner beans, arrange adjacent rows of purple and white and green sprouting broccolis or white and purple eggplants, and border his tomato "square" with okra or sesame, a rather unusual juxtaposition.

A major influence on Jefferson's gardening practices as well as the types and varieties of vegetables he cultivated was Bernard McMahon, a Philadelphia nurseryman and author of The American Gardener's Kalendar, the best American work on practical gardening in the first half of the nineteenth century. But Jefferson was perhaps most succinct in summarizing his basic gardening sensibility when he wrote his daughter, Martha Randolph, in 1809 and said, "We will try this winter to cover our garden with a heavy coating of manure. When earth is rich it bids defiance to droughts, yields in abundance, and of the best quality. I suspect that the insects which have harassed you have been encouraged by the feebleness of your plants, and that has been produced by the lean state of the soil."

U. P. Hedrick, in The History of Horticulture in America to 1860, said that "every Southerner knew of the agricultural and horticultural practices at Mount Vernon and Monticello, and followed as far as possible farm operations at these two estates." Surely Hedrick is exaggerating, but in the range of the types of fruits and vegetables and the wide spectrum of varieties, and in the progressive nature of his horticultural practices, the orchard and garden at Monticello represents the state of the art of fruit and vegetable culture in the early 1800s. Presently the garden has been restored, the orchard is mostly planted, the garden pavilion is under construction and the "great wall" is nearly completed. The restoration of the vineyards and paling fence are projected for the near future.

I have prepared a 300-page research report entitled "Fruits and Vegetables at Monticello: A Study of Varieties and Culture" in which I describe where and when a particular variety was planted, attempt to decipher Jefferson's private nomenclature, provide varietal descriptions from mostly nineteenth century fruit and vegetable literature and make an attempt at tracing the evolution of the type or variety of vegetable to the present with the hope of eventually locating it. Perhaps a third of Jefferson's vegetable varieties are described somewhere. Our attempts at locating them have been, at best, rather haphazard and are considered basically a long-term project. The orchard, on the other hand, had to be planted, so efforts toward the recovery of old fruit varieties have presently held a high priority.

At present we rely a lot on the grandchildren of old vegetable varieties still offered by some seed houses. As our efforts to locate more appropriate varieties pick up in the future, we'll become more active with the Seed Savers. If there is interest shown in the research report, perhaps I could pass on the list of the vegetable varieties grown by Jefferson for a future issue of the Seed Savers Exchange.

(The 1982 Fall Harvest Edition, pages 48-50)

THE INDIAN GARDEN PROJECT

Charles E. Hanson Jr.

Many trading posts had gardens. Some Hudson's Bay posts operated real farms. The story of this production of peas and turnips and potatoes on the frontier is a whole story by itself. The operating procedures at little posts like Bordeaux's were not so well formalized and the commissary superintendent usually wore lady's-style moccasins. The post garden at such a place was pretty sure to be the same kind of garden that would be found at any "permanent" Indian camp.

In the stories of his life as a free trader in Montana, James Willard Schultz mentioned several times the "little corn" and the pumpkins and beans grown at summer posts by the women of his family. Grown in fertile valley soil and watered from a nearby river or creek, the hardy crops thrived even in the short growing season of the Northern Plains. Most Plains Indians did gardening of this sort, the task usually falling to older people camping along a favorable stream. In 1855 G. K. Warren noted the Sioux doing this along the little spring-fed tributaries of the Upper White River (1). Bordeaux Creek is one of these little streams and it is most appropriate that the Museum Indian garden project is now preserving some of the ancient varieties grown in the same area more than a hundred years ago.

In his report, Warren stated that "The Brules and Yanktons raise a considerable amount of corn and other vegetables, and with very little attention they get a tolerable crop." However, the pressures of encroaching white settlement and the effects of the buffalo robe trade worked to accelerate the roving of the Plains Indians with a consequent diminution of any agriculture. Throughout this period the old-time crops were maintained by the Mandans, Arikaras and Hidatsas on the Upper Missouri. These village tribes depended more and more upon agriculture as the West changed. They traded corn and dried vegetables to the wilder tribes and supplied them with seeds. They also sold large quantities of shelled corn to the trading companies for employees' rations and resale to other Indians. The Fort Union Inventory for 1850 included 80 bushels of "Ree Corn" at $2.50 per bushel. In 1855, F. V. Hayden noted one Indian village selling over 500 bushels of corn to the American Fur Company in a single season (2).

These three tribes were the original seed sources for most of the varieties now preserved by the garden project. Some varieties were obtained directly from Indians of the Dakotas; many others were furnished through the courtesy of the late George F. Will of the Oscar Will Co., pioneer seedsmen of Bismarck, North Dakota. The project's beginning nearly 20 years ago was timely, for most of these varieties are no longer available. Seed houses have dropped these low-yielding, open-pollinated types, and the Indians themselves generally find it easier to buy seed from the attractive racks in today's supermarkets.

Corn is, of course, the most important crop in the garden project. The earliest variety, and one of the smallest, is Assiniboine flint. This corn was obtained from Canadian Indians and is thought to be an adaptation of Ree flint corn. The little ears are uniformly multi-colored and very attractive. The flour corns, now almost unknown commercially, were popular in primitive places because they were soft and easy to grind. Blue flour corn was grown by the Indians from Wisconsin to New Mexico, and the Mandan variety is regularly grown in the garden. It has smooth, deep blue or reddish blue kernels, and the ears grow 6 to 7 inches long. There are also similar red, yellow and white varieties of Mandan flour corn. The Indians considered sweet corn or "wrinkled corn" a special delicacy, and the garden includes an exotic red variety which was once a staple article in Mandan villages (3).

Rarest of all the corn is the "Mandan little corn" or midget blue flour corn. Ears the size of a man's thumb are borne on leafy stalks from two to three feet tall. This was the corn that produced the earliest roasting ears for the summer camp. Schultz mentioned it, old Indians of several Dakota tribes remembered it, but no specimen could be found in the 1950s. A reservation storekeeper wrote that an old Indian near Ft. Yates still raised it, but a long dusty trip to his cabin disclosed that the last seed had been lost in a drought the year before. Finally, after the search had gone on for ten years, another Indian trader gave us the name of an old Oglala woman who loved to garden as her mother had before her. She might have the seed--and did. One perfect little ear was spared for the garden project. It was enough; today there is plenty of seed to keep the variety for posterity.

Another midget Mandan crop of special interest is "Mandan tobacco" (nicotiana quadrivalis). This little species of nicotiana withstands drought and short seasons, but its product is rank and strong to smoke. The tiny leaves are rolled and cured much like tea leaves. Quickly supplanted by the better smoking tobacco brought by the trader, its use was soon restricted to ceremonies and a few traditionally-minded Indians. Today it occupies an important place in the garden each summer.

The common bean, staple food of the frontier, has not been forgotten. The original seed of the red beans grown in the garden were obtained by Oscar Will from an Hidatsa named Son of a Star, over 80 years ago. The seed of the small oval pumpkins now grown were obtained from the Omaha by Dr. Melvin Gilmore, noted authority on Indian uses of plants. The squashes include a small, round striped summer squash from the Mandans and a large blue-green winter squash from the Arikaras. Both pumpkins and squash were cut in strips and dried for future use. Strips of dried pumpkin made a tasty addition to the provisions carried by war parties and trapping expeditions.

The term "Indian watermelon" is highly controversial. Most data indicates that the watermelon is not indigenous to the New World. However, early travelers noted small round melons being grown by the Osage, Wichita, Kansa, Ponca, Arikara and other tribes of the Missouri valley (4). La Raye noted that watermelons were grown by the Arikara in 1802, and Lewis and Clark specifically mentioned the same thing a few years later. All the Indian melons were round and faintly striped, with black seeds. The Northern Indians have lost this type of melon, but the garden includes melons from seed originally obtained by Oscar Will from the Arikara in the 1880s. This little watermelon is only of fair quality, but its flavor compares with some modern varieties which are being bred more for shipping and showing than for eating (6).

With these types of corn, beans, squash, pumpkins, tobacco and melons, the garden provides a fair example of the century-old summer squaw garden on the High Plains. In recent years an interesting supplement has been developed through the addition of selected Iroquois crops.

The Iroquois nations had a tremendous influence on the fur trade. They provided the first great field of endeavor for the British, French and Dutch traders of the 17th century. They dominated much of the northwestern trade in the 18th century, and in the 19th century Iroquois men helped staff the trapping and trading brigades clear to the Pacific Coast. In addition, these tribes were expert agriculturalists and they preserved for the 20th century many primitive varieties of Eastern Indian corn and vegetables.

The garden now includes the famous Tuscarora white corn (Zea Mays anylacea) used for corn soup. It is a long-eared white flour corn with large soft kernels. It also includes the Seneca flint corn (Zea Mays indurata) for the hominy that was once a staple article of diet for canoe brigades. The third Iroquois corn is a small white popcorn (Zea Mays everta) called wata' tongwus (it bursts). There are two varieties of beans in the Iroquois collection. One is the very large spotted soup bean, the other is the fat cranberry bean, both very ancient varieties from the Seneca.

Preserving these crops and keeping the strains pure for the future has been an absorbing project for the museum staff. Inanimate relics are preserved in countless museums and collections. Game birds and animals threatened with extinction are carefully studied and receive special attention. Native trees and plants are being protected in parks and botanical gardens. The least has been done to preserve the ancient cultivated food crops which gave way first to the products of improved breeding and management, and finally to hybridization and other advances in agricultural technology. To us this has simply been another effort to accurately portray the "West That Was."

NOTES

1. Report of Lieutenant F. K. Warren, Topographical Engineer of the "Sioux Expedition," of Explorations in the Dakota Country, 1855. See Dakota Historical Collections, Vol. XI, pp. 93 and 109.

2. For much more detailed information about the corn grown by the various Indian tribes of the Plains, see "Corn Among the Indians of the Upper Missouri" by George F. Will and George E. Hyde. Originally published in 1917, this well-illustrated book has been reprinted by the University of Nebraska Press, Lincoln, Nebraska.

3. Corn was, of course, most esteemed in the green state, boiled or roasted and seasoned with bear oil or buffalo fat. The flour corns made excellent meal, and many tribes made hominy from the flint corn. Both sweet and flour corns were parched for trail rations or household consumption.

4. For an excellent discussion of Indian crops, including watermelons, see Dr. Melvin Gilmore, "Use of Plants by Indians of the Upper Missouri Region," 33rd Report of the Bureau of American Ethnology, Smithsonian Institution,

Washington, D. C.

5. In 1896 Lieutenant Wilkinson of the Pike expedition described Osage watermelons as round, the size of a 24-pound shot and finely flavored.

6. The possiblity is suggested that watermelons may have traveled up the river from the Spaniards after 1763 or along the foothill route from Santa Fe even earlier. In 1847 Robert Buist noted that Spanish melons had been introduced to the East Coast from Portugal about 1829. He described these melons as round and dark green with bright red flesh and black seeds, not as large as some others but better flavored. See: Robert Buist, The Family Kitchen Gardener, Orange Judd & Company, 1847, p. 74.

(The James Bordeaux Trading Post is located at the Museum of the Fur Trade 3 1/2 miles east of Chadron on U. S. Highway 20. Reconstructed on its original foundations, this trading post for the Sioux was active in fur trading from 1846 to 1872. A quarterly subscription to "The Museum of the Fur Trade Quarterly" is available for $5 per calendar year, from The Museum of the Fur Trade, HC 74, Box 18, Chadron, NE 69337. They will not send seeds through the mail.)

(The 1984 Fall Harvest Edition, pages 136-140)

SEED SAVING AT THE NATIONAL COLONIAL FARM

by David O. Percy, Ph.D.

We are often asked at the National Colonial Farm of the Accokeek Foundation, Inc. what practical value does seed-saving have. In reply, we note that seed-saving is the ultimate seed bank and the value of having these seed resources can be illustrated by the Southern corn leaf blight which struck the southeastern United States in 1970. This fungus (*Helminthosporium maydis*) killed about 15 percent of the corn crop, and was particularly devastating because of monocultural cultivation of maize, most of which were hybrids based on a narrow genetic base. Dr. W. Ralph Singleton, a distinguished geneticist from the University of Virginia and Director of the National Colonial Farm of the Accokeek Foundation, Inc. in Accokeek, Maryland, had been working to preserve old maize varieties, specifically Virginia Gourdseed corn. Virginia Gourdseed corn was the common open-pollinated variety of the southern colonies. When the southern corn leaf blight struck, commercial seed companies contacted Dr. Singleton. The Virginia Gourdseed corn was resistant to the blight, and using this as a parent, new blight-resistant hybrids were developed.

The National Colonial Farm continued Dr. Singleton's work in procuring heirloom varieties. Currently, we are conducting a research program to develop heirloom seed cultural and preservation techniques which can be adopted by living historical farms and others. The Farm has over seventy heirloom vegetable and field crops under cultivation. Further information about this program can be obtained by writing to the National Colonial Farm, 3400 Bryan Point

Road, Accokeek, MD 20607.

(The 1984 Fall Harvest Edition, page 152)

CORRESPONDENCE FROM OTHER HISTORIC GARDENS

Dear Kent....Firstly, I'd like to congratulate you for creating such a fine program. We are truly grateful that many of the wonderful old varieties and family heirlooms are being saved for future generations to plant and preserve. Do you know of any nineteenth century varieties of flowers, vines, shrubs or ornamental trees for plantings around two houses in Sharon Woods historical village? The plants or seeds must be nineteenth century varieties, preferably pre-1860s. We would also like to plant an area in flax. Do you know of anyone saving an older variety?....(Lynn Calaway, Route 2, Butler, KY 41006)

Dear Kent....The Association of Living Historical Farms and Agricultural Museums (ALHFAM) is currently trying to locate and then document as many sources of historical animal and plant varieties as possible for use in our member museums. Some of our members are already aware of the Seed Savers Exchange and may already participate....I think the potential for historic sites to maintain selected seed stock is great while the potential for public education is phenomenal. We introduce a lot of people to the True Cranberry Red pole bean every year. Some museums such as Kelley Farm and the Dayton Natural History Museum have seed preservation as a primary goal, while others only have that as a concern of the individuals who are employed. I would like to see more of an institutional commitment, however....One problem that museums might have with maintenance of old seed varieties is the lack of knowledge as to historical appropriateness of heirloom varieties. Robert Becker at Cornell has done some work in regards to commercially available seed....I know John Withee included historical dates for some of his bean varieties. A certain amount of the information responsibility lies in the museums, but many museums lack the horticultural expertise even though they know the historical model....(Andrew H. Baker, Old Sturbridge Village, Sturbridge, MA 01566)

Dear Mr. Whealy....Please send me a catalog of the types and varieties of vegetables you have available. Mount Vernon is interested in locating different varieties of vegetables to use on Washington's estate....(J. Dean Norton, Horticulturist, Mount Vernon Ladies' Association of the Union, Mount Vernon, VA 22121)

Dear Mr. Whealy....Cary Fowler in North Carolina recently told me about the Seed Savers Exchange. He said you may be able to tell me where to obtain certain types of old seeds....My interest in old seeds is on behalf of our state-run historic sites and particularly Historic Halifax, a village in northeastern North Carolina. At Halifax we are restoring a portion of the town to a ca. 1760-1840 historic site. Several restored buildings and a modern visitor center are already open, and we are moving into the landscape restoration phase....We hope to plant one or more period kitchen gardens....(Richard Knapp, Research and Education Coordinator, Historic Sites Section, 310 N. Blount St., Raleigh, NC 27611)

Dear Mr. Whealy....I am the gardener for the Old World Wisconsin ethnic museum. The museum consists of 576 acres and currently 44 historic buildings from the period 1831 to 1915. We are engaged in growing gardens in that era. A seed catalog from your company would significantly help us in this endeavor. (Gerard Marifke, Historic Sites Division, Old World Wisconsin, Rt. 2 Box 18, Eagle, WI 53119)

Gentlemen....We are extremely interested in information regarding horticultural and agricultural plants in use in North America, particularly the lower Hudson Valley of New York, in the period 1725-1860. If you have publications or other information, or lists of plants or seeds currently available dating to that period, we would appreciate receiving them....(Arthur H. Ode, Director of Historic Structures and Landscapes, Sleepy Hollow Restorations, 150 White Plains Rd., Tarrytown, NY 10591)

Hello Kent....I work at Sharlot Hall Museum here in Prescott, Arizona, and for some time have been talking to the Director about putting in a vegetable garden next to the "Fremont House" using the old seeds, but we didn't quite know where to go to obtain them. So when I saw and read your article I thought it was excellent. Not only a great asset to the museum but to my own personal use as well as others throughout the country....(Morey Lewis, Prescott Historical Society, 415 West Gurley Street, Prescott, AZ 86301)

Dear Mr. Whealy....I work at Conner Prairie Pioneer Settlement which is a living outdoor museum. We are looking for seeds that were used in the 1830s in central Indiana. I understand how your exchange works and I am not sure if we are properly staffed to participate in your seed exchange. But I would like your yearbook so that I can encourage our staff to help preserve and regenerate the old seeds....(Sue Cain, 252 Wallick Rd., Peru, IN 46970)

Dear Mr. Whealy....I am working on a project at the Fort Vancouver National Historic Site in Vancouver, Washington. We are attempting to reconstruct the vegetable garden as it was in the 1840s....The Hudson's Bay Company of Canada owned and operated the Fort. It was an English garden....Please send me information on your seeds, and whether or not you have any historic information on gardens and seeds at that time period. (Sally Hansen, U. S. Dept. of the Interior, National Park Service, Fort Vancouver National Historic Site, Vancouver, WA 98661)

Dear Kent....I just read about your organization in Back to Basics and would like more information on your organization, especially concerning heirloom vegetable varieties....We are developing a living history farm near Lowell, Indiana. We will be depicting the 1910-1920 time period and the 1850-1860 time period on two separate areas of the site. We will have gardens at both areas and wish to have authentic varieties....(Becky Crabb, Buckley Homestead, 3606 Belshaw, Lowell, IN 46356)

Dear Mr. Whealy....I am coordinator of North Shore Community Gardens.... which occupies land that was once part of the J. J. H. Gregory Seed Company. The Gregory Seed Co. was one of the largest seed houses in the U. S. back in the mid-1800s....In our studies we have found out a lot about Mr. Gregory and some of the seeds that he developed....He was really a very great man but not many people had ever heard about him. What we want to do is grow some gardens with some of the vegetables that he developed and introduced....We know at this time that he developed the Hubbard squash, Marblehead squash, American Turban squash and also the Middleton squash. These were all developed before 1865. Some of his cabbages were American Savoy, Marblehead Drumhead and Stone Mason Drumhead. We have a very long list of vegetables that he either developed or introduced. We have a copy of his 1874 seed catalog and are in the process of getting copies of many more of his catalogs....We at North Shore Community Gardens would very much like to participate in your Growers Network and help save some of John Withee's beans. John gardened with us up until 1979. We learned much about growing and saving beans from him. You know how he is always willing to help anyone....This garden will be a way to honor Mr. Withee. I have not told him what we are up to, but when the beans are up, we will invite him over....(Louise Bastable, Garden Coordinator, North Shore Community Gardens, 24 Acorn Street, Middleton, MA 01949)

Dear Mr. Whealy....I recently read in the January/February 1982 issue of The Mother Earth News that you are interested in listing all period gardens in the country which are maintained by living history clubs, outdoor museums, botanical gardens, etc., for a period garden directory to help such sites obtain seed varieties of the correct time period or ethnic origin....I am writing from Naper Settlement, a living historical village located in the southern suburbs of Chicago. For several years we have been involved with period flower and herb gardening, but want to branch out into vegetable and possibly crop growing. We are interested in plants and crops from the 1830s to the turn of the century that would have been grown and harvested in this region of Illinois. At this point in time, we would be unable to return seeds to the Exchange, but would hope that in the near future we would be developed enough in our garden structures to be able to save and share seeds....I think what you are doing is fantastic. We are both involved in protecting and preserving our heritage.... (Lisa Hill, Associate Director, Naper Settlement, 201 W. Porter Ave., Naperville, IL 60540)

Dear Mr. Whealy....We are preserving the Hermitage Garden at the home of President Andrew Jackson with varieties of plants that were being grown prior to 1845....Your name was listed in an article of Country Journal as a possible source of seeds from plants that were cultivated years ago....Please send me the information I need to acquire the list of seeds you have available....(William G. Randles, Horticulturist, The Ladies' Hermitage Association, Hermitage, TN 37076)

Dear Mr. Whealy....Our main museum facility has two exhibits relating to farming....one for the period 1835-1864, when our community was formed.... Would any members have some dried seed, beans, wheat, ears of corn, etc.-- that we can exhibit (instead of plant) to show what grain, beans, peas, etc. looked like in earlier periods....We have a house museum, built in 1916, that we recently opened. We might be able to plant a World War I garden each year, vegetables as well as flowers. Can you suggest which plants would have been appropriate in a northern Illinois garden during World War I? We would attempt to make seeds resulting from this garden available to others....(Dr. Jerry Musich, Curator, The City of West Chicago Historical Museum, 132 Main St., West Chicago, IL 60185)

Dear Kent....It gives me great pleasure to send you this donation for a Lifetime Subscription to the SSE in the name of the Dalton-South Holland Junior Woman's Club's Sand Ridge Nature Center Heirloom Garden....our Heirloom Garden is probably being planted as I write this letter. The gardener at the Sand Ridge Settlement is very excited about this program....The people we contacted for seeds were just fantastic. I received many letters of support and encouragement....We will have a handout available which will explain the importance of preserving plants. We intend to mention the Seed Savers Exchange along with your address, so interested parties can contact your organization....(Deborah Hart, Co-Chairman, 16458 Wabash Ave., South Holland, IL 60473)

Dear Mr. Whealy....I read with great interest the account of your Exchange in The New York Times several months ago....I am interested in old forms of common garden flowers. I am enclosing a folder about Bartram's Garden where I work on the Common Flower Garden. We are looking for eighteenth century forms of perennial and annual native and introduced garden flowers.... (Margaret Whitall Evens, President, The John Bartram Association, 54th St. & Elmwood Ave., Philadelphia, PA 19143)

(The following is from the folder that Mrs. Evans sent: John Bartram, 1699-1777, and his son William, 1737-1823, were Quaker farmers, botanists and plant explorers. John Bartram bought the 102-acre farm on the Schuylkill in 1728, enlarged the existing stone house and started to develop his garden.... Interested in plants from his youth but with little schooling, Bartram educated himself, learning Latin in order to study botanical books. His knowledge of botany brought him to the attention of Benjamin Franklin and James Logan, who encouraged him and loaned him their books. Many prominent Englishmen were curious about plants from the new world and Peter Collinson, a London woolen merchant, engaged Bartram as a plant collector. The two men corresponded for thirty-five years but never met. Their letters record the growth of the garden and Bartram's journals describe his adventurous trips in the wilderness....Bartram traveled north to New York, south to Florida and as far west as the Ohio River. The seeds and roots he brought back were planted in his garden, observed, propagated and then shipped abroad. Bartram's fame spread and in 1765 George III appointed him Royal Botanist. Linnaeus spoke of him as "the greatest natural botanist in the world.")

Dear Mr. Whealy....I am working with Kane County Restorations here in St. Charles attempting to restore a landscape surrounding an 1850 farm house. This year our goal is to plant and maintain a vegetable garden as it may have been done then....From their diary we have a fair idea of some of the vegetables they planted, but no idea of what varieties were available during the 1850s. Would your organization have varieties dated, possibly of that time period?....Also would you know of any people or organizations who collect or exchange seed for woodland wildflowers and prairie grasses and forbs?....(Roy Diblik, The Natural Garden, 38W443 Highway 64, St. Charles, IL 60174)

Dear Mr. Whealy....I am currently doing research for the Blue Ridge Institute which has almost completed construction of an 1800 Germanic farmstead. As a working farm, we intend to grow crops and vegetables that an 1800 Germanic settler would have grown in this area of Southwest Virginia....If you would know of someplace where we could obtain some Prince Albert peas or any variety of Charlton or Hotspur peas it would be greatly appreciated....(Debbie Switzer, Farm Museum Researcher, Ferrum College, Ferrum, VA 24088)

Dear Mr. Whealy....We are in the process of developing a pioneer site representing Northern Illinois in the 1840s....We are planning demonstration plots of crops of the period and I am trying to locate seeds to use at the site this year....I am interested in finding a yellow flint corn, a variety of gourdseed corn and, for use in the future, a red-bearded winter wheat, preferably a Red Chaff.... (LuAnn Bombard, Curator, Blackberry Historical Farm-Village, RR 3, Box 591, Barnes Rd., Aurora, IL 60504)

Dear Mr. Whealy....Our Department has gardens in our Missouri Town 1855 project and we would like to grow various vegetables of the period (plus produce seeds others can use)....(William L. Landahl, Director, Jackson County Parks and Recreation, 22807 Woods Chapel Rd., Blue Springs, MO 64015)

Dear Kent Whealy....I am a member of the Men's Garden Clubs of America. Several men in our local club have shared the concern you have for lost varieties. We have been contacted by a historic jail to supply plants for a garden as it might have appeared in 1859....(Timothy E. Meng, D.C., 2701 Blue Ridge Blvd., Independence, MO 64052)

Dear Mr. Whealy....Our volunteer organization supports the activities at Columbus Metro Parks' Slate Run Living Historical Farm. Opening just last May, Slate Run re-creates a family farm of the period 1880-1910....We were especially interested in your ideas in The Mother Earth News about a period garden directory....We've had some indirect contact with SSE through a couple local members and look forward to being active in preserving old varieties for the exchange network....(Sally Manifold, Garden Chairman, Friends of Slate Run Living Historical Farm, Rt. 1, 9130 Marcy Rd., Ashville, OH 43103)

(The 1982 Fall Harvest Edition, pages 51-55)

Dear Mr. Whealy....As a joint project between the National Park Service and the Concord school system, we are attempting to recreate a 1775 kitchen garden in the park. The students study old newspaper ads for seeds (Boston Gazette 1770-1775) and try to match available varieties with these descriptions. This year we plowed the garden with horses and next year we hope to do it with oxen. The children study the 1775 gardening techniques of the Capt. David Brown family, dress in 18th century clothing, help forge metal hoes and rakes at Saugus Ironworks National Historical Site, do all the gardening and answer visitors' questions....I am enclosing some copies of 18th century newspaper seed advertisements....finding the correct seeds involves comparing these seed lists with pictures and descriptions in herbals and farming guides of the past. We have taken advantage of the research done by the staff of Old Sturbridge Village, a re-constructed "Living History" village in western Massachusetts.

They have scoured the pages of early 1800s seed catalogs, and using their research findings, we can back-date certain varieties....Could you suggest a source of white flint corn, Marrowfat peas, early Charlton peas, Dutch cabbages, an old variety of cucumber listed as "prickly cucumber" and red cowhorn potatoes?....If any of your readers are interested in 18th century gardening or plants, please tell them to write me and I will send them a bibliography of books that I've found to be helpful....(Margie Hicks, Minute Man Natl. Historical Park, Box 160, Concord, MA 01742)

Dear Mr. Whealy....The Morris County Park Commission is presently developing a Living Historic Farm. It will re-create the farm of a late 19th century "Gentleman Farmer." We have accurate records of the various vegetables he planted. We are in the process of trying to locate sources for the various seeds, if still available....(Joseph R. Hovance, Director, Morris County Park Commission, PO Box 1295R, Morristown, NJ 07960)

Dear Mr. Whealy....(We are) establishing a working 17th century farm. Based upon historical and archaeological research, we are attempting to make this as accurate as possible using period agricultural techniques, livestock husbandry practices, etc. One problem we are facing is obtaining plant types that are similar to the unimproved seventeenth century varieties. While exact duplication is impossible, we hope to locate some plants that will reflect those of the period.... (Henry M. Miller, Box 168, St. Mary's City, MD 20686)

(<u>The 1978 Seed Savers Exchange</u>, page 27)

Dear Mr. Whealy....The Colonial Pennsylvania Plantation is an 18th century farm museum very much interested in the preservation and acquiring of old plant varieties found in this area during the 18th century. For many years our search for appropriate seed types has met with little success....(James D. Keiser, Box 122, Edgemont, PA 19028)

(<u>The 1979 Seed Savers Exchange</u>, page 15)

Dear Mr. Whealy....We are interested in authentic varieties of vegetables and fruits of the mid-19th century....The National Park Service is in the process of restoring Old Fort Larned. Nine historic military buildings remain from the original Santa Fe Trail fort. Once the fort is completely restored and furnished, we hope to have an authentic garden. In the 19th century army, vegetables were not provided for the soldiers. Each company of men (about 64 to 100 soldiers) often used some of their meager wages to buy seeds and garden tools to grow their own vegetable gardens. We hope to recreate an authentic army company garden....(B. William Henry Jr., Historian, Fort Larned National Historic Site, R 3, Larned, KS 67550)

Dear Mr. Whealy....The Ramsey County Historical Society runs the Gibbs Farm Museum, a living history farm....The Farm is a re-creation of an 1870s truck farm containing the original buildings and many of the implements which the family used. This season we hope to run the farm as a working truck farm ....Your Yearbook should contain much information of use to our project.... (Kurt E. Leichtle, Site Manager, The Ramsey County Historical Society, 2097

W. Larpenteur, St. Paul, MN 55113)

Dear Mr. Whealy....Pennsbury Manor is a reconstruction of William Penn's seventeenth century country plantation, and our gardening and farming programs are an integral part of our interpretation of the time period of 1681 to 1710....Locating appropriate seeds for planting in our gardens has been a continuing and frustrating job. I am hoping that your organization can help with both flower and vegetable seeds that are reasonable choices for a site that is interpreting this early colonial period. William Penn is an interesting historical personage for many reasons, not the least of which is that he was involved in the importation of English plant material for his gardens here and the exportation of native species for planting in England....I am interested in the older, smaller species that are probably the parents of the new hybrids....I need your help and possibly the resources of your Exchange to help bring integrity to the gardening program at Pennsbury Manor. I have received information and seeds from Wanigan Associates in Massachusetts who specialize in bean seeds; however, at the present time for all other crop varieties, we are at the mercy of commercial seed houses....(Nancy D. Kolb, Historic Site Administrator, Pennsbury Manor, Rt. 9, Morrisville, PA 19067)

(The 1980 Seed Savers Exchange, pages 38,39,40)

Dear Mr. Whealy....The Blue Ridge Institute has almost completed construction of an 1800 Germanic farmstead. As a working farm, we intend to grow crops and vegetables that an 1800 Germanic settler would have grown in this area of southwest Virginia....(Debbie Switzer, Farm Museum Researcher, Blue Ridge Institute, Ferrum College, Ferrum, VA 24088)

Dear Mr. Whealy....I am working on a project at the Fort Vancouver National Historic Site. We are attempting to reconstruct the vegetable garden as it was in the 1840s and need seeds of that time period. The Hudson's Bay Company of Canada owned and operated the Fort. It was an English garden....(Sally Hansen, Fort Vancouver National Historic Site, Vancouver, WA 98661)

(The 1981 Winter Yearbook, page 43)

Dear Kent....Pioneer Farm is a demonstration historical farm that provides hands-on programs for kids and adults. We have been searching for locally adapted seeds and root stock. We are currently developing grapes from the T. V. Munson collection in Denison, local peach varieties and some local corns and beans. We are also collecting native prairie seed from the few remnant patches in our locality and are using that seed to restore 12 acres to prairie. The farm is new and off to a humble start, but eventually we intend for all our cultivated varieties to be heritage varieties and hope to be a source for other gardeners in Central Texas. We invite contributions from our region and will respond by sharing what we have....(John Peterson, Assistant Supervisor, Pioneer Farm, 11418 Sprinkle Cut-Off Road, Austin, TX 78754)

Dear Mr. Whealy....The Georgia Agrirama, State Museum of Agriculture in Tifton, is on the lookout for old seeds. The living history museum, which depicts life from 1870-1899, has embarked on this project in order to make its vegetable and flower gardens as "old" and authentic as possible....In the past, Georgia Agrirama has raised only the crops which research indicates were grown in the 1880s and 1890s. Beginning this planting season, the museum grounds will feature plants from seeds which have been handed down through the generations of south Georgia gardeners....Among the particular seeds Georgia Agrirama is seeking are Jones watermelon, Henderson lima bean, Golden Prize peppers, Nixon canteloupes and Old Homestead green beans....Anyone with seeds or information about turn-of-the-century gardens who would like to assist in the project is asked to write....(The Georgia Agrirama, PO Box Q, Tifton, GA 31794)

Dear Mr. Whealy....We are in the process of developing a pioneer site representing northern Illinois in the 1840s....We are planning demonstration plots of crops of the period, and I am trying to locate seeds to use at the site this year....I am interested in finding a yellow flint corn, a variety of gourdseed corn and, for use in the future, a red-bearded winter wheat, preferably a Red Chaff....(LuAnn Bombard, Curator, Blackberry Historical Farm/Village, RR 3, Box 591, Barnes Rd., Aurora, IL 60504)

Dear Mr. Whealy....I am working with Kane County Restorations here in St. Charles attempting to restore a landscape surrounding an 1850 farmhouse. This year our goal is to plant and maintain a vegetable garden as it may have been done then....From their diary we have a fair idea of some of the vegetables they planted, but no idea of what varieties were available during the 1850s. Would your organization have varieties dated, possibly of that time period?....Also, would you know of any people or organizations who collect or exchange seed for woodland wildflowers and prairie grasses and forbs? (Roy Diblik, The Natural Garden, 38W443 Highway 64, St. Charles, IL 60174)

Dear Mr. Whealy....We are developing a living history farm near Lowell, Indiana. We will be depicting the 1910-1920 time period and the 1850-1860 time period on two separate areas of the site. We will have gardens at both areas and wish to have authentic vegetable varieties. (Becky Crabb, Buckley Homestead, 3606 Belshaw, Lowell, IN 46356)

Dear Kent....I am writing in reference to several bean varieties that I wish to grow here at Old Salem Historical Restoration. Old Salem dates back to a settlement of Germans in the 1760s. The horticultural department is attempting to recreate the gardens of the original settlers. (Joan St. John, Horticulture Department, Old Salem, 600 S. Main St., Drawer 2 Salem Station, Winston-Salem, NC 27108)

(The 1983 Winter Yearbook, pages 145, 182-185, 191)

Dear Mr. Whealy....I am an employee of the National Park Service working at Gettysburg National Military Park. Please send any information which you might have regarding vegetable or herb seeds which might have been used in an 1863 family garden. (Steve Wright, 650 Taneytown Rd., Gettysburg, PA 17325)

(The 1984 Fall Harvest Edition, page 112)

Projects

Steve Neal Displaying Some of the Potatoes Being Maintained by His "TUBERS" Project
(For More Information, See Page 69)

NATIVE SEEDS/SEARCH

(Native Seeds/SEARCH, 3950 West New York Drive, Tucson, AZ 85745, is a new non-profit organization devoted to the conservation and promotion of native, agriculturally valuable plants of the U. S. Southwest and northwest Mexico. Its directors are: Gary Nabhan, president; Barney T. Burns, vice-president; Karen Reichhardt, secretary; and Mahina Drees, treasure....I was going to write an article about their organization, but I don't think I could do better than their following report. These are fine, dedicated young people whose work you should support, even if you aren't a Southwest gardener. Associate Memberships are $10 per year. Listing of Seeds for Sale is $1.)

Native Seeds/Search Annual Report
1983 - 1984
WE'VE COME FULL CIRCLE

A child grows more in its first year of life than during any other time. Likewise, Native Seeds/SEARCH has experienced incredible growth during its first year of existence. Our goal of maintaining native crop diversity in the region has resulted in a wide number of activities that involve public education and seed conservation itself. As we become more firmly established financially, our energies are turning toward the heart of seed saving. At the present our principal means of support is by associate memberships, grants, seed sales and special contributions.

A year ago SEARCH celebrated its "coming out" to the Tucson community with a San Juan's Day fiesta held at Tucson Botanical Gardens. Prior to this event, our principal means of outreach was through letters of invitation to prospective associates, seed sales, or writeups in several national gardening publications. Since the 1983 San Juan's Day fiesta, we have also introduced ourselves regionally by participating in plant sales and fairs at the Boyce Thompson Arboretum in Superior, Arizona, the Arizona-Sonora Desert Museum Harvest Bazaar in Tucson and the Tucson 4th Avenue Street Fair. In February 1984, we sent a mailing of our seed listing and brochure to over 2,500 potential associates.

Currently our associates number 214, though this increases weekly. Nine of these are lifetime associates.

Workshops and Educational Programs

SEARCH maintains a Demonstration and Conservance Garden at the Tucson Botanical Gardens for the purpose of displaying rare Native American crops, wild relatives of crops and cultivation techniques. Esther Moore is expertly caretaking the garden on a regular basis. Two major planting seasons are observed which coincide with local Native American cultivation practices. In general, more limited seedstocks are planted here.

The first of a series of workshops was held on the topic of corn on January 21 and 22 of 1984 at Tucson Botanical Gardens. The purpose of SEARCH's workshops is to educate the public on the uses of native foods which are either not currently in common use or well known. At our January 1984 corn workshop, participants learned about the many varieties of corn and the numerous ways to prepare them. Techniques covered included roasting, grinding, lime washing and stewing.

Research Involvement

Within the past year members of the SEARCH team have become involved in a variety of research topics. Several of these topics are still in development stages, to be studied in depth when funds are available. Seed has also been provided to other research institutions. Following is a list of research topics, goals and status of projects.

1. Monitoring of wild chile populations in Arizona and Sonora. Goal: To determine whether populations are becoming endangered by human harvesting pressure in both states and to learn the economic status of cultivated and wild populations in Sonora. Status: University of Arizona Plant Sciences graduate student Cindy Baker is receiving funds from SEARCH for travel and field expenses to begin survey of these chile populations. Proposals for additional funds have been sent to Horticultural Research Institute and other foundations.

2. Preliminary investigations about the prevalence of seed-borne bean common mosaic virus in domesticated tepary beans. Goal: To screen public as well as private, or institutional accessions to isolate a virus-free seed source. Status: SEARCH grew out teparies from a wide variety of sources in cooperation with University of Arizona Plant Pathologists, who confirmed the virus presence with electron microscopy. Meanwhile, a Mexican pathologist determined that many currently available lines of teparies carry the virus even though they don't always show symptoms. We have recently obtained two lines of virus-resistant (though not immune!) teparies discovered in 1977 trials in Puerto Rico. These will be evaluated under desert conditions this coming summer. Domesticated teparies were removed from the SEARCH seed listing until a satisfactory seed source is located.

3. Cooperation with Ms. Radziah Arriffin and Dr. Charles Weber at the University of Arizona Nutrition Department in the evaluation of the chemical composition of native foods eaten by Papago and Pima Indians. Goal: To justify the consumption of these foods based on nutritional analysis. Over 40 food plants have been analyzed. Results will be published in a subsequent issue of the Seedhead News.

Seed Collection Trips

Seed collection is the activity which most clearly corresponds to the acronym SEARCH in our name. We have collected native seeds during several expeditions into Indian country, even though those trips were for other purposes. Whether the main purpose of travel is for scientific research or for arts and crafts buying, a few minutes are set aside each day for learning what the local people are growing and eating. Thus, six trips were made into Mayo and Yaqui country in 1984, where several varieties of corn, beans, squash and gourds were collected. Wild beans of Mexico were collected in conjunction with a wild bean research trip from Jalisco to Sonora. Some new seed varieties were also collected from the Tarahumara Indians in Mexico.

Seed Distribution

Our objective of in situ conservation was aided by replenishing the seedstocks of the following tribes, providing seed formerly grown by these tribes to individual farmers or to programs: Papago, Gila River Pima, Salt River Pima, Baja California Cocopa, Chihuahua Mountain Pima, Sonora Pima Bajo, Zuni and Hopi.

In addition, about 1500 seed packages were sold to the public. The Smithsonian Institution received packages of the entire collection of seeds for sale. Meals For Millions/Freedom From Hunger was supplied Mexican corn to be planted in Ecuador, and Papago Indian seeds to be distributed on the Papago Indian Reservation. During SEARCH's first year of existence we have provided seed varieties to the following researchers: 1) Dr. F. Bliss of the University of Wisconsin received common beans from various tribes to study the origin and dispersal of protein types in beans. 2) Richard Pratt of Purdue University received tepary beans from various tribes for comparison of electrophoresis "fingerprints" with bruchid beetle resistance. 3) Laura Merrick of Cornell University received wild and cultivated cucurbit seeds for study of evolution and introgression on the *Cucurbita mixta* - *C. sororia* complex. 4) Numerous archaeologists studying Southwestern crop origins were provided comparative plant material. 5. Dr. Giles Waines of University of California, Riverside received Papago 60-day corn for field growing trials.

Public Presentations

The following interested organizations have invited presentations about SEARCH at their regular meetings:

- Arizona Native Plants Society, Yuma Chapter
- Arizona Native Plants Society, Tucson Chapter
- Ocotillo Garden Club, Tucson
- Rolling Hills Garden Club, Tucson
- Tucson Organic Gardening Club

Talks by Gary Nabhan to organizations on SEARCH-related topics:

- Salt River Indian Reservation Elderly Education Program coordinated by Scottsdale Community College
- Heard Museum, Phoenix, Arizona
- Society for Economic Botany, 1984 annual meeting
- USDA Water Conservation Laboratory, Tempe, Arizona

(*The 1985 Winter Yearbook*, pages 250-252)

FRUIT SCION EXCHANGE

Over the past few years there has been an increased interest in saving and exchanging old varieties of fruit. Because an increasing number of members are starting to exchange such fruit scions through the Seed Savers Exchange, it seemed like a good idea to publish some guidelines. The following is a set of guidelines which is a consensus of the thoughts of Charles Estep, Les Hawkins, Will Bonsall and Jesse Schwartz. This should be used as a guideline for exchanging scions and budwood. In addition to the guidelines, it should be kept in mind that there are differing state regulations regarding the importing of plant material. It is each member's responsibility to know his or her own state's regulations and to not request such material from other members. Disregarding state regulations will only harm the purpose of the group.

1) Restrictions, State/Federal: All persons listing scions, fruit seeds or rootstocks should contact their County Ag. office or State Quarantine office for list of items that cannot be shipped in or out of that person's state. Do not request or offer such items in your listing.

2) Price: $1.00 per variety for LISTED MEMBERS. $2.00 per variety for NON-LISTED MEMBERS.

3) Number of scions: Two per variety (except limited quantities as listed). (All listings should contain a separate entry for Limited Quantities.)

4) Size of scion: 10-12 inches in length and 3/16 inch to 3/8 inch in diameter.

5) Age of scion: Scions should be cut from the last growth on the tree, whenever possible. Can be used from the tip where diameter is 3/16 inch where the cut is made at 6 inch length, or on down the limb up to 3/8 inch diameter. Under 3/16 inch the scion is too fragile—over 3/8 inch and it's too hard to cut.

6) Scion cutting time: For northern growers I would suggest taking scions after leaf fall (during dormancy), prior to hard freezing. For southern growers I would suggest taking scions from leaf fall to bud break, depending on the variety. For some, leaf fall is November and the beginning of bud break is January. For others, leaf fall varies during winter months and bud break runs into June. The fresher the scion the better chance of survival. (Perhaps a deadline of March 30, and later for northern growers.)

7) Storage of scions: After cutting, place scions in a plastic baggie (or old plastic bread sack). Into the baggie or bread sack put a small piece of material such as paper napkin, paper towel, toilet tissue, that has been dampened. Beware of too much moisture. A good process is to first dampen the material and then squeeze the water out of it. Too little moisture and the scion dries out, too much and the scion rots. For long storage, the scion should be dipped into a seal on the ends that are cut. Les Hawkins' system is the best I've seen. He dips his scions into a product called "Yellow-Cap Grafting Seal," made by Farwell Products, PO Box 3347, East Wenatchee, WA 98801.

8) Identifying scions: Please—mark the scions clearly with a waterproof pen or marker. Tie all scions of one variety together with masking tape and leave extra tape for name. Or, seal name of scion up in a little plastic wrap or such attached to scions. Nichols Garden Nursery, Albany, OR, carries "Sharpie Marking Pen." It is the best I've found. It is waterproof when dry and will last a year in the sun. Price: 75 cents each. Most stationery stores also carry this. In addition, soft metal tags where the name can be embossed are good to use.

9) Packaging: Wrap sacks of scions in a sheet of newspaper--to help them

maintain more even temperature and to take some of the shock of the Postal Service's rough handling. Then place in one of the bubble-filled shipping bags, a paper towel roller, or box to fit. Seal both ends of the towel roller, open up a large envelope and roll the scions up in the envelope and seal. If using shipping bags, staple end. If using box, seal with strapping tape. The sack used to wrap the scions should contain damp paper as for storage. The sack should be sealed at top.

10) Shipping time: Persons requesting scions should give preferred time of shipping. Persons offering scions should state time of year that they are willing to ship. Can be either fall or early spring.

11) Freezing scions: DON'T cut frozen scions--cut scions prior to fall freeze or after spring thaw--and store in cool place, as a refrigerator. Never store in freezer.

12) Identifying scions after grafting: Example: Use four-inch aluminum tags tied loosely to grafted scion. Use ball point pen or electric engravers to write on the aluminum tags the name, date and supplier. Then spray tags lightly with spray-enamel to preserve the tag and for easy I.D. I used yellow for the original rootstock, blue for the original tree, red for the first year's graft, light blue for the next year's graft, and this year will be orange. At a glance you can tell which ones were put on in which year. Also, spray the graft union same color as the tag to help you identify the grafted variety. In a year or two the graft will grow so you can't tell what you have unless you keep after it.

13) Apples, pears, quince are all of the same family. I have not tried grafting onto quince. I have successfully grafted apple onto pear and pear onto apple. So some members might want to experiment with this. Experiment only; this is never as good as apple to apple, pear to pear. Any apple tree that has suckers at the base with roots, the suckers can be separated from the parent plant when dormant and planted. In the spring one can graft any variety apple onto this little tree, usually at about four inches from the ground for the union. Size of tree will be same as the tree from which the root was taken.

(The 1984 Fall Harvest Edition, pages 134-135)

SCATTERSEED PROJECT

(Will Bonsall, Khadighar, Box 1167, Farmington, ME 04938)

We share your concern over dwindling germplasm pools and commercial unavailability of so many varieties, especially the open-pollinated ones. We have been organizing a small local growers and preservation network called Scatterseed Project....We are now growing a couple hundred varieties, some of our own development, but most of them are not heirlooms. In fact, many of them are available from catalogs outside the U. S. (Europe and Japan mainly).

After seeing how "together" the SSE is, we have considered whether Scatterseed would serve enough of a purpose to be justifiable, as it draws energy away from our other public service programs. After further thought, however, we think it might be better to continue at least on a small scale....Although it's extremely useful to have networks centered around an individual like yourself or John Withee, it's also expedient to have a less-centralized back-up for the long run. In fact, John is an excellent example--were it not for SSE, what would become of the Wanigan collection?

I'm also concerned about the lack of variety circulation (of course that's the basis of the problem), that is, general use as opposed to preservation. However, I can't imagine how to resolve that without functioning as a seed company, which is exactly what you're trying to avoid, and with good reason. I've wondered how it would answer to have a small for-profit seed-house specializing in odd-balls, off-beats and heirlooms, at least a wide sample of the more distinctive and/or endangered. Any opinions?

I also like the idea of a system which is simultaneously dispersed and centralized. I feel sure your Growers Network is the thing. Coupled with the SSE general membership, plus a Seed Bank, it should provide a maximum of protection.

One aspect of seed saving we've been emphasizing (and I'd like to see SSE do more of) is to teach folks how to save seeds, especially how to select for a range or set of traits. This gets a bit heavy for many home gardeners, but some crop plants differ so greatly from their forebears in a few generations as to lose some of the traits which made them unique. We've been offering a program (as part of a series on vegetarian self-sufficient living) called "Grow It, Stow It, Sow It"--covering seed-growing, selecting, hand-pollinating, viability and germination, isolation, etc. There's been more interest than I'd anticipated, and a few have expressed interest in helping with the Scatterseed Project.

We've also been trying to push some interest in "new" crops, like fava beans, edible flax (mild-flavored), seed-poppies, day-lilies, rape, etc. We'd like more varieties of all of these to use in our trials....We've also noticed a dearth of certain crop groups among seed savers. For example, while lots of people save corns, beans, squashes, tomatoes and potatoes, I've heard of little interest in alliums, brassicas and umbelliferas and would like to correspond with others who are interested in working with them....

(The 1982 Fall Harvest Edition, page 59)

(From later correspondence)--I'm building up a very formidable inventory of world germplasm sources. I'm especiallly emphasizing the NSSL and USPGS collections, and working on foreign governments' collections as well. Part of my intention is to act as a liaison between those collections and the SSE to get some of the more interesting stuff into wider circulation....The result is that I've got a tremendous amount of access and anything I select and increase can be distributed through SSE, Scatterseed Project, etc. The problem is letting everyone else know what's available so I can request it for them.

This inventory comprises many hundreds (even thousands) of pages of printout (some of it on microfiche), so it's not realistic to just send a copy to anyone who's curious. Of course I could Xerox relevant sections for anyone after a particular crop, but even that's unwieldy (just the Pullman, Washington bean collection alone is 195 pages). The cheapest I can Xerox is 10 cents/page presently....I'm just not sure how to be most effective. Of course I can procure materials for anyone who knows what they're looking for already, but what's to be had is dazzling, and I'd like to share my wide accessibility with anyone else who will use it properly. (Anyone who wants to obtain accessions through me should realize that we are expected to supply notes on performance and to be prepared to share increased seed. That's only fair.)....

I have obtained (or am obtaining) the following crops from the different facilities, and if anyone has a particular area of interest (e.g. short-season beans), they should let me know. So far I'm pursuing what I think may interest us or other members. I'm especially scanning the data for "oddballs"—onions (so far all multipliers, 17), peas (about 40 already available), squashes, corns (mostly flints so far), broccoli/cauliflowers, rapes, turnips, cabbages, Chinese cabbages, radishes, lettuces, carrots, beets, chards, spinaches, beans, soys, favas, celeries, wheats, ryes, clovers (especially early & heavy seeds for sprouting), oats and barleys (hull-less vars. of both). I haven't yet requested any suger-beets, mangels or lentils, unless I hear that any members are especially interested.... I've been asked to speak at several conferences, garden clubs, organic gardening chapters, etc. this winter, and I'll be teaching and publicizing seed-saving, as well as the SSE and Scatterseed Project.

(The 1983 Winter Yearbook, page 91)

This year I've had many opportunities to help SSE members obtain materials from the USPGS, so far mostly beans, tomatoes, carrots and cabbages. Folks have been very cooperative and grateful, and the USPGS facilities have, for the most part, been most obliging. So far members have either requested something specific (by name or description) and I've sought it out for them, or they've requested relevant sections of the catalogs, which I've sent for them to peruse (they send me 10 cents per page for xeroxing). I'm happy to continue doing this for anyone interested. I only hope people increase the materials I'm getting them and funnel it into the SSE. Part of the understanding with the USPGS is that the materials we get be propagated and redistributed. They also appreciate getting back any results or comments (via me, to keep things straight), although no one seems to insist on this. Some folks have expressed concern at requesting too much and seeming "piggy." The number is not the decisive factor; some of their requests are for hundreds of samples. The limit should be set by what the person making the request knows he can handle. If you're sure you can take care of (including isolate, when necessary), a large number of cultivars, I'll gladly request them. But we must take this privilege seriously. I'll also be requesting some varieties from SSE members to send to the appropriate USPGS facility. I hope members will be as cooperative with them as they've been with us. Even though the government collections come from all over the world and number in the tens of thousands, they are quite lacking in domestic "heirloom" varieties. This is one way we can make a significant contribution.

(The 1984 Winter Yearbook, page 115)

THE SOCIETY FOR THE PRESERVATION OF POULTRY ANTIQUITIES

(Photo by Julie Lintecum)

The Society for the Preservation of Poultry Antiquities (SPPA) is dedicated to the perpetuation of vanishing breeds or varieties of domestic poultry. "Poultry" includes standard-size chickens, bantams, ducks, geese and turkeys. A list compiled by the Society is brought up to date periodically telling what is considered to be on the endangered list. A directory of the membership is published every other year. It lists what they are currently breeding and if they sell stock, eggs or chicks. It is probably the best sourcebook for many of the rare breeds. This book is included in the membership dues along with a quarterly newsletter.

Each year a poultry show is chosen to host the annual meeting and competitive show. Many cash awards are offered to encourage the showing of these endangered poultry breeds. This show is rotated into different areas of the United States so, sooner or later, one will be close enough for you to attend.

Poultry and gardening go together very nicely. Poultry manure is one of the best natural fertilizers. And the garden can furnish part of the food for a small flock of fowl. Why not make the poultry you raise one of the rare breeds? There are many breeds and varieties from which to choose. Just as you are trying to save many of the heirloom varieties of vegetables, there are those trying to do the same with poultry. Write to the Secretary of SPPA, Marion Nash, P. O. Box 102, Murphysboro, IL 62966, for membership information.

(The 1985 Winter Yearbook, page 253)

PROPOSED NETWORK OF VOLUNTEER SEED BANKS

by John Rahart, DDS

Dear Kent....I believe that what you and the Seed Savers are doing is not only essential, but in fact it is still not enough, given the trends that are occurring in this country. After reading about the exposed state of our national seed banks and knowing you were actively involved in the Seed Banks Serving People Workship in Tucson, I got to thinking. If there is enough interest and enough seed available, let's develop a network of small volunteer seed storage facilities nationwide. These "banks" could store a good cross-section of locally adapted varieties and their diverse locations would ensure against a central bank failure. Also, as this network develops, I would like to see different groups swap samples of their collections for storage at different locations as insurance against their loss....With your central overview, you might be able to help put such a program together and organize the seed collecting mechanism. Others, such as myself, could do the actual storage (drying with our silica gel method, sealing in laminated bags and freezing), indexing, germination testing, multiplication through local growers, varietal comparisons, etc. I would like to correspond with anyone interested in setting up such a network....

I am presently in the first stages of starting a miniature long-term seed storage facility here. I intend to concentrate on native Indian seeds, especially those of Southwestern and Platt Valley and Missouri Basin Indians (drought-resistant crops)....I would like to work with any groups or individuals trying to develop cool and/or altitude-adapted strains (sub-artics and cold-hardy crops)....I would also like to hear about any studies to determine which crops or varieties are ultra-violet light-tolerant....I am willing to advise anyone who is interested in setting up a small storage facility or to answer any questions concerning our silica gel seed drying technique....

Biodynamics is actually a highly refined organic approach....Earlier this century a man named Rudolf Steiner was making some amazing predictions and statements....He was asked, "Why is the world so aggressive?" He replied, "It's a matter of nutrition." This surprised most people because we don't generally think of war-like hostility resulting from a root cause of some nutritional variance....In Europe at that time the large farms were beginning to use the same non-husbandry of the soil techniques that we have seen grow into modern American agriculture. Many farm families, who had been growing and saving wheat and oats for a century, all of a sudden were having trouble getting their seeds to germinate properly. The situation was deteriorating rapidly and they were having to buy seed from foreign sources, which was very disturbing to them. So they came to Steiner again to ask him the cause of this. He laid out the concepts of bio-dynamics at that time. He essentially advised them how to get back into building their soil using his directions. Many of his concepts seemed to have no previously known scientific justifications or background, but within a few years these same grain farmers were some of the most highly prized grain sources in all of Europe....It's becoming exceedingly obvious that our concept of agriculture in this nation appears to be way off base....We are losing topsoil at an incredible rate, our crops are becoming more nutritionally hollow and increasingly harder to grow....All of these things together give me a keen desire to find out if we have indeed jumped the track....

(Dr. John Rahart, PO Box 1186, Saratoga, WY 82331. The writings of Rudolf Steiner (and those of Dr. Ehrenfreid Pfeiffer, the principal bio-dynamicist in the U. S.) are available from: Bio Dynamic Farming & Gardening Assn., PO Box 550, Kimberton, PA 19442. Send a self-addressed stamped envelope for a list of the titles they have available....)

(The 1982 Fall Harvest Edition, pages 57-59)

A LAYMAN'S EXPERIMENTS WITH TEOSINTE
(From Correspondence with Tom Ackley of Minnesota)

There are about seven varieties of annual teosinte all located in southern Mexico and Guatemala, along with one tetraploid perennial and one diploid perennial. The tetraploid has been known to exist for about 100 years, but because it has 40 chromosomes, it cannot cross with maize (although tetraploid synthetics have been developed by, among others, Cornell University and the University of Illinois). In 1979 the diploid perennial, which has 20 chromosomes as does corn, was discovered and sparked off excitement and debate on its potential uses....

For about the last century, much controversy has surrounded the origin of corn and the place where teosinte figures in. Recent evidence and deduction fairly lays it to rest and teosinte is now the accepted ancestor of corn, though the fight in academia was a hard one and perhaps bordered on the ridiculous. My own breeding work has convinced me of this, without the help (or interference) of the "experts." Because of their genetic similarity, the chromosomes match up easily, cross readily and produce fertile seeds. The "hybrids" are relatively higher in protein, and yields can increase in the second and third generations as much as 100 percent. Because of the perennial nature of Zea diploperennis, the possibilities in planting style could be revolutionary, to say the least. One representative of a major seed company told me that their company wouldn't touch it because "We'd be cutting our own throats."

Now here's the hitch and one of a suspicious nature to me. At a recent seed trade convention I heard Dr. Lambert, University of Illinois at Urbana, who has studied the feasibility of the use of teosinte extensively, declaim the use of teosinte in the future, citing too many unknowns and other "difficulties." This pessimism extends everywhere I look. Dr. Alexander, Professor of Plant Genetics, wrote in a recent letter to me, "It does not seem likely that corn breeders will use the teosintes in an improvement program on a very large scale in the near future. In short, I see very little to come from the use of teosinte in corn breeding." He made it clear that little research was expected. Such a turnaround and in only a short period of time leaves me wondering at their actual intent, especially since the pessimism's nearly universal.

The only one I've met or heard of who is serious about using teosinte (openly that is) is a Chinese scientist doing a tenure at the University of Minnesota. He was quite impressed at the results I have obtained. Quite a shot in the arm for me because, though I am no scientist, I think I know corn. And if that particular aspect is really being left unpursued, I'm going for it....

The use of teosinte in corn breeding programs has been attempted since the turn of the century with a variety of "success"....Recent statements concerning the success of these early attempts are clouded (I've read old, old material about this) and paranoia suggests the intent to enrich a few pockets by gaining time....Last year I had plants that had up to 40 ears and that, I think, justifies itself....

I first obtained annual teosinte from Dr. Beadle, a noted heavyweight in genetics, and have since been in regular contact with him. That year, 1978, some crossed....Intrigued, I devoured as much related material as I could and now find myself knowledgeable enough to do something about it....

Last year I got seed of Zea diploperennis from Dr. Iltis and Dr. Doebley, its "discoverers." This year I will begin to climatize them....Because of its own peculiar physiology, the teosintes must be climatized with a short day treatment before they will flower at my northern latitude....After such treatment, the seeds will produce plants which will produce seeds which will produce plants without the need for the treatment, just like any other corn....Dr. Galinat of the University of Massachusetts offers a day neutral teosinte developed from Guerrero teosinte and northern flint, although one needs not cross with a corn to get a day neutral teosinte, only to apply short-day treatment (covering the teosinte with a light-proof material for an hour in the a.m. and an hour in the p.m. will suffice). This applies to the perennial also....

Climatizing plants, if required and depending on the complexity of the plant, has become more or less elementary. Plant manipulation in terms of cell biology and genetics has grown, in my opinion, from a science to a philosophy....what used to be stepping stones to understanding have become understanding itself....Corn is inherently selective and exhibits such a reproductive plasticity that it has been climatized to nearly every habitable spot on the globe. Teosinte, being an expression of corn (or corn, being an expression of teosinte), is an admirable tool in the "undeveloping" of corn. Its half-breed children exhibit all the variants available to corn from the discarded to the exploited. Teosinte is extremely sensitive to photo period (day length) and must be kicked out of that condition, which is really pretty easy to do. This year I am doing that with the diploid perennial, as my seed came directly from Mexico. Once done, teosinte is on easy street, for Zea mays is highly adaptive (plastic as the experts so lovingly say), and that info is retained genetically. My annual teosinte now needs no more attention than ordinary corn and the crosses are treated <u>as</u> corn....

This year I've started a larger program with the assistance of a local farmer, one Reid Volkman, whose land is ideal and whose outlook on life and attitude are similar and compatible with mine (I am an Indian and have always used the holistic attitude and approach to life and propagation). On his land we intend to introduce teosinte into the wild, hoping to establish a wild population of both the annual and the perennial, to use as stock when we want it. Teosinte is essentially a weed found in the wild and able to propagate itself, its fruit being similar to that of a head of grass. In all other respects, it is nearly indistinguishable from the corn plant. We are confident it will work....On my farm we will do the crossing.

(From later correspondence)--....Much has happened regarding my teosinte--all good....I've been breeding this for three years and studying it even longer.... I've learned a lot since I began; more, I suspect, than if I'd followed a more traditional mode of study. Obviously, sorting out all the "expressions" of teosinte hybrids into manageable categories would take any one person longer than three years, and I don't pretend to have done that. What I have done is develop the viability of using teosinte as a food service plant, or, rather, bridging the gulf between teosinte and corn--the one being a weed, the other a food. A bold statement, perhaps, in light of claims to the contrary. But it is in truth relatively easy to do. It's just a matter of approach....

This is my third year in combining these two plants with the emphasis placed on sorting out usable and prolific lines. Barring any belligerent storms, I anticipate all the success I could desire. I have 1,000 plants, half of which sport from six to twelve medium-sized ears. The next few years will be devoted to retaining this quality through inbreeding. I also will do much the same with the diploid perennial, but keeping these lines separated from each other (about a dozen or so) will require time and space about one magnitude larger....

(Still later)--....just learned of a seemingly ominous development--one that might in time relate to diploperennis, but not at all confined to it. I will quote from a USDA publication dated August 6:

"....In a move to speed government scientific discoveries into commercial use for the public, the USDA will grant some exclusive licenses for private companies to manufacture and sell products patented by the USDA. The aim is to get a higher payoff for publically financed research....Anson R. Bertrand, USDA director of science and education, 'Companies often feel that without some license protection, they can't afford the development and startup costs to produce and sell a product competitively. Instead, they will put their limited resources into their own patented or licensed products that promise higher returns than a product marketed by many companies....I expect more and more companies will seek exclusive licenses to USDA inventions....'"

The article goes on to say that the first company to receive such a license is Pennwalt Corp. of Philadelphia for the control of nematodes. It takes no great imagination to include seeds of new varieties. Question: Will the USDA grant such licensing to companies based in or owned by another country?....

(Still later)--....Just a note to say my plants are in and I'm cataloging results. I am very pleased and excited about next season....

(The 1981 Fall Harvest Edition, pages 50-51)

HENRY DOUBLEDAY RESEARCH ASSOCIATION

Henry Doubleday Research Association (National Centre for Organic Gardening, Ryton-on-Dunsmore, Coventry, England CV8 3LG) is an organization of organic gardeners in Great Britain which has more than 6,500 members (including a group in India and three in Australia). The Association carries on Henry Doubleday's work on the medicinal, nutritional, agricultural and horticultural uses of comfrey--the fastest builder of vegetable protein yet known. During 1978 they had over 500 garden experimenters working and reporting back on various porojects, including: standard organic gardening, no digging, shallow cultivation, long vs. short shoot potatoes, French intensive gardening, survival gardening, potato ellworm trial, green manuring trial, companion planting, clubroot trial, comfrey and perennial vegetables.

They publish a quarterly newsletter, maintain "the best organic advice service in Britain," sell two valuable booklets called The Vegetable Finder and The Fruit Finder, maintain The International Fruit and Vegetable Library which catalogs all varieties and their known sources, and will send you brochures containing descriptions and prices for their many publications, guides and experimental results. They literally lay a wealth of knowledge at your feet. Annual membership in their organization is eight Pounds Sterling (same for overseas membership).

Of interest from their newsletter: "....We are fortunate in having a fine collection of fruit varieties at the National Fruit Trials, Brogdale Farm, Faversham, Kent. Here they hold almost 2,000 varieties of apples, 450 varieties of pears, 250 varieties of cherry, over 500 plum trees and a large collection of bush fruit....The University of Manchester has a gooseberry collection, and a rhubarb collection exists at the National Vegetable Research Station...."

Last year the Association experimented with a method of planting potatoes - "....plant your seed potatoes with at least 4 inch long sprouts on, so that these start putting out roots from more joints underground giving the tubers a flying start...." First reports back from their experimenters seemed to average about a 30% increase in yield, but their own experiment at the Trial Ground failed to prove this out. So there will be another experiment next year to try and isolate the factors which may have produced up to a 50% increase in some cases.

(The 1979 Seed Savers Exchange, page 35)

****

From a letter from Director Lawrence D. Hills -- The situation over here has continued developing and the Gene Bank is now in operation and is making collections of nearly all the varieties that are still commercially available. The Gene Bank is open to all plant breeders....I am also trying to get a grant to get a lady Ph.D. archeologist who has great experience to travel in Greece and collect the peasant varieties....

I have enclosed our latest account of the Vegetable Sanctuaries, which is what I have gotten started. But we still do not have any money and these four sanctuaries depend on the good will of the people running them. I think if you have some cash behind you and you are getting something like our seed sanctuary system raising them, you are well away....Since you have funds enough to do what you have already done, it sounds to me as if you are doing the very best you can....

If you computerize your commercial inventory and all the private holdings, you have really got hold of something that should keep seeds going in America ....Regarding your underground seed vault, how are you fixed for technical information on this? I know Professor Hawkes in Britain who invented the cool storage system, and I could get his advice, if this would help you....

I get tremendous demands from people who want seeds contemporary with the Mayflower or Thomas Jefferson, as well as the various periods of English history. These people are interested historically, but not genetically....

I do not think I would have gone back and done anything differently. I did what came naturally, went for maximum publicity and this produced the Gene Bank, though the people who run it tend to ignore us and freeze us out....What I need now are funds to get a man with knowledge of the seed trade to help run the sanctuaries....

You are, of course, in contact with the International Coalition for Development and Action--the people who produced Seeds of the Earth. I am trying to get them to produce a popular version and will probably be going with them to the FAO Conference in November, perhaps taking along our man who has started a vegetable sanctuary in Sri Lanka....

Our groups in Australia have all got separate seed libraries, because you have only one seedsman in Australia--Messrs. Yates--and if they do not stock something, you cannot get it anywhere. They are still fighting the legislation....

We cannot handle exchange of seeds with large numbers of people. We want to make certain that our stuff is true to name and being adequately grown and we need retired seedsmen to go round and see what is being grown and how well. I think you might do well to get someone with long experience, ideally one of the many seedsmen driven out of business by the multinationals, who would be willing to help for the sake of genetic diversity....

(The 1981 Fall Harvest Edition, page 57)

BLUE RIDGE SEED SAVERS

Kim Domeratzky and Friends Form Local Seed Exchange Group in Virginia

King Harvest has surely come....peaches, tomatoes, cukes, corn, okra, beans, elderberries....they're all happening at once here. The garden is in full bloom and very colorful....I have been interested in growing and gardening since I took a course at University of Massachusetts in Boston taught by Garrison Wilkes, called "Plants and Human Affairs." He made me aware of the loss of genetic diversity in our major food crops and the extinction of local vegetable varieties since the Green Revolution. Since then, my life has focused towards agriculture. I have worked on an organic pick-your-own farm. I do orchard work in the fall and plant trees for the forestry in the spring. I am currently working in a local greenhouse....

You asked me to write again about our local seed exchange....I've decided to call it "Blue Ridge Seed Savers." Our first meeting was in late September of 1980. I put up posters in the local country stores, but most people showed up by word of mouth. It was a warm, sunny afternoon when people started arriving at Free Creek Farm in Batesville, Virginia. Free Creek is a small goat farm with chickens and ducks and a wonderfully fertile raised-bed organic garden owned by Jim Bryan....the people (young and old, children and grandparents) socialized with each other for the first hour and a half. Then we set up benches of apple crates and boards....

Amy Rice talked about the propagation of herbs and gave a demonstration and shared the herbs she brought with everyone. Then Mr. Bedall, an elderly fruit grower in our community, spoke about the various techniques of grafting and budding fruit trees. He demonstrated with ease the different techniques and then let us go up and try our hands at it. It was a hands-on demonstration with time enough for many of us to sharpen our skills. Alberto Vasquez then talked on the different ways of preserving seeds. Finally I spoke on the importance of saving seeds and keeping old varieties alive. Then people who had brought seeds and plants took them into the greenhouse to distribute. Alberto (our greatest contributor) brought about 20 varieties which he made available to people....About 25 people passed through at various times that day....Farm refreshments were served--goat's milk, herb tea, salads, home-baked bread and homemade wine. Music was played until after dark....

The second seed exchange was held this spring, March 17, 1981, at Loiland Orchard. My friend Loi Patkin owns the orchard, which is now in its fourth year. It is a small orchard located in the mountains of Greenfield, Virginia. Much of the work is done by women....Loi has been trying to preserve and maintain the old varieties of apples. We budded several onto rootstocks last year. She sells locally and people come to the orchard to buy apples and make cider on her old apple press. Sometimes they come from the city and bring their kids and a picnic lunch. There is a creek full of crawdads and minnows for the kids to entertain themselves. She bought a huge old copper apple butter kettle and last fall we had The First Annual Loiland Applebutter Day. We peeled apples all morning and cooked and stirred and played old-time music all day. It turned to butter just about dark....

Anyway, once again we were blessed with a nice sunny day. Again I publicized with flyers in country stores and I sent out cards or called the people who had come to the previous one. This time when the people arrived, they just sat down on the applecrate benches and immediately started trading seeds. I supplied masking tape and pens, and wax paper sandwich bags for people to put the seeds in....

One couple brought seeds saved by a great-grandmother in Kentucky--two nameless corns, one a white dent corn that was good for cornmeal and the other a white dent with a red fleck. It is what I've heard called Bloody Butcher. A friend of mine grew Bloody Butcher one year and it was really a beautiful crop. Most ears were red and white, although sometimes there would be a solid red or a solid white one....one neighbor of Loi's distributed seed collected while studying in Mainland China--a Chinese cabbage and a Chinese pepper. Another woman brought Scarlet Runner beans which she grew here and saved seed from....other seeds exchanged were luffa, cucuzzi, many flowers and herbs, seed from 40 lb. crookneck pumpkins, seed of a 9 ft. tall red okra, purple potatoes, big birdhouse gourds, Roaster squash, West Indian gherkins, currant tomatoes and yellow plum tomatoes from Alberto (Alberto brought two sixpacks of jars filled with seeds). I brought some white pine trees and people dug up some Damson plum seedlings. There were also Chinese cabbage plants, comfrey, Jerusalem artichokes, multiplier onions (potato), and one person cut budwood from some of Loi's vintage apple trees....

There was hot cider, bread and stew, raspberry applesauce, cookies, homemade beer and honeysuckle wine. Before we split, I spoke about how to save different types of seeds and handed out a four-page pamphlet on various techniques. I also explained the problems associated with seed patenting and explained why it is so important to preserve old types of seeds....Once again about 25 people attended; half of them had attended the previous one....

Our next seed saving meeting is tentatively scheduled for November 7, 1981. A local botanist, Jeff McCormack, who has corresponded with you, may lecture and I might also ask Peter Hatch from Monticello to speak....I am somewhat undecided about expanding the size of the group, yet I would like to spread the word. A larger group would mean more variety and diversity, but it could lead to disorganization....most of us are new to seed saving. So I plan to hold a group discussion on our experiences this year....share with each other how the seeds obtained from the exchange grew (some didn't), how to improve next year's seed saving such as tagging the early tomatoes so that they don't get eaten, planting enough beans for eating and seed saving, methods of hand-pollinating corn, and how to keep beans from sprouting in the pod when it rains. I'd like to share tips on things that can be planted in the fall (garlic, spinach, lettuce and dill) to come up in early spring, and things that will reseed themselves. Also greenhouse gardening, cold frames and season extenders (cloches and Dutch lights)....

As much as I dislike organized business meetings, I'd like to see this group have a correspondence secretary at some point in the future, so that we as a group could be a member of your Seed Savers Exchange nationally....((What a great idea - Kent))....Is it necessary to know the variety name to participate in the Seed Savers Exchange or programs that you have set up?....((No! Many fine old varieties have lost their names just because they have been kept so long or have changed hands so many times. That doesn't make them any less excellent. Just call them "No Name" and give as complete a description of their history and characteristics as possible - Kent))....

Correspondence is not one of my better traits (I am amazed at the volumes of correspondence that you are able to handle personally), but I do the best I can....I want people to understand the importance of preserving our variety of genetic resources, but I am not a very good speaker....((Neither am I, but I know it's necessary for me to get much better at it and the only way to do that is just to keep getting up and talking. Read what Forest Shomer said about this very thing in his speech at the Tucson Conference. I marveled at how easily Forest spoke before a group. It just flowed. But he's been getting up before groups and giving seed talks at least once a month for seven years. If there is one thing that I came away from the Seed Conference with, it's knowing that we must all realize just how much we can educate and affect those around us - Kent))....I want to try to reach more of the old-time farmers and gardeners in the area who may be saving heirloom seeds. I plan to advertise with flyers in country stores and in some local newspapers....

Kent, thanks for being such an inspiration....You have my admiration and respect for starting the Seed Savers Exchange with all the hard work and correspondence that involves....Feel free to mention my name to other Virginians to contact me about common interests....(Kim Domeratzky, Box 106, Batesville, VA 22924)

****

(Alberto Vasquez, SSE Member from Virginia)--....I've been visiting Kim Domeratzky of Batesville, Virginia (about 25 miles away) a good deal this year. She and Jim Cary are growing a good garden with quite a few SSE vegetables. They are also accomplished stringband musicians, The Master Batesville Band, and I've been sitting in with them some. Kim is really ambitious and enthusiastic and her local Virginia Seed Exchange is quintessential....((which means - "the most perfect embodiment of something" - Kent))....If every state had a few local seed exchanges, think how much consciousness would be raised. I am really looking forward to the next Batesville exchange.

(The 1981 Fall Harvest Edition, pages 52-53)

****

(Later mail from Kim Domeratzky Cary)--

I finished my course in Advanced Basic Programming. For my last project, I wrote a program to set up a database of Blue Ridge Seed Saver members and print out mailing labels. The program creates a code similar to the one that you use. It uses the county for the first two digits, then the first two letters of the last name, then the first letter of the first name. I have it on disk and hope to use it with our next mailing....I would like to set up a database of plant varieties that are offered in our exchange. Can you send me an example of the file structure that you use in your inventory?....I'm sorry that I haven't gotten back to you sooner. Sometimes the correspondence seems overwhelming....I am thinking of having David Via speak at our fall meeting. He held a Tree Crops Conference in Charlottesville last year. I would like to learn more about planting nuts from seed.

(I think what the Blue Ridge Seed Savers group is doing is fantastic. Active local groups can really wake people up and also do a thorough job of searching a region for heirlooms. I truly hope that groups like this will develop around enthusiastic individuals or in garden clubs all around the country -- Kent.)

(The 1982 Fall Harvest Edition, page 65)

ORNAMENTAL CORN IMPROVEMENT PROJECT
(by Mark Widrlechner, Director)

June 1, 1981--Dear Kent....I read with great interest the article about your efforts to coordinate the preservation of vegetable varieties that appeared in the June issue of Organic Gardening. I'm very glad to see that you and the Seed Savers Exchange are performing this important work. It is always good to have alternatives to the national seed storage and germplasm facilities....

One of the goals of the Ornamental Corn Improvement Project, Inc., is to breed better types of decorative corn for the Upper Midwest. In conjunction with that work, we might be able to cooperate with the Seed Savers Exchange by increasing varieties....

June 25, 1981--....The catalog is most fascinating. My wife and I have spent hours pouring through it....OCIP was incorporated in April with the purpose of providing continuity and protection for work that Sherry and I have been doing in earnest since 1979. Actually, my interest in ornamental corn goes back farther than that--from 1977 to 1979 I was a research assistant under Jack Harlan at Urbana. It dawned on me at that time that there are many plant breeders working on corn: field corn, sweet corn and popcorn; but the restraints of marketability do not let those breeders take advantage of the marvelous variation present in the species.

I believe in a flexible, diverse agriculture. Since ornamental corn varieties have multiple uses (aesthetic, corn meal, popcorn, roasting ears, feed, etc.), I think that it's worthwhile to examine ornamental types carefully....

OCIP is a non-profit corporation with a three-member board. We have four major goals: trials, breeding, education and historical research. Sherry is putting together a slide presentation for garden clubs and we've written an article for the state horticultural society. Our trials and breeding program are relatively small. Most of our material has come from three sources: local varieties, seed companies and the Plant Introduction Station at Ames. Next year we will be cooperating with a couple of public breeders, too. We keep our eyes open for striking varieties that also have good plant characteristics, such as little lodging, few earworms and smut resistance. We are not presently searching for particular varieties. If any of your members have been selecting and improving decorative types for a number of generations, those types would be the most useful to us....our growing season is shorter than yours. We can handle 110 day types okay, but 120 is stretching it....My plans for 1982 are reasonably well blocked out, but I could be more flexible in 1983....

July 11, 1981--Your suggestion about using 1982 to locate sources from your members is probably the best way to go....As for particular types, I'm most interested in 12 to 16 row flints (and flours) that will silk in 65 to 80 days. Preferably ones with attractive colored ears, that have been carefully selected for a number of generations. Even from seed companies, I've seen varieties that could use a lot of work....

As for helping you by increasing varieties, I think that it would be smarter for us to to 1/2-sibs for your increase rather than selfs. If we increase for you by selfing, we'll eliminate a good deal of the diversity in the population and run the risk of inbreeding depression. Culling obvious outcrosses (off-types) and then doing controlled 1/2-sib crosses would be a lot better....

You asked about hybrids vs. OP "discovering." I look at OP varieties as sources for inbred lines. I'll self for at least three generations with selection. Then I'll do test-crosses. Depending on the results, either the lines will be selfed to homozygosity (or close to it) and hybrids will be tried, or else the S3 lines will be combined into synthetics. The synthetics will be treated like OP varieties and will be selected for uniformity. In other words, after sorting out the "junk," new OP varieties would be built.

I'm reading a most fascinating book that might be of interest to SSE members. Corn Among the Indians of the Upper Missouri by George F. Will (Oscar's son) and George E. Hyde was first published in 1917. Bison Books of the University of Nebraska Press reprinted it in 1964. It's worth a few hours....

August 24, 1981--....Just thought I'd let you know that the Ornamental Corn Improvement Project, Inc. has received 501 (c) 3 and public foundation status from the IRS. It looks like we'll have a pretty good harvest, too. What else can we ask for?

(The 1981 Fall Harvest Edition, page 49)

THE LINGUISTIC ATLAS OF THE GULF STATES

(The following letter was received recently from Susan L. McDaniel, Associate Editor, The Linguistic Atlas of the Gulf States, Department of English, Emory University, Atlanta, GA 30322.)

In a recent issue of the Christian Science Monitor, I read an article about your work in preserving heirloom seeds. Old-fashioned fruits and vegetables formerly grown in the South and Southwest are of considerable interest to me in my work, so I am writing to ask if you could be of any assistance to us.

"The Linguistic Atlas of the Gulf States" project is a study of dialects in the states of Tennessee, Georgia, Florida, Alabama, Mississippi, Arkansas, Louisiana, and Texas. Our data base consists of tape-recorded interviews with 1,121 natives of those states. In the course of the interview, fieldworkers routinely asked about certain fruits and vegetables grown in the area.

Of course, the varieties most commonly mentioned are those that can be found in any standard seed catalog--Blue Lake beans (which my husband grows) or Elberta peaches. But in a number of instances, old farmers and others from rural areas spoke of varieties that were grown in their childhood, or other varieties that we have not been able to verify, though we have consulted several agricultural extension agents.

The Monitor article caught my eye because it mentioned Moon-and-Stars watermelon, one of the very varieties I had been looking for. If you or your members could supply verification (or at least correct spellings) for some of our "mysteries," we would be very grateful.

The following is a list of the vegetables and fruits that are presently unknown to us. All spellings are guesses, based on the informants' pronunciation. We would appreciate any information we could get on any of them:

All Heart watermelon, Augberta peach (different from Elberta), Big Hook lima bean, Bird-eye pepper, Blaber pea, Blue Boy tomato, Blue Shell lima bean, Boones Irish potato, Boss Talbott watermelon, Breezina potato, Bruce Reds potato, Cow Pumpkin, Cowhorn okra, Crouch peach, Cubas sweet potato, Cue (or Q) melon, Cue (or Q) pumpkin, Dorritons watermelon, East India potato, Garris (or Garrett) Gray watermelon, Georgia Boys peaches, Gray Stone watermelon, Gray Strand watermelon, Ground Roomers peanut, Grunyea Italian lettuce, Hester watermelon, Hickory Chickens (mushrooms), Hog berry, Honey Ball melon, Honeycomb watermelon, Joan of Arc sweet potato, Le Tang blackberry, Little Boy tomato, Little Butts onion, Long John lima bean, Marlon melon, Medallion bean, Moody sweet potato, Mountain Sprout (or Sprite) watermelon, Muscogee grape (not Muscadine), Musgwine onion, Musk Rose sweet potato, Panancy potato, Parker berry (like huckleberry), Patch berries, Pindalini (small Italian tomato), Piufa Gray watermelon, Pogrom mushroom, Radagoos tomato, Red Goose (or Late Goose) bean, Reyuda Wonder watermelon, Rhubibs (like shallots/rhubarb?), Rosin berry, Seven Sisters (small onion), Shab bean, Snider watermelon, Squabbly squash, Star melon, Steamboat bean, Sugar Town watermelon, Sulfur bean, Sweet Valikia watermelon, Watkins Green watermelon, White Autaud peanut, White Planks (onion).

(The 1984 Fall Harvest Edition, page 155)

CHARLES STEVENS MEMORIAL SEED LIBRARY OF APPALACHIAN HERITAGE CROPS

(The following form letter was received from R. Otto, Banner Elk, North Carolina, along with a list of the old-time varieties he has already located. This is exactly what needs to be happening in different regions all around the country. I really admire what he's doing and hope that all of you who are in his area will support the Seed Library - Kent.)

Dear Friends and Seedsavers....Many of you have sent us some very worthy varieties for the Seed Library that we will be propagating this year to increase the seed. We are still on the lookout for as many of these varieties as we can find, so if you know of anyone who has seed of old-timey varieties of fruits, vegetables and field crops, please either secure for us a sample or get your friends to send us a sample of their seed. Please be sure to label the name, area and soil type on each sample of seed. Also a history of the variety will be very useful for compiling the future catalog of these seeds.

The way the Seed Library works is that anyone who wants to join the Library as a patron has only to send us a $5 membership fee....In return we will send you a copy of Seeds for Tomorrow, a book that we will be printing in the fall of 1982 which will tell you how to grow and save your seed in a way that is not against the Laws of Nature. We also will send to you a sample of the seed you wish to "check out," and the only requirement you have to follow is to return double the amount in the fall.

This will enable the Library to have more seed to make available to our patrons and allow us to grow a "blending patch" of seed from each of your gardens. This will help us to blend all of the acquired genetic traits of the variety into the resulting seed with the end result of developing the strain that's adaptable to the many different soils, climatic variations and disease problems that we can encounter here in the Appalachian Mountains.

Until we are able to acquire a grant or some other type of funding, we are dependent on your support and the sale of this book to be able to meet the expenses of this most important work. I am supporting it as much as a small farmer can and hope that I will soon be able to establish a permanent base of operations. Thank you so much for your interest and support. -- Charles Stevens Memorial Seed Library of Appalachian Heritage Crops, c/o R. Otto, Route 3, Box 658-A, Banner Elk, NC 28604.

(I must admit that I was bothered a bit by the term "blending patch." I had visions of, for example, a half dozen heirloom cantaloupes being planted together, crossed and lost in an attempt to develop one adapted strain....When I wrote and asked, I found that I was very much mistaken....- Kent):

Further correspondence from R. Otto --....To explain the "blending patch" idea, I will describe the work of a geneticist friend of mine, Dr. Darrell Langham. He pioneered this concept under famine pressures in Argentina during WWII due to a submarine blockade of the country. Seed was needed to enable the people of the various climatic zones to be able to produce food locally, as all imports had been cut off. He collected seed from all over the country and planted all of the varieties together in one patch. Then through careful selection he chose his seed. The result was a variety that was successful in the full range of climates in just one season. He then went on to refine and stabilize it over several years.

My application of this idea would be to limit it to one variety grown in several different locations for at least three years. Then the pick of these crops would be planted together in a blending patch and a thorough mixing of pollen accomplished. This would be continued for three more years and the seventh year would be the "scattering of the seed" to various maintainers who would be part of a rotation, so the seed would move at least biannually and not become lop-sided again. This rotation is a very old European custom (probably introduced by the Druids) that I came across while studying Bio-Dynamics in England and Europe.

Since I am expecting some rather trying changes in climate in the next two decades, I feel that it is very important to work on the quality of seed and its survivability. The older varieties carry old genetic "memories" of past climatic conditions. However, I believe that much of this memory is dormant due to continuous cropping in one climatic zone. The blending patch is an idea that I believe can break this dormancy. As Luther Burbank once said, "Nature sets no limits on the length of time through which a submerged character may be transmitted."

Once the dormancy of genetic material is broken, we will see some startling changes in once-familiar varieties. It is then that we must keep to the guide-lines of Burbank: 1) Select elements or characteristics, 2) Combine chosen elements, 3) Bring combinations to quick bearing, 4) Select one out of 500 that is perfection and then select again until stabilized. Once we have a plant able to express its inherited traits, we should help it to keep that creativity flexible by using the "rotation of maintainers"....

What I need most are memberships to help defray the costs of mailings, seed storage, seed location and collection, etc. Also, any donations that can be found can be channeled through the University which is a quasi-co-sponsor of the Seed Library. I am a part-time instructor with Appalachian State University in the Earth Studies Department where I teach Biological Gardening and Homesteading and Bio-Dynamic Agriculture....

We are also currently testing commercial fertilizers, both organic and chemical, doing variety testing for our bio-region, and testing seven breeds of chickens for their value in a home feeds, self-sustaining mountain small farm....I am seriously looking for pre-1940 cane-fruits--raspberry, boysenberry, loganberry, dewberry, blackberries and strawberries. Also, pre-1940 small grains--wheat, barley, oats, spelts, etc. What I would like to get for next spring are some Hulless Oats and Barley....

(The 1982 Fall Harvest Edition, pages 56-57)

(I continue to be very interested in this project, but a recent letter sent to R. Otto at the address in Banner Elk was returned stamped "Undeliverable as addressed." If anyone has any information about R. Otto or the Charles Stevens Memorial Seed Library, I strongly urge them to get in touch with me - Kent.)

SOUTHWEST TRADITIONAL CROP CONSERVANCY GARDEN AND SEED BANK

The Southwest Traditional Crop Conservancy Garden and Seed Bank is sponsored by the Meals for Millions/Freedom from Hunger Foundation. The project offers assistance and seeds for free, but has a priority system for distributing the rarer seedstocks, since many are now endangered. The project also aids Native Americans in the United States Southwest and northern Mexico in solving their seed storage problems and recovering seed varieties that were once grown in their communities.

Gary Nabhan, agricultural ethnobotanist and collector of most of the seeds, is the project manager. Gary is a former member of the Seed Savers Exchange and has helped me tremendously with various projects during the past year. Gary spent three days in my home this last summer and we brainstormed and bounced ideas off each other until we were both nearly exhausted. Then we went swimmin'. Took the family to Nine Eagles Park to try to beat the heat wave. I can't ever remember spending a more stimulating three days in my life.

This last summer Gary and others at the Conservancy Garden grew: a Pueblo Waffle Garden with a rock wall surrounding the growing area and a foot of sand spread over the surface to simulate Hopi growing conditions and to permit deep planting; a Sonoran Kitchen Garden which allowed them to show the intercropping of beans (teparies) with corn (chapalote) for support; and a Papago/Pima runoff-supplemented field which was protected on one side by a living ocotillo fence. These demonstration gardens of traditional crops and growing methods were at the Tucson Botanical Garden's Porter Gardens.

Although they are distributing seed, they wish to stress that they are not a seed company or a seed give-away. (That sounds familiar.) They distribute seeds as their supplies allow and according to the following priority system: to individuals of the culture from which the seedstock was derived; to researchers willing to maintain the seedstock and/or return a portion of their increase; to organizations or communities of other cultures in a similar environment willing to try to maintain the seedstock; and to seed exchange participants or small seed companies willing to increase the seedstock and distribute it to others.

Members of the Seed Savers Exchange are among those selected to receive small starter samples of the following seedstocks: red dye amaranth of the Hopi; Pueblo flour corns which the Hopi used as grinding corns; I'itoi's onion (a small, wild, desert, bunching onion brought into cultivation by the Papagos); Sonoran panicgrass—the Warihio Indians of Sonora-Chihuahua ground its small seeds for a cereal; a grain amaranth also from the Warihio; and Mexican dye indigo of the Mayo—a perennial leguminous shrub, frost-sensitive, leaves used for making wool and cotton dyes.

Obviously, SSE members living in the desert Southwest will be the most useful to this project. Also when requesting these samples from Gary, you may want to send a small donation to aid their project. Seed distribution always incurs postage and packaging costs.

(The 1981 Winter Yearbook, page 58)

(Ed. note: For more information about the current projects of Gary Nabhan and friends, see the article on Native Seeds/SEARCH earlier in this chapter.)

NAFG SEEDS FOR SELF HELP PROGRAM

by Will Raap, Membership Director, National Association for Gardening

Through recorded history, the giving of seeds has been a sign of friendship and good will. They have been bought and sold, given as precious gifts, borne in pockets and saddlebags, sailed in ships' holds and jounced over mountain and desert in wagons. Thus plants native to every country have been tried in many other countries, and many have readily adapted to their new locations.

We in North America, seeking ways to show our feeling of friendship with our neighbors to the South, cannot do so in a more natural, more acceptable way than by sending them seeds from our gardens to help them in the present difficult times.

For centuries, people in Central America and the Caribbean have had an agriculture-based life, self-sufficient and reliable. In recent years the old way of life has lapsed. People are now removed from the habit of growing their own food. Long-used growing techniques are forgotten. Locally-adapted seed varieties, once passed religiously from parent to child, have not been propagated and are lost forever. Food is imported, and costly.

Costa Rica is one of the countries faced with collapse of its food system....Far-sighted leaders in Costa Rica, realizing the country hasn't enough money either from natural resources or industry to cover the necessities of life for its people, see a return to home food-growing and small farms as a solution....Many in Costa Rica can't afford to buy seed or can't find adapted varieties. We can send seed that will be given them free....We have found a place that will grow out any seed we send, to make sure that it is adapted to one of the three growing regions of the country, then send it to the appropriate area. We are calling our program National Association for Gardening (NAFG) Seeds for Self Help.

If you are a gardener who saves your own open-pollinated (non-hybrid) seeds, please send us seeds from your favorite varieties. They must be clean and dry, and you can send them in envelopes (marked for hand-cancellation), a little plastic bottle or film canister, or whatever is handy that will protect them in the mails. Also, we must have the following information turned in with each seed variety you submit: 1) your name and address, 2) U. S. growing zone (if known), 3) crop, 4) variety, 5) days to maturity, 6) source (if known), 7) any comments (special qualities, disease problems or resistance, drought tolerance, cultural needs or requirements). Mail to: Seeds for Self Help, National Association for Gardening, 180 Flynn Ave., Burlington, VT 05401.

We will sort, grade, package and ship the seed to the Centro Agricol Contonal in Turrialba, a major regional center for agricultural research and assistance operated by the Costa Rican Ministry of Agriculture. The Center in Turrialba will set up a seed bank program for all of Costa Rica, involving several other regional centers. Each center will receive seed varieties most suited to its area. The seeds will be planted, the plants and production will be evaluated, and the most successful varieties will be identified. In a few years enough seed from the best varieties will be available to be distributed free to home gardeners and small farmers throughout the region.

(<u>The 1982 Fall Harvest Edition</u>, page 60)

OTHER PROJECTS

Kurt Christiansen, PO Box 875, Ben Lomond, CA 95005--I am an apprentice at the University of California Santa Cruz Farm and Garden Project. We are trying to establish seed gardens at both the Farm and Garden sites. Our goal is to make locally adapted seeds and information on growing seeds at home available to the community. The gardens will also serve to provide a place for growing seeds now endangered by selection and the recent patent laws pertaining to seeds, both of which threaten to diminish the existing gene pool of plant varieties drastically. Regionally-based seed reserves serving the surrounding communities would be a welcome step for all concerned. Anything which you or your members could share about growing plants of all types for seed, the need for doing so, or anything that could help us document the critical need for more seed gardens of this nature would really be of great help....

Keith F. Keating, Windstar Gardens, Snowmass, CO 81654--I have just begun a garden project for the Windstar Foundation based on the teachings of Alan Chadwick's system--French Intensive, Bio-Dynamic. I studied with Alan for two years. The focus of our garden is high altitude gardening, as we are 7,200' in the Rockies....Also a focus of mine is the continuation of special varieties that will grow in an arid high altitude climate with a 2 1/2 month growing season....I wish to express my support for what you are striving to do.

(<u>The 1981 Winter Yearbook</u>, page 6)

Beni Adom, 3262 Williamsburg, Ann Arbor, MI 48104--I do not have any heirloom or endangered seeds myself, so I am really interested in your Growers Network....Since I have a finite amount of space, I would prefer to be growing heirloom or endangered varieties exclusively....I live in a 427-unit low-and-middle-income housing cooperative on the edge of town. We are allowed to plant our yards however we wish (in my case and a number of others, we have ripped out all the grass and turned everything into a garden), and the co-op has a policy of trucking in manure and top soil all spring for anyone to use....In addition to almost 300 gardeners, the co-op has an active 4-H club, which is almost exclusively garden-oriented....

My hope (or, at this stage, my fantasy) is to see what <u>I</u> can do with several heirloom varieties and then publicize what I'm doing to the rest of the co-op. We are the single most active (and successful!) ecology and environmental organization in the area, and I have rather high hopes that there will be a number of folks here to whom your program will appeal....It is fortunate that there are a few people like you around to keep things going, and I sure appreciate what you are doing....

(The 1981 Fall Harvest Edition, page 62)

Ronald Goodman, Wanikiya Cooperative, Box 65, St. Francis, SD 57572--Wanikiya Cooperative is a group of people here on the Rosebud Sioux Reservation who have started to garden and pray together towards a new/old unity. We want to resume growing the vegetables which were evolved by the Native American people of this general area. We are especially interested in beans and squash varieties....It is our intention to inherit and honor the agricultural work of those ancestors who related to Mother Earth with so much dignity and precision and gratitude....

(The 1982 Winter Yearbook, page 89)

Robert J. Brooks, Chief of Program Development, State of Connecticut, Dept. of Corrections, 340 Capital Ave., Hartford, CT 06106--....We have organized a group of prisoners to work with endangered and uncommon plant species. We would like to obtain fruit and vegetable varieties that are no longer being generally grown....The work you are doing is quite extensive and obviously important. We would like to be one small link in the network and see that our efforts here encourage other prison systems to enable men/women to do likewise. When you think about it, that's an army of half a million!....

(From a later letter)--....Yesterday we had the dedication of the endangered plant species project among the prisoners here....Six acres of woodland on the Correctional Center's grounds have been designated the Cochegan Wilderness Lands and set aside as a habitat for rare or endangered plants. Inmates will raise the heirloom vegetable seedlings in a donated greenhouse....Your spring issue is great. It has been dog-eared by many of us pouring over all the contents. Five of the prisoners at the Montville facility have ordered seeds for our group to plant this summer....Your interest in us is most helpful....

(And still later, reported in The 1984 Winter Yearbook, page 77)--....The Cochegan Wilderness Lands, a prisoner project, is thriving and is now into its third year. The project is located at the Montville Correctional Center (PO Box 980, Uncasville, CT 06382). The developments include a generously sized greenhouse obtained by the participating prisoners as a donation from a local nurseryman. The greenhouse was moved, pane by pane. In an open area near the greenhouse an heirloom apple orchard has been started. This now includes 10 antique varieties of apples. During the past growing season, the men began growing several heirloom vegetable varieties and these are now being offered through the SSE. The project was begun with a small private donation, but it is now self-sufficient through income generated by the weekend sale of plants, largely ornamentals, to prison staff and visitors. The major need is for one or more gardening volunteers to assist the men and provide continuity as members leave at the expiration of their sentences.

Michael Siesel, Director, Flagstaff Community Solar Greenhouse, 320 N.

Aztec, Flagstaff, AZ 86001--....We are a community based, non-profit organization promoting alternative energy, self-sufficiency and organiculture, and hold many popular varieties of garden vegetables in a non-official seed bank....We are interested in testing vegetable varieties which are open-pollinated here in Flagstaff. Our altitude is 7,000' and the growing season is a scant 90 days. Because of our short season and relatively low photo period, as compared to the northern latitudes, many vegetable varieties developed in the subarctic seed stations do well, but I feel there may be seed, perhaps from the Appalachian region, which may do better....I would like to commend you on your foresight and dedication. It is the "least worn path" that leads to our fulfillment as individuals, and you have clearly chosen that trail. We at the Solar Greenhouse applaud your efforts and hope that we can assist in some small way to help you achieve your goal....

Craig A. Siska, 905 Goldsboro Ave., Virginia Beach, VA 23451--....I am a professional gardener/landscape architect and a former apprentice of Alan Chadwick and his Bio-Dynamic/French Intensive system of horticulture....We have ten 50' beds under cultivation already this spring (2,500 sq. ft. of growing space) that are sown with alfalfa and clover for soil improvement this year. All work is by hand/labor intensive....Paramount to our project is to become a seedbank for excellent heirloom, open-pollinated varieties of vegetables, fruits, herbs, flowers, shrubs and trees. We will not grow for production this year or next year. We will grow exclusively for seed. In 2-5 years we hope to have an extensive seed bank and then make those seeds available to members of your Exchange!....Thank God for your Exchange!

Country Kids of Armanda, c/o Sherry Watkins, 77111 Cryderman Rd., Armanda, MI 48005--I am head of the Gardening Program for the "Country Kids" 4-H Club in Armanda,Michigan. I am also a certified Master Gardener, a national program which is part of the federal Extension Service....This year's national 4-H theme is Our American Heritage....When I read your article, it seemed like a project in perfect keeping with our theme. I also feel it could be a valuable learning experience that these children will carry with them throughout their lives....My goal is to teach our 4-Hers the total significance of gardening, not just the pride of putting a few vegetables on the table. Our garden is the nucleus of many of our programs, and anyone who is in our cooking, baking, preserving or outdoor cooking classes must be a member of the gardening class, because we use as much produce from our own garden as possible. We also provide the produce for the club's annual 4-H picnic. This year we hope to have enough extra produce that we can sell it at the flea market and use the money to purchase some fruit trees and bushes, which will be an Advanced Gardening Class....We are very proud of our gardening program and feel that the Seed Savers Exchange could become an important part of it....

(The 1982 Fall Harvest Edition, pages 61 and 120)

Jack E. Rice, Rt. 5, Box 157-C, Laurinburg, NC 28352--I live in the southern part of North Carolina which, to me, is the finest lima and cowpea growing area in the U. S. (I'm sure some folks will take issue with that statement.) In the South, "limas" and "cowpeas" are not common words to gardeners, especially the older ones. They will tell you that they grow "butter beans" and "field peas." Why fight tradition! (The experts are now even trying to replace the term cowpea with "protepea." There is no other vegetable crop that is surrounded with more confusion in classification and varietal nomenclature, and this just adds to it. They are generally divided into 11 basic groups and you hear names like cowpeas, field peas, Southern field peas, Southern peas, black-eyes, table peas, crowders, etc., etc....and now protepea. I don't even like the word, so they'll always be cowpeas to me)....

The SSE has growers that specialize in squash, tomatoes, beans and corn. I would like to specialize in limas and cowpeas for the SSE. I would like very much for everyone in the SSE to send me a small sample of any lima or cowpea that they are keeping (along with any information on origin, history, culture, development, range, soil, planting, harvesting, etc.). One day in the near future I would like to start writing a book on limas and cowpeas, and I have been gathering this information for that purpose....My goal is to maintain a collection of limas and cowpeas, to develop an ongoing trial garden for them and to always share my surplus through the SSE....My collection now stands at about 200 limas and about 50 cowpeas, but this is very misleading. They have many localized names, especially the Florida Butter, Sieva and the little Dixie Butterpea. A large percentage of the ones I receive now, I already have by another name. But all of the names will be kept and cross-filed as proven. A name should never be dropped, as this is part of the tradition of seed saving. The old localized names are the original names, not the modern seed catalog names or the names we have given them....

It feels great to be involved with a group of people who enjoy growing and preserving our heritage of heirloom vegetables.

(The 1983 Winter Yearbook, pages 118-119)

Jon K. Carrington, Silver Hills Institute, RR #2, Lumby, British Columbia, Canada V0E 2G0--We sponsor a three-year Medical Missionary Training Program in a holistic lifestyle (physical/mental/spiritual). This is a Seventh-Day Adventist 800-acre organic farm teaching health and health reconditioning, building, farming, greenhouses, general country living, and how to help others and be a problem-solver with true Christian principles.

(The 1984 Winter Yearbook, pages 197 and 233)

Paul Bredberg, Box 66A-1, Route 2, Rogers, AR 72756--I am a staff member at Realife Ranch, a Christian orphanage with an emphasis on training the children in every practical area of life, because many will be in missionary positions in the U. S. and abroad. Two former staff members are in Zaire and Guenea Beasau right now and could really benefit from growing their own seed and vegetables....I've been gardening for several years and last year we had approximately five acres and about 3,200 sq. ft. of raised intensive beds. Our seeds are donated to us each year. However, I feel we should move away from dependence on the seed suppliers. Also, if all the children learn how to save and grow their own seed, they can carry that skill with them wherever they go....

(The 1984 Fall Harvest Edition, page 35)

Seed Companies

"Hey Dad, how do you spell swamped?"
(Tracy Whealy at Kent's Desk, Answering a Few Letters)

SEEDS BLUM

(Jan Blum, Idaho City Stage, Boise, ID 83706)

When Seeds Blum first became a seed company in 1981, I had a lot to learn about how seed companies work. Being avid seed gardeners, I envisioned us growing all the seeds we offer in the catalog. This seemed to be simply a matter of planning carefully and then working our tails off. That first spring enlightened me considerably! Fantasy blended imperceptibly into reality as we felt ourselves to be elfin jugglers straining to keep several huge balls in the air. What balls were we balancing? - Order filling and customer correspondence; spring garden planning, planting, watering and weeding; and the third ball took in all the loose ends (things like eating, sleeping, planning and recording business "strategies," not to mention washing clothes, cleaning house and totally forgetting socializing....).

We "klutz-juggled" through that first spring and summer, surviving narrowly. Sprinting into the fall garden chores (which includes a lot of seed gathering and cleaning) we collided headlong into another "time competitor" -- catalog writing and printing! The truth began seeping into my enthusiasm. Two people just can't run a seed company as if they are Burpee's and remain sane, if even survive at all. I turned desperately to my fellow seedspeople for answers. How do other small seed companies do it?

Each time I asked this question, I bumped into the same fact: They buy their seeds. But from whom? Although some companies are seeking other sources, everyone directed me to the same place--the wholesale seed companies, of course! Perhaps I could have accepted this answer and rejoiced at such a simple solution, had I not read so many Seed Savers Exchange booklets, or grown so many superb heirloom varieties. After all, the whole reason for starting Seeds Blum had been to reintroduce these great seeds into the mainstream of American gardening and to keep them alive and available. I didn't have a lifelong dream to do a seed company. I had simply had a love affair with heirloom seeds and now found myself married to a seed company that was proving difficult to live with. How could we continue to grow heirloom varieties and still have time to run the business and make them available to other gardeners on a consistent basis?

Suddenly I began to understand why there is a seed crisis. If all of us (large seed companies, smaller "alternative" seed companies and, in turn, gardeners) essentially get our seeds from the same source, then of course that limits the type of seeds available. Everyone gets the type of seeds that the wholesalers offer. Even that fact has been palatable, as long as the seed wholesalers offered a sprinkling of open-pollinated varieties and seeds with traits attractive to home gardeners. But things have changed drastically in the last ten years. The profit incentive has swung breeding directions to hybrids and seeds that suit the needs of large mega-farmers. Now those are the seeds offered to everyone! A few enterprising seed companies offer varieties that still appeal to home gardeners by buying seeds from wholesalers in foreign countries; but the situation is changing worldwide, and it is only a matter of time before even those seed sources are no longer available. It was beginning to look like duck soup for all of us.

The next few weeks watched a real struggle going on here at Seeds Blum. Our whole reason for existence as a seed company was to offer heirloom varieties and other seeds suitable to home gardeners. It was painfully obvious that we couldn't possibly grow all the seeds we wanted to offer. If we relied on the traditional wholesale seed sources, we would have to drop many varieties each year because they would be unavailable to us. (About this time, we began having nightmares of ourselves, as angry gardeners, charging up the steps of Seeds Blum's offices, and meeting ourselves, as seedspeople, helplessly trying to soothe the situation.) Indeed, we wanted to expand (rather than shrink) our offerings of these varieties that had spiced up our own gardening so dynamically. The more we cogitated about the whole situation, the more determined we became that at least we had to go down fighting. As the greenhorn seedspeople on the scene, we certainly couldn't blame other seed companies for the course they had taken. When they had started, many of them ten years before us, things had been quite different in the seed world. Open-pollinated varieties had been plentiful, even from the large wholesale companies. And how well we knew what a welcome provision buying seeds can be when only a handful of people are working long hours just to fill orders, do trials and get information out. Bound by all the same limitations of time and money, we stared blankly at all our stated alternatives, then turned instinctively and cut loose, running blindly in an altogether unknown direction. At least the cold air felt good.

When we first stumbled onto the idea of paying small-scale growers, including backyard gardeners, to grow seeds for us, we immediately sat down and dashed off letters to everyone we thought would benefit from "the idea." Dead silence. Then a few snickers. Then a few pretty strong arguments about how impractical the idea is, especially our "outrageously high" pay scale for growers. Finally, a few people said they had been toying with similar ideas and did we have specific plans on how to get going. Before we had time to even think about "how to get going," we found ourselves plunged in neck high. Publication in the Seed Savers Exchange of a letter in which I had written about many of these things brought a whole flood of responses--ideas, folks who wanted to be growers, seeds and suggestions that we carry certain heirloom varieties and why. And so the growing season of 1984 yielded unexpected harvests. Not only did more than 50 experienced, good quality growers nurture seed crops that have been bought by Seeds Blum, but we have all grown in other ways, as well. We are all learning to plan ahead better and communicate more directly. We are learning to clean seeds faster. And, beneficial to everyone, our Seeds Blum 1985 catalog will offer many "new" heirloom varieties suggested and grown for us by these innovative gardeners.

Not in dramatic sweeps or with clarion bugle calls, but step by quiet step, a garden revolution is marching across America. Although it is 1984, and this struggle for independence is of a slightly different nature, the Boston teaparty keeps popping into my mind. Frankly, it's the most exciting thing I've ever been involved in! It puts me in touch with the type of people I've always admired most and allows me to form close working relationships with them. I learn a lot from others and am also respected for what I know. It enables us to make a living at what we love doing, and to pay others for doing what they enjoy. In time, as we need larger quantities of seed, and as other seed companies implement and succeed at similar grower programs, we will open up viable options

Jan Blum Digging Potatoes

for formative or floundering small farm operations. This whole thing is fun, breathtaking and deadly serious. Not only have we staked our future here at Seeds Blum, but many of us believe our survival as a human race depends upon our success (and/or the success of others also trying to save our genetic diversity). Let's say the "small grower" idea is a challenging and hope-filled plan. So, what exactly is this idea and how does it work?

We need small scale seed savers with two basic qualities: conscientiousness and honesty. These special gardeners divide into two groups: experienced seed growers and beginners determined to learn. The experienced seed savers (saved seeds and grown those seeds out for at least three seasons, and preferably more) have a ready market for their seeds now. Fall and winter is the best time to plan for seed crops the following year, but we can do some planning through March 1. (After that, be forewarned, letters start to pile up as we put our full energy into filling orders and getting our own gardens planted.) Inexperienced seed savers can be a part of the action now by joining one of our other Grower Programs and beginning to save seeds of varieties they want to offer. We are happy to answer questions and make suggestions along the way. There are a few crops that are ideal for beginning seed savers. Experience is very helpful but it is just like anything else; determination, dedication to quality, the ability to be honest and ask questions when there is a problem, almost eclipse whether you have experience.

We have several different kinds of growers, and have room for more of each. There are gardeners who grow only one variety for us, and backyard gardeners who grow small amounts of many varieties (from 2 to 30 different kinds!). Some gardeners are "seed multipliers." They increase small lots of seed to large amounts so others can grow them as "growers." Some folks are seed hobbyists and they build specialized collections of certain varieties, such as Shaker, American Indian, Russian, Amish, tomatoes, okra, beans, etc. They say it's not fair to have favorites, but I must admit that I always choose a favorite of everything. And my favorite seed savers are the gardeners who have been saving family heirlooms or neighborhood "had around" seeds, and they want to grow that variety for Seeds Blum. It is a matter of policy here that anyone who offers us family or neighborhood heirlooms has the first opportunity to be the "seed grower" of that variety. Only if they decline, do we look for another grower.

We have a grower brochure that tells what we expect of growers and what we provide. It also reviews selective breeding, how to avoid inbreeding depression, isolation distances, seed harvesting and cleaning. We provide forms for record keeping and a pamphlet that tells the quantities of various types of seeds that we need and how much we pay. We've run into a lot of snags over the money. Not from growers, mind you. But we've been warned that we are being much too extravagant in the prices we offer.

First of all, no seed company, no matter how much we want to, can pay much higher prices than we pay and survive. Lest dollar signs dance in your eyes while you quickly write to inquire about our Grower Program, let me assure you that most growers couldn't afford to grow seeds for much less than we offer. Our prices are very high as seed companies go, but not so high when you work all season to save a few cups of seed. We offer higher prices than others for two simple reasons: 1) we ourselves are also growers and we know what goes into seed saving, and 2) we want consistent, good quality growers who are happy to grow for us year after year.

Our experiences with our growers have been very good so far. We are listening to their comments, and we give them feedback so they know where we stand. It took a season of working out some of the lumps and bumps, but you should see the seeds fly now! We are discovering that there is a lot of unused energy hiding in backyard gardeners and many growers write to us when they realize we are interested in paying folks to grow seeds for us. Sometimes we write to customers who seem to have a real eye to choose the unusual varieties from our catalog, or seem to have a real interest in one type of vegetable. Wherever we meet them, we have come to really appreciate every one of these seed growers who are joining with us to revolutionize what seed companies are.

In addition to our Seed Grower Program, there are other interesting gardener participation programs that help us meet more gardeners and get familiar with their interests. Our Trial Garden Program is very popular, and we have sent out hundreds of free packets of seed for gardeners to test and tell us what they think of these varieties we are considering carrying in the catalog. In Grassroots Garden Research we work with gardeners who are active researchers and like to keep track of what they plant when and how things do in their specific microclimate. Each participant gets a copy of "Pleasure Gardens," an annual brochure that shares many ideas for harvesting more pleasure from the garden, and a report of othér gardeners in the same microclimate. Gardener's Choice brings participants a "Mail Order County Fair" newsletter. The Gardener's Choice portion of the newsletter compares gardeners' favorites with a "new to them" variety that they grow on the same basis as trial gardeners--the difference being that they compare it with their favorite variety. It's pretty interesting reading, in addition to all the recipes, stories and even poetry that are interspersed among the pictures. It is a way for our customers to meet each other, and learn new ideas at the same time.

1984 has been an excellent year! Our business has tripled as new folks have discovered us. We have built rooms onto our garage and the house that give us much-needed space. But the most important progress we have made has been accomplished in cooperation with the many folks who raised seed for us this year. This has been a real dream come true and it encourages us to keep pushing forward, realizing we can be the seed company we want to be. We can be truly independent from the large chemical-owned wholesalers and support small-scale agriculture at the same time. On a personal level, we still enjoy cleaning seeds in the autumn evenings, while the wood stove takes the chill off the crisp fall air. A great big thank-you to all the Seed Savers Exchange folks who have reached out to us and made this adventure possible, and more than that, successful. We have a long way to go (many more varieties to grow out and offer), but we're well on our way! We welcome your comments and suggestions, varieties you would like to see us carry or ways you think we could improve our service. We are here to serve you, and our company sprouted after I had read and learned from Seed Savers Exchange members what is actually happening in the seed world. It is really the Seed Savers Exchange that has given our company such a dynamic difference in the world of seed companies, and we are most happy to work with Seed Exchange members in any way we can. We are working very hard to print our own packets with specific growing instructions on them for 1986. The 1985 catalog includes much more growing and varietal information because you requested it.

I want to end this letter on a personal note. Not only has the business prospered because of your participation, but we are much richer as individuals, as well. Many times I feel completely overwhelmed by all the negative forces in the world. But then I think of all the heartwarming letters we receive from gardeners, all the seeds people have unselfishly sent us and all the work our growers are doing because they believe in positive change. And I have strength and courage--because our lives have touched. Thank you!

(Seeds Blum catalog is available for $2 from Seeds Blum, Idaho City Stage, Boise, ID 83707.)

(The 1984 Fall Harvest Edition, pages 89-95)

SOUTHERN EXPOSURE SEED EXCHANGE

(Jeff McCormack, P. O. Box 158, North Garden, VA 22959)

In 1982 a number of my personal and professional interests were synthesized together in the form of Southern Exposure Seed Exchange, a seed company specializing in heirloom, traditional and modern varieties of open-pollinated varieties of vegetables.

I envision the seed company as a participating member in a network of organizations working to promote sustainable agriculture based on ecological principles. The main objectives of the company are to : 1) encourage cooperative self-reliance in the mid-Atlantic region, 2) promote seed saving and exchange, 3) promote the use of disease and insect resistant varieties, 4) conserve and locate rare and endangered varieties, and 5) specialize in varieties for special applications, such as solar greenhouses and varieties which are regionally adapted.

The seed company is viewed mainly as an instrument for meeting these objectives. As the seed company becomes more successful, profits will be used to conduct research and to offer educational services of: 1) integrated pest management, 2) plant breeding and variety introduction, 3) development of a non-profit regional seed bank, 4) development of a pre-1950 catalog collection for use by agricultural historians, 5) research on the interplay between culture and agriculture, and 6) publications based on literature and agricultural research.

We are well on our way to meeting these objectives. It has not been easy, but it has been enormously challenging as a personal adventure, and sometimes just plain fun. We are gaining an understanding (we hope) of the basic economics of the seed business, and have begun gathering resources and developing contacts necessary to develop the long-range objectives outlined above. Our first catalog offered 65 varieties of vegetables. In our second year we offered 140 varieties including herbs. Now entering our third year, we offer an expanded selection of vegetables, herbs, and have added flowers, books and supplies. Our facilities have expanded with our offerings. Initially the seed company office consisted of a hallway closet filled with jars of seed, and a desk outside in the hallway. We soon expanded from the closet to half of the living room and portions of the bedroom, kitchen, utility room and attic. We now have an addition to our home which contains an office/lab, better storage facilities, and are working on completing the solar greenhouse. Our growing area has increased from 2,000 sq. ft. to several small gardens and a few participating growers. My wife, Patty Wallens, is now an active partner in the business. And we also have a three-month-old "sprout," Timothy, who is very active in the fertilizer business.

Jeff McCormack and Timothy at
The 1985 Campout Convention

Some of our specialties include a large and diverse collection of multiplier onions (potato onions and shallots) from foreign and domestic sources, and heirloom tomatoes and peppers. In addition to these specialties, we offer a wide range of traditional varieties, some several hundred years old. Our catalog is developing its own character with detailed cultural information and variety descriptions, historical information and botanical illustrations. We are including seed saving resources and information such as isolation distances for seed savers in the mid-Atlantic region. We charge $2 for our catalog/gardening mini-manual (free to previous customers).

(The 1984 Fall Harvest Edition, pages 100-101)

BUTTERBROOKE FARM SEED CO-OP

Shortly after the Fourth of July last summer, my family and Glenn Drowns had the good fortune of meeting Tom and Judy Butterworth and their ten-year-old son, T. J. They had been visiting Tom's parents in Wausau, Wisconsin. They set up their tent at a local campground on the Upper Iowa River and came on over to the house. We barbecued chicken that evening on the patio and sat around and talked until late.

Tom and his family live on a four-acre place in the Quaker Farms section of Oxford, Connecticut. Tom is a professor at Western Connecticut State University in the Biology Department. Although his specialty is microbiology, he also teaches courses in Horticulture, Nutrition and Biology & Society. Tom and Judy share an enthusiasm for gardening with nature, self-reliance and cooperative efforts.

When they bought their four acres in 1976, they finally had the elbow room they needed and the chance to start raising nutritious, good-tasting, chemical-free food on the scale they'd dreamed of. Tom built a passive solar-heated growth chamber to extend their growing season through the winter months. He gradually converted his garden areas into raised beds and by using French Intensive methods he doubled and then tripled his yields. Their smallholding, which is known as Butterbrooke Farm, also includes a Christmas tree planting, berry patches, fruit trees, chickens and ducks, a pond stocked with bass, and a wildlife area. Getting a maximum yield from a minimum of space is a concept that Tom and Judy both practice and preach.

As Tom and Judy's gardening and yields increased, they began giving away extra produce. They soon found that, except for their immediate neighbors, folks that they didn't know always wanted to pay for things. So they started putting up signs advertising cheap prices on whatever was ripening in their quarter-acre garden at the time. As more and more people started dropping by Butterbrooke Farm, they used the opportunity to introduce folks to some of their favorite vegetable varieties which they had discovered through the Seed Savers Exchange.

Tom was a member of the Newtown Food Coop for nine years. The fact that it recently folded makes him sad. It was there that his mail-order seed business was born. One season he left a box of extra vegetable seeds at the coop for distribution. Members were delighted. Tom decided to set up a seed cooperative to buy seeds in bulk from family-owned seed companies in New England. All of their seeds are open-pollinated, chemically untreated and rapidly maturing strains suitable for growing in New England gardens. Substantial savings are passed on to his members by buying cooperatively, packaging the seeds by hand in plain envelopes, printing a plain pricelist and advertising by word of mouth.

Several hundred gardeners now participate as members in the seed cooperative. Others simply take advantage of the opportunity to obtain a few packets of seed. Tom has to buy several 50-pound bags of some of the beans and peas. He buys at least pounds of everything else. 1984 was the seventh year for the Butterbrooke Farm Seed Co-op. Seed is offered in two different-sized packets. "A" packets are a standard-sized packet containing a generous quantity of seed. "B" packets contain three to four times that amount. Prices are an unbelievably low 35 cents for A's and 75 cents for B's. Several gardening guides are also available. Anyone can buy seed at the above prices just by writing and asking for the "Seed List."

For those wishing to become more deeply involved, Tom invites you to become a member of the Butterbrooke Farm Seed Co-op for an annual fee of $6. Members receive: 1) a 15% discount on all orders, 2) a subscription to the quarterly Butterbrooke Farm Seed Co-op Newsletter, 3) an organic gardening advice service, 4) a role in seed variety selection, 5) an opportunity to purchase rare or heirloom seeds from a supplemental seed list available only to members, and 6) notification and invitation to meetings, workshops and potluck dinners. (And for $15 you can become a Sustaining Member of the Co-op, which entitles you to additional benefits.)

From talking with Tom, I came away with a sense of his deep commitment to the cooperative movement and also to self-reliance. As he said, "I really don't understand why more people don't band together to make their seed purchases, instead of continuing to pay for the color pictures in catalogs....Most people buy the same varieties anyway." I don't know why either. Makes perfect sense to me. If you are interested, send a SASE to Butterbrooke Farm Seed Co-op, 78 Barry Rd., Oxford, CT 06483.

(The 1984 Fall Harvest Edition, pages 98-100)

GIANT WATERMELONS

Late in 1979 I was sent one seed from a watermelon which officially weighed in at 200 pounds even. It was grown by Ivan and Lloyd Bright of Hope, Arkansas. I traded a few letters with Lloyd and found out that Hope is the giant Watermelon Capital of the World. There have been annual Watermelon Contests there since the 1920s, and 100-pound watermelons are common in the area. Almost 50 years ago a 195-pound watermelon was grown there, but it wasn't until 1979 that the 200-pound barrier was finally broken.

Lloyd sent me a unique 58-page booklet entitled Producing Giant Watermelons ....an Arkansas Tradition by Lloyd Bright. It's loaded with pictures of melons and their growers and their fields during the 1920s contests and up through the years. It's fascinating and had me grinning most of the way through. The fun and rivalry among the local growers really shines through—the Laesters, the Middlebrooks, the Brights and many others. There are chapters on Watermelon Contests, Giant Varieties, Breeding Watermelons, Seed Care, Feeding Watermelons, Little Extra Tips, Soil Preparation and Tilling, Disease and Pest Control, Handlng Giant Watermelons and much more.

Lloyd writes, "....We grow only two types of watermelons and both are large, sweet and have good taste. We've developed our seed lines over the last seven years by selecting only the largest melons for planting. These varieties grow large because of their inherited size and do not require wicking, glucose feeding or any special feeding. They do need adequate fertilizer, weeding, pruning and watering....Our O. F. Blue Rind has a blue-green rind, but O. F. Lloyd claims he named them after his mule Blue. It produces nine possible rind and color patterns. We planted two acres out of four small seed melons we liked and the largest weighed 154 lbs. on Saturday. I think it may be through growing. These are similar to the extinct Jumbo Triumph line, but these taste better than the Jumbo....Our Carolina Cross is a long, light-striped melon that often grows in excess of 150 lbs. It germinates slowly in cool soil and produces large numbers of odd shapes. No other variety produces larger melons. We have won the Hope Contest six out of the last seven years with this variety."

Richard & Jason Bright—World's Champion 200 lb. Hope Watermelon (Photo by Clyde Davis of Hope, Arkansas)

Lloyd said their seed prices will vary, and the books will be $3. If you send a self-addressed, stamped envelope (to Giant Watermelons, PO Box 141, Hope, AR 71801), Lloyd will send you a sheet describing what's available and you can choose the rind and color patterns you want. What's the date of the Watermelon Contest, Lloyd? I bet the weigh-in is something to see.

(The 1982 Winter Yearbook, page 91)

JOHNSON SEED CO.

(Jim Johnson used to be the Historic Gardener at Old World Wisconsin. His projects there in 1980 included: an 1897 Finn orchard, a 1915 Finn orchard, a 1915 Finn potato field, a 1915 Finn vegetable garden, a large 1860 Pomeranian German flower garden and vegetable garden and potato patch, an 1880 Pomeranian German orchard and potato patch and vegetable garden, an 1850 Norwegian vegetable garden, 1860 oat and corn and barley fields and five large 1880 flower gardens. He has done a tremendous amount of research since 1975 tracing back the introduction dates of garden vegetables, fruits and flowers.... Jim printed the first catalog for his Johnson Seed Co. in 1983. In 1985 he hopes to be offering 150 varieties of heirloom vegetable seeds. To receive the catalog, send two "loose" first class stamps to: Johnson Seed Co., 227 Ludwig Ave., Dousman, WI 53118....The following is from a talk we had during the summer of 1983 in Dousman, Wisconsin.)

I started my seed company as a reaction to certain trends within the seed industry: smaller seed packets, higher prices, slower order filling, lack of gardening information in the catalogs. And all this at a time when a lot of new people were going into gardening. The garden seed companies are staying completely within their traditional formats. The only revisions being made are to modernize the appearance of their catalogs. There are no revisions in terms of techniques or methods. They're doing things exactly the same as they did ten years ago or twenty years ago. There's just no change, except for a decrease in service. My catalog will be black and white. I might put some pictures in next year. There'll be some drawings. The cover is hand-drawn and very attractive. In fact, it's frameable. The size of it may change next year. This year I used an 8 1/2" x 14" sheet of paper folded in half....I want to publish a four-year cycle of gardener hints (repeat it in four years). Some, that I consider key ones to success in growing a certain plant, will be carried each year. But you're limited in space....I intend to keep the prices moderate. I'll put more seed in the packets and try to build genuine credibility with the catalog. I've gotten a lot of compliments on the techniques that I'm using....There are three categories in the catalog: modern, heirloom and unusual. It will include many older commercial varieties, pre-1900....

There's a monastery not too far from here. I've been there many times during the past few years, doing little odd-jobs for them. They've got about a 15-acre field that's back in the corner which they would like to rent out to a farmer. But they've never been able to find a farmer who wants to rent it. They've got an 80-acre field that they do rent out, but no one wants to mess with the smaller ones. There's some scrub brush starting to grow in it. I've got a tractor and equipment. Next year, time permitting, I hope to start growing many of the unique seeds that I offer. The idea would be, whenever possible, to grow things on a five-year cycle and then freeze the seed. Grow five years' worth as best you could figure, so you're not growing the same things every year. Then the next year grow a different set of plants. That way you'd get the maximum effect for your personal labor....

Years ago I was a county agent in Spooner, Wisconsin, which is a small town in Washington County near Superior. I used to do a radio program and I'm well-known in that area. I'd like to go up there with about 100 varieties of incredibly odd and unusual vegetables, display them and sell the seed at the County Fair. I think it would be a wildfire sort of a thing. How could a gardener resist buying? Just go up there and have a lot of fun with it and see what would happen....I could take my black carrots, purple carrots, red carrots, yellow carrots, white carrots, three-foot-long carrots....the red sunflower....green radishes, yellow radishes, black radishes, the giant radishes....there's all kinds of odd tomatoes and squash....large bushel gourds and various gourds people aren't familiar with. White cucumbers....Moon & Stars watermelon. A rainbow of beans, red brussels sprouts, red celery, pink celery....and there's all shapes and sizes and colors of potatoes....and some really different types of corn. I figure I could come up with about 100 varieties that would stop gardeners in their tracks.... That's half of all of this--just the fun of it. You could get to be a traveling medicine show! You could hit county fair after county fair. If you had varieties like that to display and had the seed available too, in a season or two you could make your name in the whole area. It would be tremendous publicity. All the local newspapers would write you up. They all would, everywhere you went. And you could give your interviews, on the one condition that they publish the name and address of your seed company, and they'd willingly do it. The radio stations would interview you, too. I don't see how you could lose on it. You'd have your seed business going strong after just one summer....

I've done a lot of work with tracing varieties back to the earliest date that they are mentioned. An excellent and little-known source of such information is State Fair Premium Books. Wisconsin developed and published such hardcover volumes, and just about any library in the state has them. The State Fair in Wisconsin has been going for 130 years. These were the premium entry lists. They include the names of the winners and the towns they were from. Some of the descriptions are so very general that they're worthless, but others are quite good. So it is a very nice source of varietal information. Not every variety of the time, but the major varieties of the time are in there. These were published in hardcover by an agricultural society that's still in existence. It's under a different name now, but it was the Wisconsin Horticultural Society. They had an annual meeting and all of the speakers were published verbatim. These are big books. They'd have the State Fair premium lists along with the top three winners in everything. I don't care if it was horse pulling, the best chicken breed and so on. It was all in there....

We all have that one vegetable that we regret not grabbing when we had the chance....I was in the Peace Corps in Nigeria from 1966 to 1969. There is a squash in Nigeria that is grown just for the seeds. You eat the seeds. The seeds are silver dollar size and kind of flat, but they've got some thickness to them. You take the seeds and broil them. Then the outside skin becomes real loose and can be taken off easily. Each seed is a mouthful. And you don't grow it for the meat at all. They would just cut the thing green, take the seeds out of it and throw the rest away. To be honest with you, I cannot remember what the squash itself looked like. But I ate some of the seeds over there and they were quite good. And I've wanted to get some of that. I'm not worried about it going extinct, because it's common there. But I don't know how to locate someone who could get it for me....

There are hardly any foods that are native to West Africa, which is actually the reason that the region developed so slowly. I thought when I went to Africa that it would be a great adventure--exotic berries and nuts and things. Forget it! There's a lot of things that grow prolifically there now, but they've all been brought in. There's a type of yam which is a very poor food source that grows wild there. The Africans basically don't even bother with it today. There's a wild nut that grows on a vine. I've eaten that and I can see why it's not very popular. There's also a native breadfruit there, but that's a super-poor-man's food. And that's about it. There's just nothing there to eat. The tribes settled along the ocean, where they could fish. There are parts of inland Nigeria where Christian missionaries didn't even arrive until 1920....

I was working with protein crops, but I was only there for 10 months until the Biafran War. And then I was transferred to Ghana....In Ghana I also worked strictly in protein crops and planting techniques. They improved production so tremendously....I got a letter back from a guy after I left and was gone for a year or two. He told that many farmers were growing with the techniques that I had brought them....

(The 1984 Fall Harvest Edition, pages 96-98)

ABUNDANT LIFE SEED FOUNDATION

Abundant Life Seed Foundation (P. O. Box 772, Port Townsend, WA 98368) is a non-profit, tax-exempt foundation dedicated to propagating and preserving the plants and seeds of the North Pacific Rim. It was founded and is directed by Forest Shomer. Forest and friends hold quarterly seminars on such topics as winter gardening, starting nurseries, saving seed from biennials, commercial herb growing, seed growing and collecting and the self-sowing garden. They also conduct seed collection and plant identification tours of coastal British Columbia and Washington State. With a sliding fee scale from $4 to $10 according to your means, membership will bring you their quarterly newsletter, a book list, and also their annual catalog of garden vegetables, herbs, trees, shrubs, wildflowers and ornamentals.

Over the years Forest has offered almost 100 locally adapted heirloom vegetable varieties through their catalog. He is also carrying six tomato varieties from the collection of Czechoslovakian tomato breeder Milan Sodomka. These are the Czech varieties that were lost from the collection of Ben Quisenberry last year and I didn't know until just recently that Forest also had them. I urge all of you to first send for their catalog/newsletter subscription. Buy the garden seeds from them that you would normally order from the mail order giants. Forest has a great selection of vegetable varieties, and the fine work he is doing deserves your support. Then after you have seen what his organization is all about, I hope you will support him with an annual membership.

(The 1981 Fall Harvest Edition, page 59)

GOOD SEED COMPANY
(Will Ross, Box 702, Tonasket, WA 98855)

In 1980 I went public with my private collection hobby and formed the "Good Seed Company," a national mail-order venture. My gamble is that "old-fashioned" varieties are still in demand because of their fine qualities which include excellent production, flavor, nutrition and beauty, as well as sentimental and academic values. My goal is to collect and produce a diversity of high quality "old-fashioned" seeds. Since many old varieties are presently being abandoned by their new multinational owners, the possibilities for collectors are momentarily staggering. But this opportunity will soon pass....My personal vision is to offer an inventory of strictly heirloom seeds, ignoring modern breeders' triumphs. I honestly feel that if my reselecting from these old varieties is attentive enough, their "modern triumphs" are going to face some stiff competition from the very same items they discarded....

The Good Seed Company is a "collectors" company. My goal is to collect definitive arrays of cultivated Native American food crops (not wild plants) and representative arrays of Old World food crops. Also, my old ornamental herbiculture background has me selling bulk wildflowers on the erosion control market to help make ends meet....I am not in the breeding business. I'm not a plant scientist. I'm a garden artist and a rare seed collector! I know the limits of my energies and can only do so much at once. I'm very good at roguing out undesirable specimens and selecting seeds from good parent lines. As for crossing and mixing and playing Luther Burbank, well at the moment I'm too busy collecting and selecting to play matchmaker....

We have moved back to Washington and leased two acres as our 1982 production and trial grounds. (We're shopping for 20-40 acres of farmground to develop as our permanent home.) We are at 2,000 ft. elevation, nine miles east of Tonasket. We're very far north, only 24 miles from Canada! Long days ....Are you familiar with the Pacific Northwest climate differential? From the California redwoods to Alaska it's rain, rain and more rain. But east of the Cascades it's dry with only 10" of rain each year or less. The rain shadow is narrow, and east of us towards the Rockies it's much wetter and colder. So this niche is a weird northern desert. I like it lots! The dryness gives us a different range of crops than the populated wet side, so my working here will complement Forest Shomer's. I am the only seed company east of the Cascades (he's got three other small competitors on the wet side). I'll always be grateful to Forest for his example and encouragement....

Our emphasis this year is on "old-fashioned" tomatoes and beans, especially pioneer varieties from the 1800s. Future directions for the Good Seed Company include research into superior homestead varieties, production of cover crop seed and a product line of useful farm and garden inoculants, implements and irrigation supplies....Our annual catalog of seeds and supplies, published each December, is available for $1....

(The 1982 Fall Harvest Edition, page 115)

TAOS PUEBLO NATIVE SEED CO.

(At the Tucson Conference I had the pleasure of meeting Jim Reyna and John & Marie Kimmey. They were giving out samples of Taos Pueblo Blue Corn and had huge almost black ears of it on display. They also had pictures of beautiful fields of it growing and also pictures of their grandfather, who is over 100 years old, who is supervising the planting. The following is taken from correspondence and literature concerning what they are doing)--

We, the Reyna family, are pleased to present to you a strain of seed which has been handed down and tended annually for hundreds, perhaps thousands of years. It has never seen fertilizers other than that which nature provides in our special earth and water. Our source of irrigation is a high mountain lake, Blue Lake, held sacred by our people as the source of all spirit. The earth is part of the highest mountain in New Mexico.

This native blue corn is indigenous to the Taos Pueblo Tribe. As a member of the Tribe, my partner Jimmey Reyna has reclaimed a family field which is known historically for its high yields of corn for many generations. This past spring Jimmey, along with his father who is one of the elders still actively farming, restored some antique harness and plows which were used around the turn of the century by most of the Taos Indians. With a well trained horse, the Reynas have planted several acres this year.

We have been reading recent reports from scientists and agricultural experts regarding the need for commercial corn producers to utilize the tough, disease-resistant strains of native corn to build up the genetic strength of our presently weak hybrids. It is our belief that corn of the toughest varieties, grown under the conditions of its origin, can serve this important function of re-enriching our nation's staple crop. It is this market which interests us most....but as this is our first year of offering this seed commercially, we are also interested in exploring all markets for our product.

This Taos Pueblo Blue Corn is a staple of our people and makes excellent cornmeal for piki or cornbread. It can either be cooked, dried and ground for atole, or can be ground raw for tortillas and tamales. We feel sure that you will find it of the highest quality and extremely nutritious.

This seed is an ancestral strain which has only been grown by family members until now. It should be planted about four inches deep in reasonably rich soil and will take about 100 days to mature. This corn does not require a great deal of irrigation, as it has survived many droughts.

We hope to be able to offer other native seeds next fall such as squash, beans, pumpkins and chiles. We intend to contact the Rio Grande Pueblos, Hopi and Zuni in order to get the best selection of seeds while helping the presently dismal economies of these ancient villages.

Taos Pueblo Blue Corn is available in 1 lb. bags at $6 per pound, 5 lb. bags at $4 per pound and 10 lb. bags at $3 per pound. These prices include shipping via UPS. Write to John Kimmey and Jim Reyna at Talavaya, P. O. Box 9289, Santa Fe, NM 87504.

(The 1981 Fall Harvest Edition, pp. 55-56)

SIBERIA SEEDS

We are Ron, Cynthia, Sarah and Angela Driskill. We make up the family seed company called Siberia Seeds. As you may know, Siberia Seeds bounced into the limelight about two years ago when articles began to appear in major gardening magazines throughout the continent about the Siberia tomato. Gardens For All started it first, with an article in the May 1983 issue, and Harrowsmith, Country Journal, Farmstead Magazine and Countryside followed thereafter. Several newspapers ran a story on Siberia (which they are still doing) and the UPI wire services picked it up. Since then we have added Glacier tomato to the list—which is another early-bearing, cold-resistant variety—and several other heirloom tomatoes have been tested, with promising results.

In addition to being an SSE member, we are also conducting our own seed exchange program, and we will gladly send trial samples of either Siberia or Glacier to SSE members for any cold-natured variety of tomato, beans, peas, squash or pumpkin that is no longer being sold. We are also interested in foreign varieties and those which are exotic or unusual in some way.

Now for the most important part. We need growers and test gardeners. We We want approximately ten people per state and ten people per Canadian province to test our seeds for us, who preferably represent the various climates found in each domain. In that manner, we know how well a certain variety is going to do in each place before we sell it, and that helps us to plan our publicity, as well as to help gardeners choose the tomato they want to try.

GROWERS are our main concern, however. We need people who are willing to grow tomatoes and/or other vegetables for us. We have received several good heirloom tomatoes that are better suited for warmer climates, and we need to find people who are willing to grow them, strain them, separate the seeds, dry them and ship them to us. The going rate among the big seed companies that do this is $40 to $50 per pound, and that is what we would be willing to pay our growers. Our growers need to be living in a growing season of at least 140 days (preferably).

Thanks very much. If you would like a catalog, write to SIBERIA SEEDS, Box 3000, Olds, Alberta, Canada T0M 1P0. Send 50 cents and a self-addressed long envelope.

(The 1984 Fall Harvest Edition, pages 102-103)

GLECKLER'S SEEDSMEN ARE BACK!

Merlin W. Gleckler participated in seed exchanges and research with over 100 foreign countries, and the results were clearly evident in his unique catalog. He had a severe stroke in July of 1980 and some members told me that Gleckler's was going out of business. But his kids have started the company back up and I recently received their new catalog.

It's still all there! Corn from Zululand, 3-foot-long Imperial carrots from Japan, pink celery from England, pink cress from India, Roselle--the Jelly Plant (the okra family, India) and tomatoes from Greece, Belgium, Puerto Rico, Japan, India and Italy which include: Peron Sprayless, Giant Belgium, Watermelon Beefsteak, Dutchman, Evergreen, White, Egg, San Pablo, Napoli, Ruffled, Thessaloniki, Goldie (150-year-old pioneer variety), White Beauty, Russian Red, Angora and many others.

I called George Gleckler, Merlin's son, and he has been unable to find any pure seed of his father's Cob Melon. Also, he would like to obtain seed of their Pink Grapefruit tomato, which some of you may have. They are asking 50 cents for their catalog to cover printing and postage (send to Gleckler's Seedsmen, St. Route 120, Metamora, OH 43540).

I urge all of you to support them with as much of your seed business as you possibly can. If you don't realize that we all have a stake in seeing that they make it during this recession, then you've missed the whole point of the SSE.

(The 1982 Winter Yearbook, page 90)

(From a later letter from the Gleckler family) -- Our father is now in a rest home....It really gladdens his heart and mind knowing that his business is being carried on....Next year we will be offering quite a few new unusual seeds, as we have expanded the countries we import from. A few of the seeds are: White Broccoli, Edible Burdock, Chinese Giant Pepper, Large Bamboo, India Rubber, Coffee, Violet Eggplant, King Purple Broccoli, Golden Beet, Armenian Yard Long Cucumber, Celtuce, Yellow Tree Tomato, Passion Fruit and yellow Baby Seedless Watermelon....Thanks to the request you put in your last catalog, I am again growing my father's Cob Melon, which I had lost....So far none of your members have been able to come up with his Pink Grapefruit Tomato....our catalog is still 50 cents....

(The 1982 Fall Harvest Edition, page 117)

Invocation

Plant a Tree

Because you came into a fruitful world that your foreparents planted for you. So will you do for your children!

Because the pattern of white stars on a shimmering red apple is some kind of secret pressing forth.

By way of weaving a green mysterious carpet of love all over the dry land.

As a deafeningly, noiseless reply to the headlines in this morning's newspaper.

As a symbol of all the unselfish, constructive forces in this country working in the creation of the true commonwealth.

Because once in the obscure trauma, rapture and illusion of an instant you felt an opening through and beyond the conditionalities of merely individual existance. You felt Life—

Living Tree Centre, an educational association devoted to furthering aesthetic and cultural evaluation, is near the town of Bolinas which is located amidst spectacular beauty on the northern California coast. Their nursery specializes in preserving and propagating antique apple varieties such as Cox's Orange Pippin, Irish Peach, Motherlode King, Skinner's Seedling, Winter Banana, Calville Blanc d'Hiver and over 30 others. From their Experimental Garden come unique seeds of proven vigor and hardiness, including: the legendary Escondido Gold Melon, Red Russian Kale, Red Orach and several others.

Living Tree Centre offers a three-month apprenticeship program for those who wish to learn and practice the art of propagating fruit trees. Subjects being offered include: managing a small nursery; designing and planting small orchards; techniques of grafting and budding, including "tongue and whip," "T budding" and "chip budding"; topwork and frameworking of mature trees; fruit exploring in ancient orchards (locating and collecting scionwood from promising cultivars); and pruning and maintaining fruit trees.

Their catalog/newsletter....includes detailed variety descriptions and step-by-step cultural information. The price is $4.50 from Living Tree Centre, P. O. Box 797, Bolinas, CA 94924.

(The 1984 Fall Harvest Edition, pages 132-133)

SEEDPEOPLE NETWORK FORMS

With grant support from the C. S. Fund of Freestone, California, the Seed Foundation was host to the second Cascadia Bioregional Seed Gathering, August 18-19, 1984, at Port Townsend: a networking meeting for seed producers and distributors.

Preceding the meeting, Pamela Lee of Down Home Project, Missoula, Montana, presented a slide show, "Agriculture in Nicaragua and Nearby Countries," based on her early 1984 travel and work experience, from Costa Rica to southern Mexico.

Meeting sessions took the form of group discussion on a series of issues. Mushroom Alan Kapuler of Peace Seeds opened a session on Genetic Diversity with an explanation of the Five Kingdoms (an expansion on the old animal-vegetable-mineral understanding), commenting that "everyone's neighborhood is a plant collection and one can map a gene pool, put it in families." Michael Pilarski of Friends of the Trees Society suggested forming caretaker groups around plant families, such as the Actinidia (kiwi) group that he helps network. Another session, devoted to Seed Politics, touched on concerns about plant patenting legislation, consolidation of ownership of seed companies, and the November 1983 FAO International Undertaking on seeds. John Schneeburger of Garden City Seeds noted that "plant breeding is offering choices which provide technological solutions that may overlook concerns such as building soils." During a discussion on Networking Seed Grow-outs among small companies, Jan Blum, of Seeds Blum, suggested that "we pool our efforts on the more common varieties to allow ourselves the chance to grow more specialty varieties. I'm less interested in bulk buying than in building our networks, supporting growers." Two sub-groups spun off, to network vegetable seed grow-outs, and to develop inventories of regional species. Other sessions discussed Organic Standards for Seeds, led by Tilth Producers Cooperative, and Advertising and Publicity (turning "advertising" into "outreach"), from which emerged a proposal to collaborate on a travelling slide show or video, depicting the work of regional seed companies, showing what we are doing to produce seed and inform the regional public about the world seed situation.

This session ended with a sailing cruise on Admiralty Inlet, but not before an organizational discussion that carried over into the next morning, with the intent to form a SeedPeople Network, a non-profit educational project to carry on the work of this annual meeting on a regular basis. Proposed articles and bylaws were reviewed and a subsequent meeting was scheduled for the Tilth Jamboree, September 28-30 near Ellensburg, Washington, with the hope of ratifying corporate instruments at that time and getting on with our collective work.

The second day of the conference provided views of seed production at the Seed Foundation gardens, followed by a visit to Dungeness Organic Produce near Sequim, Washington, where seed production complements the main purpose of raising market food, herb and cut-flower crops. Group feeling was so positive by this time that most participants stayed a full day beyond the meeting to network, play volleyball and generally enjoy the opportunity to be together.

Conference proceedings were recorded and are now being edited. For information on cassette tapes of discussions on Genetic Diversity; Seed Politics; Networking Seed Grow-Outs; and commentary accompanying Agriculture in Nicaragua: send a self-addressed envelope to The Foundation.

(Reprinted from the Fall 1984 issue of the Abundant Life Seed Foundation Newsletter, P. O. Box 772, Port Townsend, WA 98368.)

(The 1984 Fall Harvest Edition, pages 145-146)

OTHER SEED COMPANIES

Marapa Nursery and Plantsmen, Bob Howard, 1275 Cherryvale, Boulder, CO 80303—We are interested in developing a seed catalog for retail sales by mail order. We presently collect and sell tree seed for the Japanese market and wildflowers from the Rocky Mountain region. I also have seed developed by Alan Chadwick and his apprentices. We plan to grow these plants for seed production for those interested in bio-dynamically (or organically) grown seed. We would also like to carry old varieties in our catalog. Can you recommend any Seed Savers who would be interested in supplying seed for a mail order business?

Bill Wilkerson, Rt. 3, Box 73-C, DeFuniak Springs, FL 32433--....We grow chufas....Cyperus esculentus....There are five varieties in the U. S. I'm not sure which we grow but it is the one used in Spain as a food....We grew 12 acres in 1981 at a yield of 3,000 pounds per acre. So there is no shortage of chufas; however, few people seem to know where to obtain them....Presently there is a small market for chufas for game food plots for wild turkey, and turkey hunters keep us in business. However, people used to plant chufas for a snack food and for their hogs (before 1940)....It would please us to see more households with chufas in their garden....Send a stamped envelope for a price list and brochure....

Fern Hill Farm, John Gyer, Jessup Mill Rd., Clarksboro, NJ 08020--Enclosed is a picture of the Dr. Martin pole limas we raise here at Fern Hill Farm....The largest beans are typical of fully mature produce. However, the smaller are the size we prefer for table use. In both cases the skin is green and tender, the flesh is smooth-textured and the flavor is pleasantly nutlike. Three beans per pod are typical, but four per pod are common under very good growing conditions....They are the only pole lima bean listed as "excellent" in the book Burrage on Vegetables....grown under state certification....germination tested, and hand selected for uniformity and quality....growing suggestions are included with each order....Send SASE for order form; we have no catalog....

The Charles C. Hart Seed Co., PO Box 9169, Wethersfield, CT 06109 (Note from Kent)-- I have finally located a source for Green Mountain Potatoes. They never appear in the Hart seed catalog because the company never has any firm prices until after their catalog goes to press. I have been told that if you write them, they will send you a list of potatoes and their prices later. They also said that they carry a large selection of potatoes and do make retail sales in quantities of five pounds or larger.

Hastings - Seedsman to the South, PO Box 4274, Atlanta, GA 30302 (Elizabeth Whittle, Catalog Research)--....Many of our customers (mostly home gardeners from all over the Southeast) are requesting "rare" and "unusual" and "old-fashioned" seeds....If you know of any varieties that are again becoming popular or are particularly good for the South, please let me know....Keep up the good work....I am a home gardener myself and am very interested personally in learning more about the Exchange....(Note from Kent: When I read this letter, it made me wonder if our efforts are having a larger impact than we can imagine....I want to thank Elizabeth and the other fine folks at Hastings, because they are the only seed company in the country that is systematically sending me seed (of the varieties they are discontinuing) to be maintained by our Growers Network. It's too bad that the other approximately 120 companies who were asked aren't as concerned....Hastings carries a really fine collection of vegetables. About a third of everything they carry is either unique to their company or specifically adapted to southern climates.)

St. Lawrence Nurseries, Bill MacKentley, R. D. 2, Route 56-A, Potsdam, NY 13676--Twice in the past week, Mr. Whealy, your name has come to my attention from totally different sources, so I guess our karma is running closely....In 1974 I was on the N.Y.S. Committee for Protection of Endangered Plant Species. My raison d'etre was simply that our genetic heritage is being rapidly depleted! That was my feeling eight years ago and lately the problem is becoming much, much worse....

In 1972 (after college and teaching) I apprenticed with Fred Ashworth, a 75-year-old internationally known plant breeder. For over 60 years Fred had sent and received plants from all over the world. As a self-educated "Hereditarian," Fred touched all aspects of agriculture from vegetables to flowers to fruits and nuts (he was a close friend of J. Hershey and J. R. Smith - Tree Crops)....During the 1940s, Fred did breeding work with the Irish potato, crossing Solanum demissum (collected from Central America for him) with existing cultivars. They grew over 4 million seedlings, and the world's first blight-resistant potato was developed. His initial goal was to develop a potato for Iceland which has similar daylength requirements to S. demissum....

I think it is a good idea to have Allard's Principles of Plant Breeding as well as the Yearbook of Agriculture for 1936 and 1937 (especially 1937) under your belt, if you are in the seed business or saving old seed....

Our nursery has two major goals: 1) to offer all northern hardy fruits and nuts able to be grown in our climate and 2) to promote tree crops and offer selections of tree foods not normally thought of as having potential food value. We presently have over 100 varieties of apples plus much other material that we are not offering until we can get sufficient stocks propagated. Enclosed is an abbreviated list of our offerings....

(Note from Kent--Send a self-addressed stamped envelope for this list. What immediately caught my eye were several unique nut trees, including an American Chestnut that seems to be exceptionally blight-resistant.)

(The 1982 Fall Harvest Edition, pages 114, 116-118)

J. L. Hudson, Seedsman, A World Seed Service, Box 1058, Redwood City, CA 94064--....Our Seed Exchange has worked out very well so far. I've received many seeds that are not commercially available. One lady sent a species that was desperately needed by some researchers; they had written all over the world for it for a year with no luck. She had a plant growing in a corner of her garden....Concerning low vigor seed--two years ago I had several ten-year-old seeds of a very rare plant I wanted to preserve. The first season's plants were about 1/10 the size they should have been and were weak. Seed from these produced much healthier 3/4 size plants this year, and I expect they will have regained their full vigor in a season or two. I have had similar experiences in the past. It seems that old seed, though weakened, retains the potential for vigor that the parent plants had in the genes....We will not be adding any more F1 hybrids to our catalog. We feel that there are many true species far more worthy of cultivation than these endless hybrids....The cultivation of genetically uniform crops can have disastrous results....I really enjoyed reading the Seed Savers Exchange. It is very interesting and well put together and the Hodgepodge section was most enjoyable....(Catalog--$1)

(The 1979 Seed Savers Exchange, page 37)

Saving Garden Seeds

(Editor's note: This entire book is made up of material printed in The 1984 Harvest Edition or earlier issues. The only exceptions are the articles on pages 292-328 of this chapter which appeared in The 1985 Harvest Edition. We have been publishing a Seed Saving Guide in almost every issue since 1977. But it makes no sense to reprint any of our earlier guides which now seem incomplete and outdated. By including these recent articles, we have made this chapter as current and complete as possible. It contains more information on "how to" and the "need for" saving seeds than most books on the subject.)

Kent Whealy, Founder and Director of the Seed Savers Exchange (Photo by Alan Buie)

CONSERVING VARIABILITY WITHIN COLLECTIONS
by Gary Nabhan

"If only more people could have had the chance to see Wilfredo Salhuana's slides at the Seed Banks Serving People Workshop," Dr. Roos said pensively, "it would have helped them observe what is otherwise an abstract concept associated with probability statistics." Dr. Salhuana showed slides of the variability in a corn population which, for several years, had been grown from 500 seeds planted per year. Then, for just one year, he reduced the population size to only 100 seeds planted. The next year, he planted 500 seeds from the progeny of those 100, but the genetic variation had been drastically reduced as the result of the one-year, 100-seed bottleneck. You could clearly see the lessened diversity. If those seeds were the only ones left to represent that population, it would mean that part of the original variation within that corn collection would be permanently lost.

Through the above (paraphrased) conversation with Dr. Eric Roos of the National Seed Storage Laboratory, Kent Whealy and I were impressed by the fact that how seed savers plan and implement their growouts--particularly in how many seeds are grown and how they are selected--makes a great difference in how much genetic diversity we preserve.

Dr. Eric Roos has worked with this problem for years, attempting to improve coordination between the National Seed Storage Laboratory and its growouts at USDA Regional Stations. Since he is a bean enthusiast, as are many SSE participants, it may be illustrative to use beans as an example of how to follow his guidelines. Often, a collection of beans made from a Mexican market, an Indian farmer or a county fair is quite variable--four to six color variants are not uncommon in one batch, and some similar-appearing beans may differ in other traits. For instance, there might be two isogenic lines of white beans--one, early maturing; the other, late maturing--that are visually identical in the same batch.

To maintain as much of the variation in the bean collection's population as possible, it is important not to grow out too few beans. For instance, if you had eight different lines within a mixture that totaled 100 beans, and you reached into your bean bag and took only 16 seeds to plant, there would be only a 30% chance that your sample would include all eight components. If you planted 32 randomly-selected seeds, you would have an 85% chance of including all eight components. It would take at least 64 seeds planted to reach a 99% chance of having at least one of all eight components. (Dr. Roos has published notes regarding conservation of bean variation in the Seed Banks Serving People Proceedings, as well as in several volumes of the Annual Report of the Bean Improvement Cooperative, a voluntary and informal organization to effect the exchange of bean information and materials. These reports are maintained at the Dept. of Seed and Vegetable Sciences, New York State Agricultural Experiment Station, Geneva, NY 14456.)

Dr. Roos suggests that people should grow out a sizeable sample of a few collections one year and others the next, rather than an inadequate sample of every collection every year. With better storage conditions, the interval time between growouts can be increased, thereby reducing the number of times a bean population is exposed to potential crop failure or sampling error. However, it is important to do periodic germination tests, to ensure that the percent of viable seeds in storage does not get too low, or else variation will be lost that way. He also encourages growers of mixed batches of beans to remember the "hidden traits" as well as the colors--if yours includes those two lines of white beans, one early and one late, grow and harvest them in conditions where both will be renewed. Growers who harvest just once at the season's end may miss early beans whose pods have already popped. Gardeners in areas with short growing seasons may lose the late-maturing beans due to early frosts, possibly wiping out that "hidden line."

For those of you living near good science libraries or bookstores, a recent semi-technical softbound book describes these concepts in more detail and relates them to conservation efforts: Conservation and Evolution, by O. H. Frankel and Michael E. Soule, 1981, Cambridge University Press, 32 E. 57th St., New York, NY 10002, 327 pages.

(The 1982 Fall Harvest Edition, pages 23-24)

GUIDELINES FOR MAINTAINING PURITY IN PEPPER VARIETIES

by Jeff McCormack, Ph.D.

It is strange that the cultivated pepper is a major food crop yet there has been very little investigation of the relationship between natural cross-pollination (NCP) and isolation distance. In 1941 Odland and Porter wrote that, "Plant breeders and seedsmen disagree considerably in their opinions relative to the amount of natural cross-pollination in the cultivated pepper....In seed production, knowledge relative to natural crossing is a great aid in determining the isolation necessary in the seed plots" (4, p. 585). The lack of information is due partly to the number of variables affecting NCP such as location, time of year, changes in insect populations and climatic factors. Also, NCP is more difficult to study in peppers than in some other crops.

Isolation distance recommendations are based on the intended use of the seed. In other words, what degree of purity is sufficient for the intended purpose? For example, for certified seed production, very high purity is required (virtually 100%). Many large commercial growers of pepper seeds isolate sweet bell peppers by 1/4 mile, and hot varieties are isolated from sweet or hot varieties by one mile (1,7). For home gardeners wishing to save their own seed, the recommended isolation distances are much smaller. Several seed saving guides have recommended distances ranging from several feet (5) to 50 feet (2) to separation by the length of the garden or as far as practical (2,8). In these cases, the seed is being saved primarily for one's own use or for exchange among a few individuals, not for commercial purposes where there is potential for widespread distribution. In preservation efforts, such as that being undertaken by the SSE, there is always the potential for widespread dissemination of the seed: therefore a greater attention to seed purity is essential.

The purpose of this article is to recommend an isolation distance suitable for preservation efforts, and to present some of the considerations involved in determining isolation distance.

I have grown out some pepper varieties from the SSE and elsewhere only to find that some of the varieties were impure. Impurity is defined as an unacceptable percentage of plants containing excess genetic variability. A high degree of genetic impurity is not necessarily undesirable. Again, it's a question of intended use. If the intent is preservation of varieties true to name, then genetic variability is to be discouraged.

Because so many gardeners are now involved in preservation efforts, and because 1/4 mile isolation distance may be difficult to achieve in practice, there must be some middle ground where the trade-off between purity and practicality is small. What is needed is an isolation distance recommendation which will give a reasonably high degree of purity (at least 98%, preferably 99% or better) for small plantings.

A misconception exists among a number of individuals who save their own seed. The misconception is that no crossing has occurred if the fruit and foliage of the first generation (F1 generation) or subsequent generations appear no different than the parental generation. This misconception can be illustrated by performing the following test. The test consists of growing a row of hot peppers next to a row of sweet peppers, both varieties having approximately the same shape and color of fruit and otherwise similar in appearance. The seed from the fruit of the sweet variety is then saved and planted. When the fruit of this planting (F1 generation) is eaten, a high percentage of these peppers will be found to be hot (due to the presence of a dominant gene received from the hot variety). One enthusiastic bite into a hot sweet pepper will illustrate that similarity in appearance does <u>not</u> mean absence of cross-pollination. Incidentally, the hot trait will not disappear in the next generation (F2 generation) or subsequent generations unless the hot plants are rogued out (each plant would have to be grown in isolation to do this). Instead the genes will "move around" in the plants of the subsequent generations. This experiment demonstrates the obvious results of cross-pollination, but the manifestations of NCP between two similar varieties will be less obvious, especially where recessive genes are involved. These recessive genes are not so easily removed from the population of plants resulting from the NCP of two different varieties, partially because these recessive genes are not readily identified.

What percent of NCP can be expected of two varieties of pepper grown side by side and how does isolation distance affect the percent of NCP? In a recent series of experiments (6) conducted over a two-year period in five commercial fields in New Mexico, test plants of one variety were transplanted into commercial plots of another variety. Tester plants were placed 12" (30 cm) from adjacent plants of the commercial variety. The average NCP was found to be 42% with individual plants having a NCP value as high as 91%. This percentage may be unusually high because the commercial cultivar outnumbered the tester plants, and it's not usual procedure to plant two different varieties in this fashion. In a similar series of experiments by Odland and Porter (4) it was demonstrated that NCP values ranged from 9 to 38% depending on the variety of pepper tested. In my own experience I have had peppers appear to cross at 6, 15 and 25 feet with barrier plantings in between. The pollinating agents in most of these examples are honeybees, bumblebees and halictid bees such as sweat bees. None of the studies available to me investigated NCP as a function of distance. The authors did not recommend isolation distances possibly because there are so many variables involved whose effects are only partially understood. Listed below are some of the variables involved and their effects on isolation distance.

Variables Affecting NCP	Effect on Recommended Minimum Isolation Distance: Increase distance	Effect on Recommended Minimum Isolation Distance: Decrease distance
Greater number of plants	+	
Greater number of varieties	+	
Variety characteristics (e.g. hot versus sweet)	+	
Large number of pollinators present	+	
Presence of one or more barrier crops		-
Presence of alternate pollen sources		-
Staggered blooming times (seed collected from first blooms of first blooming variety)		-
Collection of seed from center of block plantings		-

If you are growing only two or three plants each of two varieties, you may be able to get by with a smaller distance, say 50 feet plus barrier plants. Hot peppers as a rule of thumb may require approximately four times as much isolation distance as do sweet bell peppers. Flowers of hot peppers have style and stigma (female reproductive structures) which protrude further beyond the anther cone (male reproductive structure). This situation is called stigma exsertion, and whenever the stigma is exserted, the flower is more likely to be cross-pollinated.

The Lend-Lease Act of World War II states that for peppers, 1/4 mile isolation distance is desirable between two varieties, but the distance should not be less than 150 feet plus a barrier crop between two varieties (1,3). Evidently the 150-foot distance plus barrier crop represents a compromise between purity and practicality. In the absence of hard evidence on NCP as a function of isolation distance (but using the information from the Lend-Lease Act), I am proposing the following guidelines for SSE members which should give 98% or better purity for small plantings. Here the intended use is for preservation efforts where fairly high purity is desirable, where the seed may not be grown out each year and where dissemination is limited. On this basis the following isolation distances are recommended for the SSE:

****

Between sweet bell varieties — 150' plus barrier crop*

Between hot and sweet varieties
(or between two hot varieties) — 600' plus barrier crop

****

*Note: Additional distance is desirable between long-fruited sweet varieties and sweet bell varieties, or between two long-fruited sweet varieties.

If necessary, modifications of these distances can be made by consideration of factors presented in the preceding Table of Variables Affecting NCP. If you cannot achieve the recommended minimum isolation distances of 150 feet and 600 feet for sweet and hot peppers, there are other alternatives for keeping the varieties pure:

1. Grow only one variety, and be sure to check the distance to peppers in neighboring gardens.

2. Protect the flowers with tape before they open and then hand pollinate with a camel hair brush.

3. Cage each variety under an insect-proof material such as cheesecloth or window screen. (Note: Be careful to screen all entrances since bees are quite adept at finding small openings.)

4. Isolate varieties in time rather than space.

As a final note, I welcome any comments from people working in this area. I will revise these guidelines as necessary as more information becomes available.

LITERATURE CITED

1. Borchers, Ed., Director, Virginia Truck and Ornamental Research Station, Virginia Beach, Va. 1984. Personal communication.

2. Bubel, N. 1978. The Seed Starter's Handbook. Rodale Press, Emmaus, Pa.

3. Hawthorn, L. R. and L. H. Pollard. 1954. Vegetable and Seed Production. Blakiston, N. Y.

4. Odland, M. L. and A. M. Porter. 1941. A study of natural crossing in peppers (Capsicum frutescens). Amer. Soc. Hort. Sci. Proc. 38: 585-588.

5. Rogers, M. 1978. Growing and Saving Vegetable Seeds. Garden Way Publishing, Charlotte, Vt.

6. Tanksley, S. D. 1984. High rates of cross-pollination in chile pepper. HortScience 19 (4):580-582.

7. Villalon, Ben., Texas Agricultural Experiment Station, Weslaco, Tx. 1984. Personal communication.

8. Whealy, K. 1983. Seed Saving guide. In: The 1983 Winter Yearbook. The Seed Savers Exchange, Princeton, Mo.

(Jeff McCormack operates Southern Exposure Seed Exchange, a new seed company which is now in its third year. They specialize in seeds for the mid-Atlantic states, greenhouse varieties and a growing list of heirlooms. Send $1.00 for their catalog to: Southern Exposure Seed Exchange, P. O. Box 158, North Garden, VA 22959.)

(The 1984 Fall Harvest Edition, pages 24-27)

CONTROLLING SEED-BORNE DISEASES IN TOMATOES

by Mike Courtney, Ph.D.
Excerpts from Lecture and Slideshow
"Seeds for Yesterday and Tomorrow" Conference
Toronto, Ontario, February 11 and 12, 1984

You can harvest fruits and collect the seeds from them any time that fruit shows color. If you see a pink blush on the fruit, then you can harvest it at that time and extract the seeds and the seeds will be viable. We like to wait until the fruits are very ripe, almost overripe, because it makes it easier to clean up the pulp....(Next slide.) Take a knife and cut the fruit, just like you were cutting it for tomato slices for a sandwich, except we're going to have two halves here. (Next slide.) You can see the seed cavities within the fruit. Each one of those is filled with jelly, and embedded in that jelly are the seeds. You've got to get that gelatinous coat off, because it does contain germination inhibitors and you have to get rid of it in order for the seed to germinate. (Next slide.) The easiest way is just to get yourself a glass jar, squeeze the tomato, and as you squeeze, scrape off the seeds into the jar. If you're short of seeds, occasionally you have to go fishing down in the cavities with a spoon to get all the seeds out. But in most cases just a good, healthy squeeze and they'll all come right to the surface and you can scrape them into the jar. (Next slide). We actually take the cardboard tag that we used to identify the plant and stick it into the jar with the seeds. We figure that we've got less chance of losing it, if it's in the jar. You have to be careful when you do that, because some of the tags that you can buy commercially will tend to disintegrate in the jar. We've tested a large number of brands and we know which tags to use and which pens are really indelible....

In some cases where you're very short of liquid from the fruit--in this case we only had the one fruit--you may want to add a little water. But this tends to retard the fermentation process. So in most cases, if you've got a good supply of liquid in the jar, don't worry about adding anything else. You also want to keep an eye on the jars so that they don't dry out. If it looks like they're going to dry out, add a little water to it. You let this sit for three or four days. As you'll see in the next slide, you get a very vile fungus growth all over the top of it. It's not pleasant to look at and it's not pleasant to smell. The fruit flies absolutely love it. So if you've got someplace (preferably warm, but isolated) where you can do your seed extractions, it's probably for the best. This fungus is what's doing the work for you. It's eating away that gelatinous coat on the seed. It's not doing any damage to the seed at all. In fact, it's doing some good things. The fungus produces a whole series of antibiotics and those antibiotics help you to control seed-borne bacterial diseases. These include diseases like bacterial spot, speck and canker. Canker in particular can be very troublesome in a home garden because it's very unattractive on the fruit. It causes a canker on the fruit itself, but once it gets onto the seed it also infects the seedlings and the seedlings can be killed quite quickly. So fermentation is the only reliable method of controlling these bacterial diseases. (Next slide.)

You will need a strainer of some sort, hopefully with the holes smaller than the tomato seeds that you're going to be straining. Add some water to the jar, swirl the whole mixture around, pour it into the strainer, hold the strainer under a running faucet, sit there with your fingers in the running water, swish the seeds around, the pulp will just melt right away and go right through the strainer, and you'll be left with the seeds. This next part is a little hard to demonstrate, so I'll tell you about it and then show you the end result. Scrape those seeds, once they're clean, right down into the middle of the strainer. Lay a paper towel out on the counter somewhere and then very quickly grab the strainer, bring it overhand and slam it right down onto that paper towel. It looks horrible when you see somebody do it. It makes a lot of noise. But, as you see in the next slide, the seed is deposited in one very small pile exactly where you want it on the paper towel, if your aim is good. Believe me, if you try to sit there and pick those seeds out of that strainer and smear them onto the paper towel, they'll stick to your fingers and you'll soon see the logic of this forceful method that we use. I guess most people figure that they're going to miss the paper towel, but after awhile you get good at it. So the seed is deposited on the paper towel, and we'll stick the same tag on there too. Some people like to write the information from the tag onto the paper towel, as just an extra back-up.

We have a special seed dryer which has fans in it that force warm air through a big box and the seeds dry either in paper towels or on special screens that we have. In the normal weather that you get in late summer and fall, the seed will be dry in a day or two. We don't worry about any special drying precautions. You can dry them over silica gel to get them down to a very low moisture content if you wish, but we don't....(Next slide.) After the seeds are dried, we just scrape them off the paper towel and into a storage envelope. We stick the same tag in with it. (Next slide.) Finally, on the envelope we write the information that we think is pertinent....write the name of the variety, the date that the seed was saved, and any other information that might be helpful to you in the future. For instance, why you saved the variety in the first place. There's nothing worse than coming through seed storage and finding a tomato variety that has no notes. Nothing. You have no idea why it's there and you're faced with a problem. Do you grow it out again and maybe find out that you didn't like it in the first place? If so, you've gone through twice the labor and still get nothing out of it. So put some sort of reason on the envelope as to why you saved the seed in the first place....Then you can seal the envelopes of dry seeds in an airtight jar and put them in the refrigerator, or better yet in your freezer where you know they're going to stay frozen and not freeze and thaw, freeze and thaw. Then you ought to be in good shape.

QUESTION: Are there any seed-borne diseases that your fermentation process doesn't control?

Nothing seed-borne that I know of, at least none that we have trouble with. There has been some question recently as to whether tobacco mosaic virus can be seed-borne. Even if it is, I doubt if fermentation would take care of it. Hot water treatment in that case would probably be needed.

QUESTION: How long do you let the fermentation process go?

Usually until we see good mold growth. Then we stop it at that point. Certainly no longer than five days. The trouble you run into is that the seed will start to germinate in the mixture and then you can lose it. But within three to five days, you're usually safe.

(The 1984 Fall Harvest Edition, pages 38-39)

VIRUS DISEASES IN POTATOES

by Bob Coffin, Ph.D.
Excerpts from Lecture and Slideshow
"Seeds for Yesterday and Tomorrow" Conference
Toronto, Ontario, February 11 and 12, 1984

Now you people are interested in saving seed of potatoes and that is a very big chore and let me tell you why. We're confronted with a major headache with that problem even under the best of conditions at the Cambridge Research Station....What do you see (pointing to slide being shown) when you look down that middle row of potatoes? Some of the plants look nice like this big healthy one here. This one here has curly leaves, and see how some of them are a little bit lighter color? Those are symptoms of virus diseases. Virus diseases are the most serious diseases of the potato crop that exist. They are responsible for the degeneration of varieties, if you try to keep them for a long time. Now under some conditions you can keep them for a long time, as long as you can keep the viruses out of them. But some lines of potatoes, if you tried to grow them and save your own seed, after five to six years you wouldn't get back any more potatoes than you planted, simply because they've been run down by virus diseases.

Now, we'll tell you a little bit about the virus diseases and the way that the seed growers try to minimize it and the way that it might be feasible for someone to try to maintain some potato lines. (Next slide, please.) Now these are extreme symptoms that I'm showing you. The virus diseases are often named by letters of the alphabet. There's A, X, Y, M and so on. This one happens to be virus Y. It's a very debilitating virus. (Next slide, please.) What do you notice about that plant at the end of the row there? The leaves are curling. That virus disease symptom is called leaf roll. Very serious virus disease. Now, how are these viruses spread? (Next slide.) Aphids. The primary vector for virus diseases is aphids. All aphids are not necessarily infected with viruses. But if they do get viruses in them, the winged types of aphids will move from plant to plant or from field to field. The way the professionals grow seeds is that they try to locate their farm in an area that has a relatively low incidence of naturally occurring aphids. So the places where they grow the best seed potatoes in Ontario are way up north, up by New Liskeard, Thunderbay, places where the incidence of aphids, especially the Green Peach aphid, is quite low. They also monitor the populations of aphids with aphid traps, so they'll know when they show up. Generally they reach their peak populations in late August. Sometimes if the aphids don't reach their peak until later, you can get a good crop of potatoes. But sometimes they arrive early and we try and kill down the tops of the potatoes so that they just can't transmit. And, like it or not, there's a lot of insecticides used. Some people put systemic insecticides on the potato plant. The potatoes take up the systemic insecticide. The aphid sticks his feeding mouth part in there and takes one sip and falls over backwards. But that still doesn't stop the total transfer of viruses. The mere fact that that aphid landed on the plant and took one little suck before he got killed could have transmitted viruses to that new plant. So what has been done here by Agriculture Canada to try to furnish good quality seed to growers is an elaborate program of certified seed. And we'll briefly touch on that. (Next slide, please.)

The three tubers on the left are infected with spindle tuber viroid. A viroid is even smaller than a virus. This is a healthy tuber on the right. Sometimes we don't see the symptoms in the top parts of the potato plants, so we have to use indicator plants. (Next slide.) These are tomato plants and we took leaves of a healthy potato plant and rubbed it on this tomato plant. Nothing happened. Then we took leaves of a spindle tuber infected potato plant and rubbed it on the leaves of the other tomato plant. As you can see, it's very susceptible. So sometimes we have to use other crops as an indicator line to detect the presence of viruses or viroids. (Next slide, please.)

The one thing we do in our seed plots is we rogue. By roguing I mean we pull out and destroy any diseased plants, so hopefully the insects will not transmit the virus from the infected plants to other ones in an adjacent row. (Next slide.) Another way that we try to check out our seed is by eye indexing. This is done in the greenhouse during the winter. We have lines of potatoes which we wish to maintain as free of viruses as possible. So what we do is we scoop out one eye from the tuber and we grow that one eye in the greenhouse. (Next slide.) You can see that we plant the eye and we also number the tuber that it came from. Now let's take, for example, #3, there. We'll take the eye of #3, plant the eye and put the #3 tuber back in cold storage. Then we'll grow that little plant that comes from that eye (next slide) in the greenhouse and take a look at the plants. (Next slide.) Here we've got a healthy plant on the right and a virus-infected one on the left; if the one on the left happened to be #3, next spring when we're cleaning out our storage we would throw away the #3 tuber because we know it has viruses in it. So the only tubers we plant in the field are those which show no symptoms in the greenhouse eye indexing test. Now, this is an old-fashioned method and it works pretty good....Agriculture Canada and other researchers have come up with more sophisticated chemical techniques that they can take a little piece of a potato and analyze it chemically for the presence of viruses. And there's going to be a lot more of those labs built....

(The 1984 Fall Harvest Edition, pages 40-41)

ISOLATION DISTANCES FOR TOMATOES

by Jeff McCormack, Ph.D.

Most seed saving guides lack specific information about the minimum isolation distances for predominately self-pollinated crops such as tomatoes. Some specific guidelines are needed because frequent natural cross-pollination (NCP) of tomatoes may occur when two or more varieties are grown side by side in a garden under certain conditions. Even a small percentage of NCP over a number of years could eventually cause the loss of one or more characteristics which are unique to a particular open-pollinated variety.

Prompted by this concern, I am proposing some guidelines for isolation distances specifically suited to home gardeners who wish to keep their varieties pure. In preparing these guidelines I reviewed the scientific literature on tomato pollination, talked with three tomato breeders in different areas of the country, talked with people who have saved their family heirloom seed for many years, and have made my own observations of bee pollinator activity on tomato blossoms. The guidelines I propose are for both modern and old varieties of tomatoes, and further refinements may be necessary pending the results of experiments I have planned using genetically marked tomato lines.

Isolation requirements may be different for commercial growers than for home gardeners saving their own seed. Commercial growers or breeders of tomato seed may separate varieties by as little as 10 feet, primarily to avoid mechanical mixing of the seed crop. Such plantings are often made in areas which are bee-poor due to pesticide use or lack of suitable habitat. NCP may further be reduced by planting the seed crops in large blocks with a barrier crop in between. Furthermore, seed may be collected from plants in the center of the blocks and not the edges.

In contrast to large block plantings, many home gardeners tend to plant row crops of many varieties in a small space. These crops are frequently visited by wild bees (halictid bees, such as sweat bees) and bumblebees in search of pollen. This situation may contribute to a high frequency of NCP in bee-rich areas in crops that are primarily self-pollinated. Factors leading to higher NCP are explained below.

The amount of NCP of tomatoes is a function of a number of variables: (1) wind movement; (2) variety characteristics such as style length; (3) environmental variables affecting style length such as light intensity, day length and carbon-nitrogen ratio; (4) type of bee pollinator and its behavior on the blossom; (5) isolation distance; and (6) presence of other pollen-producing plants in the area of the seed crop.

Wind movement of the blossom tends to increase the amount of self-pollination. Even though tomato pollen may be blown some distance, NCP by this method is probably of little significance.

Tomato varieties having long styles (pollen-receptive organs) are more likely to be cross-pollinated by bees than varieties with short styles. If the length of the style exceeds the length of the anther cone (pollen-producing organ), NCP by

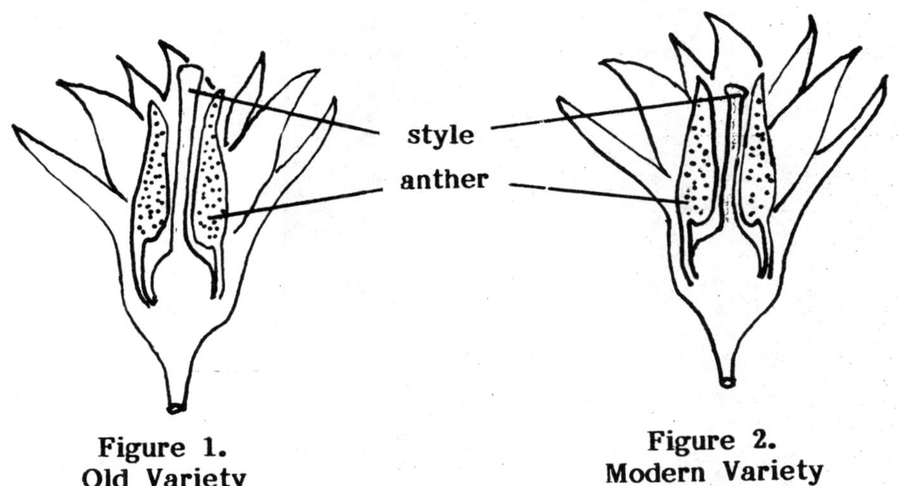

Figure 1.
Old Variety

Figure 2.
Modern Variety

bees is more probable, and probability increases as style length increases (see Fig. 1). Gardeners attempting to preserve old varieties need to be aware of this point because many older varieties generally have longer styles than modern varieties. Most modern varieties have styles equal in length, or shorter than, the anther cone (see Fig. 2). Our modern varieties were derived originally from wild tomato ancestors (primarily from Ecuador and Peru) which relied on bee pollination to a large degree. As these wild types were transported out of their center of origin to new geographic areas, the absence of their usual bee pollinators resulted in selection for variants which had shorter styles and an increased capacity for self-fertilization. Although style length is genetically determined, environmental conditions may cause style length to increase, thereby affecting the probability of cross-pollination.

Another factor affecting NCP is insect activity. Generally, tomato flowers are unattractive to bees if other pollen sources are available; however, in some bioclimatic regions of the U. S., bee visitation of tomato flowers may be quite common even in the presence of other pollen sources. Such a situation exists in regions of California and parts of the Mid-Atlantic region. For example, in parts of Virginia I have observed and photographed bumblebees (Fig. 3) and halictid bees (Fig. 4) such as sweat bees collecting pollen from tomato flowers. Bumblebees tend to vibrate the flowers while halictid bees appear to chew the anthers to get at the pollen. In terms of their behavior and position on the flower, halictid bees seem more likely to cause cross-pollination than the bumblebees, but this has not been fully investigated.

Close interplanting of two tomato varieties may typically produce 2-5% NCP; however, factors such as long style length, frequent visitation of tomato flowers by bees and suitable environmental conditions may produce much higher NCP values. Various studies have reported values of 12, 15, 26 and 47% NCP values in interplanted tomatoes. The wide range of results reflects the influence of different methods and variables used in these studies; however, it is clear that NCP values can be high under the right conditions.

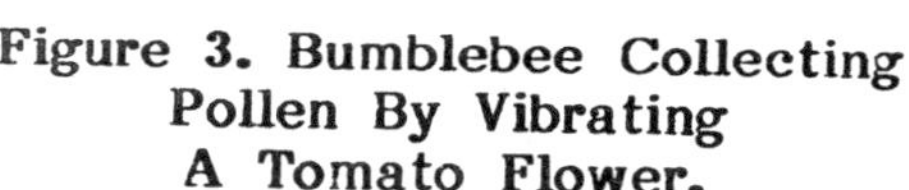

Figure 3. Bumblebee Collecting Pollen By Vibrating A Tomato Flower.

Figure 4. Wild Bee (Halictid Bee) Collecting Pollen From A Tomato Blossom.

What does all this mean for gardeners wishing to save their own open-pollinated tomato seed where there is high bee activity on tomato blossoms? Modern tomato varieties (style length equal or less than anther length in most cases) should be separated by a distance of approximately 10 feet to give a high degree of purity. Older varieties may require a 20 to 25 foot isolation distance. In both cases, a tall barrier crop or pollen-producing crop such as squash should be planted between different varieties of tomato to attract the bees away from the tomatoes. (Squash blossoms will present a lot of pollen to bees in the morning, but not in the afternoon.) In most cases, these recommended isolation distances will give <u>average</u> purity values of approximately 99 to 99.5% or better. Because occasional outcrossing may occur at large distances, plants used for stock seed may require an isolation distance of 50 to 150 feet or more in some localities.

The relationship between isolation distance and NCP is geometric, not linear. Thus as isolation distance increases, the amount of NCP falls off rapidly. A study by Currence and Jenkins (1942) illustrates this point very well (Fig. 5). Therefore it is evident that even a separation of a few feet between varieties in a small garden will greatly reduce NCP of tomatoes even though minimum recommended isolation distances cannot be achieved. Also NCP can be reduced or eliminated by taking advantage of different blooming times of early and late varieties.

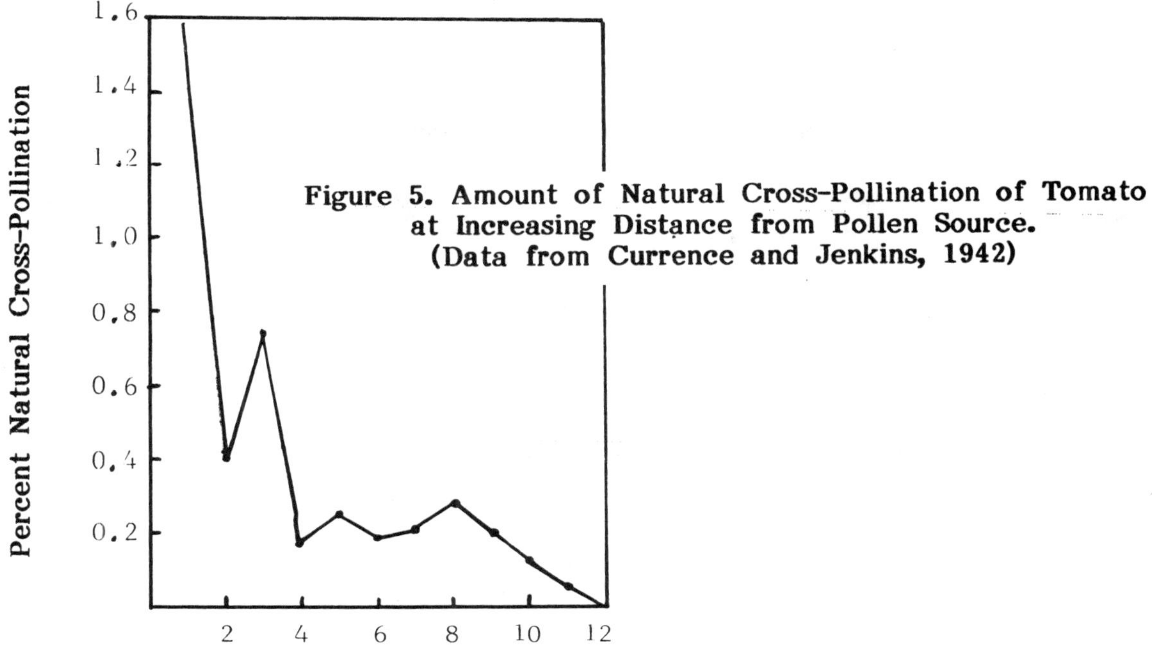

Figure 5. Amount of Natural Cross-Pollination of Tomato at Increasing Distance from Pollen Source. (Data from Currence and Jenkins, 1942)

Number of Rows (Approximately Six Feet Apart, 1.8 Meters) from Pollen Source.

Gardeners should not be discouraged from saving their own seed because of a small amount of NCP. A small amount of NCP could eventually improve a variety, but it could just as easily cause the loss of quality of a variety. If you are trying to preserve a variety in its purest form, then isolation distance becomes very important. Although a small amount of NCP may not be a problem one year, its effects may be additive and detrimental to preservation efforts in the long run. The goal is not just to save the variety from year to year but for generations to come.

Selected References

Bennet, J. 1983. A tomato blossom for all seasons. Horticulture 61: 53.

Currence, T. M. and J. M. Jenkins. 1942. Natural crossing in tomatoes as related to distance and direction. Proc. Am. Soc. Hort. Sci. 41: 273-276.

Rick, C. M. 1949. Rates of natural cross-pollination of tomatoes in various localities in California as measured by the fruits and seeds set on male-sterile plants. Proc. Amer. Soc. Hort. Sci. 54: 237-252.

Rick, C. M. 1950. Pollination relations of *Lycopersicon esculentum* in native and foreign regions. Evolution 4: 110-122.

(The 1984 Winter Yearbook, pages 247-250)

INSECT CROSS-POLLINATION OF OKRA
by Ted Gibbs

Okra, Abelmoschus esculentus (formerly Hibiscus esculentus) is self-pollinating; however, bees are active in cross-pollination. Some crossing in our heirloom okra varieties has been observed. This observation set this gardener in motion to seek advice from the professionals. These professionals would be the first to agree that this is not the last chapter in this story. Much scientific investigation needs to be done on okra and its environment.

A survey of distance requirements to isolate okra seed crops from insect cross-pollination provided the following results:

University of Missouri, Department of Horticulture	1 mile min.
Plant Breeder from Florida	1/2 mile or preferably 1 mile
U. S. Dept. of Agriculture (1905)	1/4 mile
State of Oklahoma,* Registered or Foundation Seed**	1,320 feet
State of Oklahoma,* Certified Seed**	825 feet

*Oklahoma is one of the few states with a standard for okra seed production. Texas has a standard (unable to get figures for this publication) which is more lenient than Oklahoma.

**The commercial seed certification classes include: Breeders Seed, Foundation, Registered and Certified. Breeders Seed comes from a small increase that the breeder feels is a true genetic representative of the variety. Foundation Seed can produce Registered, and Registered can produce Certified. All classes are controlled and protected.

A survey of data on okra cross-pollination under field conditions provided the following results:

University of Missouri, Department of Horticulture	4% to 18%
U. S. Department of Agriculture, estimates as high as	42.2%
Other reports from U. S. Department of Agriculture	4% to 32%

Unless a gardener has a small, isolated island for each okra variety, isolation from insect cross-pollination by distance is not feasible for the heirloom okra gardener.

A basic understanding of sexual reproduction in okra is needed before presenting our workable insect cross-pollination isolation method. Ideally, okra is self-pollinated and fertilized within the blossom without outside (the blossom) interference. This is a very complex process which I hope to simplify. The following photographs will hopefully aid in the identification of the principal blossom parts in this reproductive process.

Style - Cylindrical-shaped part in center protruding up from the blossom base or ovary where the female egg is located. The style is white, approximately 1/4" in diameter and 1" in length.

Stigma - Female sexual part located on the upper end of the style. Very dark maroon color. It becomes sticky on the surface when receptive to pollination.

Anther and Pollen - Anther is the male sexual part (not readily visible to the naked eye); however, the bright yellow pollen held by the anther is visible around the outside length-wise center of the style. At maturity (Anthesis) the anther ruptures, distributing the pollen, and automatic self-pollination may occur.

Open Blossom

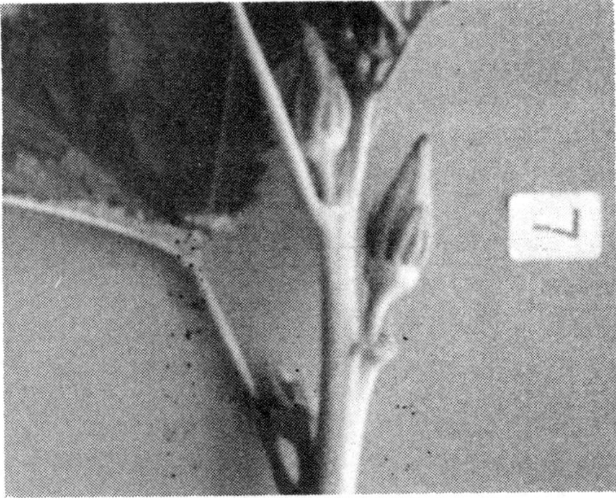

Blossom Bud

Some Petals Removed To Show Style

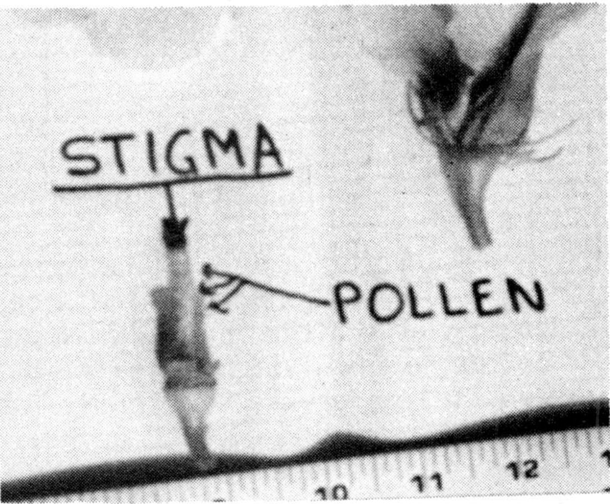

Stigma And Pollen (On The Style)

A simplified version of events at sexual maturity (Anthesis) is as follows:

The okra blossom opens, the stigma develops a sticky surface and the anthers rupture releasing the bright yellow pollen. Pollination occurs with the arrival of pollen on the sticky stigma. The pollen grains may be numerous and may have come from the same blossom or a different blossom of the same variety (via flying insect) or a blossom of a different variety (via flying insect) or all of these. As pollen lands on the stigma, a tube growth is initiated from the pollen grain down through the style to the female egg in the ovary at the blossom base. While many pollen grains could be present on the stigma and even initiate tube growth, fertilization (male parentage) will be accomplished by the single pollen grain whose tube completes the journey to the egg first and transfers its genetic make-up via the tube. If the successful pollen grain is from the same flower or variety, the desirable self-pollination and fertilization occurs. If the successful pollen grain is from a different variety, undesirable cross-fertilization (hybrid seed) occurs. This fertilization affects the current generation seed only. As these seeds are grown out the subsequent season, the genetic make-up of the seed may be apparent in the plant characteristics. If you fail to identify the hybrid plants during their first grow-out (generation), they will probably be apparent in a subsequent grow-out.

Heirloom okra seed can be preserved by isolating the blossoms of selected seed plants from insects, thereby assuring self-pollination and fertilization. There are several methods to accomplish insect isolation, but none appear so appealing as "bagging" practiced by our very helpful friend at the West Kentucky Research and Education Center (University of Kentucky) in Princeton, Kentucky. A 4" by 5.5" bag made of very light, thin synthetic material with drawstrings at the top for closure, it resembles those used for certain (roll-your-own) tobacco several years ago. We don't think these bags are available commercially, but they should be easily made or some improvised cover could be substituted....(Small muslin bags with drawstrings are available from Crystal Springs Packaging Co. -- Kent)....Avoid cotton material that becomes heavy with dampness. We want to simply cover the okra blossom before it opens up to keep insects out while minimizing loss of sunshine, air circulation and visibility of the bud blossoming. (Kent suggested discarded ladies nylon hose.... think it over). We need to bag or cover the okra blossom bud the day before Anthesis (blossoming). When you see the bud blossomed inside the bag, mark that seed pod by loosely tying a string or ribbon around the blossom stem below the bag. Leave the bag on for the next full day, and then the following day remove the bag or cover from that blossom while ensuring that the seed saving marker is in place....the process is over.

If you miss bagging a bud before it opens, simply add that pod to your food pod harvest in six to eight days....that goes for any pod without a marker. Pollen is no longer viable after one day at room temperature....there could be some pollen on the removed bag....be safe and let the bag rest one day in the house (or equivalent) before reuse. About five weeks are required from blossoming to the production of a dried pod with fully viable seed. Therefore, count back five weeks on the calendar from your area's last average frost-free date. Stop bagging on this date and use all new pods for food harvest. Save all marked pods for seeds.

Missouri University Department of Horticulture advises that if we methodically isolate our seed from cross-pollination/fertilization and systematically eliminate rogues/off-plants from our heirloom okra grow-outs, we should see results in two years. The continuation of this program should produce Breeder Seed Quality from our heirloom okra gardens. I'm sure that each of you agree that anything less is not acceptable. Some of the okra variety characteristics which could be recorded and monitored are as follows:

--Pod color, diameter***, length***, axis straight or curved***, number of ridges/angles/suture, vitamin C content***, yield per plant number and weight***.

--Blossom petal blotch***, base color***, venation color***, stem color***.

--Stalk height***, branching habit.

--Leaf stem color***, degree of lobing.

--Calyx or sepal (blossom protective cover) color***.

--Seed color, weight (seed per gram).

--Over-all plant: resistance to drought, rain, insects, disease; degree of spines or smoothness; sensitivity to day length; sensitivity to pod seed saving; annual or perennial behavior.

--Time required from planting: to blossom buds, to blossom opening***, to first pod production***, to last pod production.

***Identified by the U. S. Department of Agriculture in 1978 as an inherited characteristic in okra. I am not suggesting that you watch all of those characteristics. I do want to give you a choice. Some characteristic changes are much easier to see than others. You will be limited to your records of characteristics unless you are able to improve them.

SELECTED REFERENCES

Welch, James R. 1981. Fundamentals of Plant Genetics and Breeding. John Wiley and Sons, New York.

Beattie, W. R. 1905. Okra: Its Culture and Uses. U. S. Department of Agriculture. Farmers Bulletin No. 232.

Martin, Franklin W. and Ruth Ruberte. 1978. Vegetables for the Hot, Humid Tropics Part 2. Okra, Abelmoschus esculentus. Science and Education Administration. U. S. Department of Agriculture.

Calvin, Clyde L. and Donald Knutson. 1983. Modern Home Gardening. John Wiley and Sons, New York.

The Staff of Organic Gardening Magazine. 1978. The Encyclopedia of Organic Gardening. Rodale Press, Emmaus, Pennsylvania.

Minnich, Jerry. 1983. Gardening for Maximum Nutrition. Rodale Press, Emmaus, Pennsylvania.

Glimn-Lacy, Janice and Peter B. Kaufman. 1984. Botany Illustrated. Van Nostrand Reinhold Company, New York.

SELECTED TECHNICAL DATA CONSULTANTS

Dr. Art Gaus, Vegetable and Fruit Specialist, University of Missouri, Columbia, Missouri.

R. Mark Hurley, Research Horticulturist, University of Kentucky, West Kentucky Research and Education Center, Princeton, Kentucky. Member, American Society for Horticultural Science, Alexandria, Virginia.

Tom V. Williams, Plant Breeder, Homestead, Florida. Member, American Society for Horticultural Science, Alexandria, Virginia.

(The 1984 Fall Harvest Edition, pages 28-32)

LONG TERM STORAGE OF GARDEN SEEDS

by Bruce Bugbee

By following simple principles, seeds of all species can be stored for several years with no loss of germination and only minimal loss of vigor. Vigor can be defined as the ability to germinate rapidly with good resistance to disease. The vigor of a seed is lost in stored seeds well before the seed dies completely.

The two greatest enemies of stored seeds are high temperature and high moisture. A rule of thumb is that the storage life of a seed is doubled for each 10 degrees F. decrease in temperature and/or for each 1% decrease in seed moisture. If seeds are stored in moist, warm or fluctuating environments, their vigor and germination will decrease rapidly.

The best way for a homeowner to store seeds is to get them as dry as possible at a low air relative humidity and then put them in a moisture-proof container. There are only two common materials that are completely moisture proof, metal and glass. A glass jar with a rubber gasketed lid screwed on very tight makes a good storage container. Another good container is a coffee can with its heavy plastic lid taped on with electrical tape or several layers of masking tape. Lightweight plastic bags are not moisture proof and do not make good storage containers.

After moisture has been excluded from the seeds, they should be kept in a cool place. The colder the better, but the temperature should remain relatively constant. The very best place is in a freezer; the next best place is in a refrigerator. Dry seeds will retain their vigor in a freezer for several decades; in a refrigerator they will stay healthy for about ten years. Since most home storage will be for less than five years, the coolest spot you can find is quite adequate. If the container is really moisture proof like it should be, it can be kept in moist soil underneath a house or in the root cellar where the temperature will fluctuate as little as possible.

Without protection, seed vigor will begin to decrease long before the seeds die. But with proper storage, seed vigor will be preserved about five times over the natural storage life.

(Bruce Bugbee, Department of Vegetable Crops, Hunt Hall, University of California at Davis, Davis, CA 95616, is a graduate student in vegetable seed physiology at UCD. Part of his research has been with seed vigor testing and long-term seed storage. He's a friend of and co-author with Ken Kern, and I thank Ken for putting me in touch with him. Seed Savers Exchange is better because of Bruce's interest in us and I thank him for his help. The following are parts of his letters including some questions and answers he agreed to do with me for your information:)

BRUCE: Glad you can distribute this seed information. I am often disturbed by the mistaken advice that gets published....You should purchase Principles of Plant Propagation by Hartman and Kester. It costs about $16, but is very complete....I think you're doing a really valuable thing by encouraging people to experiment with seed production and new varieties. The big seed companies would probably take what you're doing lightly because of your lack of technical expertise, but to my mind when your heart is in the right place, you can pick up the technical information....A book you would find useful (if you can get ahold of it) is Vegetable and Flower Seed Production by L. L. Hawthorn and L. H. Pollard, published in 1954; it is now out of print....

Q. - Can seeds be dried in an oven with no damage?

A. - No. Damage begins to occur above 96 degrees F. and it is very difficult to keep an oven below 100 degrees F. The commercial companies dry seed at 90-95 degrees F. Drying on a screen in the sun when the temperature is less than 96 degrees F. is a good method. The hygroscopic material will do the rest. If seeds are dry enough to break rather than just bend when folded, they are ready to be put in storage.

Q. - Does freezing ever damage dry seeds?

A. - Yes. If the seeds are not dry enough, water expands and breaks cell walls. However, if seeds are at the proper moisture for storage (less than 8% seed moisture) they can be stored in liquid nitrogen (-197 degrees C.; -322 degrees F.)....

Q. - I read in one of Rodale's books that unusually strong plants can result from planting really old seed. Is this true?

A. - No. The best seeds at harvest time will keep the longest in storage, but their vigor never increases. Vegetable seeds are at their prime as soon as they reach maximum dry weight on the mother plant.

Q. - If seed of low vigor is grown, will the seed kept from such plants have full vigor again?

A. - No. Proper storage only maintains the vigor that was present when the seeds were harvested.

Q. - How well can people here expect seeds from other countries to do?

A. - Seeds will do fine, if kept dry and cool. Plants, however, are very sensitive to changes in daylength.

(The 1978 Seed Savers Exchange, pages 25-26)

Powdered Milk vs. Silica Gel--Their Effectiveness as Drying Agents

(Information obtained from Bruce Bugbee and Lauren Hastings, Plant Science Department, UMC 48, Utah State University, Logan, UT 84322)

To research this question, we dried equal volumes of powdered milk and silica gel in an oven at 70 degrees C. (160 degrees F.). After 48 hours, the samples were weighed and set on a shelf in our lab which was at 30% relative humidity. After four days, both dessicants had reached equilibrium with the air moisture: the powdered milk absorbed 14.3 mg water/cm cubed (.235 g/in cubed); the silica gel absorbed 165 mg water/cm cubed (2.71 g/in cubed); the silica gel was 11.5 times more powerful in absorbing moisture than powdered milk....We also calculated how much dessicant would be required to dry a one-ounce packet of seed from 15% to 5% moisture (a typical requirement). The answer: almost a full cup of powdered milk, but only a tablespoon of dried silica gel. To dry 16 one-ounce packets of seed, a gallon jar of powdered milk would be required. The same task could be accomplished with only a cup of silica gel....When we did the whole experiment again to be sure that we had all the numbers correct, we got the same results. Powdered milk is a poor seed drying agent. Also powdered milk taken right out of a new box before drying was already half full of water and could absorb only 8 mg water/cm cubed....

Silica gel (SG) is an excellent moisture absorber and is available in camera shops, drugstores and hardware stores. SG is normally white, but most SG is sold with a dye treatment (cobalt chloride) which is what changes color with the humidity. It is a deep blue when completely dry and gradually changes towards light pink as it picks up water vapor from the air. It can be redried indefinitely at 200 degrees F. in about eight hours. Occasional stirring speeds up the drying of a large batch. Less time is needed for small batches (less than one pound). It will turn black if dried at too high a temperature....Without going into exact calculations, I estimate that for already dry seed (4-7% moisture) only a small quantity of SG is needed in a jar (equal to about 5 or 10% of the seed). For less dry seed (7-12% moisture) about equal parts of seed and silica gel should be put together. Since this gets expensive, it is better to dry seeds below 7% moisture before storage in a jar....

(The 1982 Winter Yearbook, page 88)

SEED DRYING: AN EXPERIMENT
by John Rahart, DDS

Here is the initial report from the research that we have been doing. Dr. Bruce Bugbee from Utah State did the analytical moisture determinations. The research is by no means complete, but at this stage we have enough information to design a simple cookbook-type recipe which should provide people who do not have access to sophisticated scientific apparatus to dry and package seeds for storage, with enough information to get a longer and more uniform shelf life. It should not be unreasonable to hope for 20-30 years reliability from this method. The system is simple and easy to use.

We utilized gallon mayonnaise jars (available from restaurants, school cafeterias and other places where institutional cooking is done), but any other airtight containers will do, such as canning jars, etc. We put our seeds into paper coin envelopes (make sure the envelopes are not the moisture-proof type like those used by most commercial seed companies). Then we determined the total weight of the seeds and packets which were to go into the container and also weighed out an equal amount of silica gel. The seed packets and silica gel were then sealed in the gallon jar.

It appears that both the small seeds (tomatoes, peppers, etc.) and also the large seeds (peas, beans, corn, etc.) reach moisture levels that are excellent for storage after 7-8 days. At this point the seeds are dry enough to substantially increase their storage life without getting into the potential dormancy problems experienced when large-seeded legumes are dried to under 5% seed moisture.

The dried seed packets are then removed from the drying chamber jar and placed into another moisture-proof container (canning jars, heat-sealable laminated bags such as the National Seed Storage Lab uses, etc.) and stored <u>without</u> any desiccant. Since we are not sure how fast the very dry seeds will reabsorb moisture, it would be best to minimize the amount of time the seed packets are exposed to room air during transfer.

Dr. Eric Roos from the National Seed Storage Lab has suggested that it is a good idea, when removing seeds from storage which have been dried to low moisture levels, to let them stand in "room air" for a few days before planting. By allowing them to slowly pick up moisture, we lessen the shock they would receive if we took them from very dry storage and immediately put them into a very moist planting environment.

Our major interest in designing and researching this project is to provide the home seed grower with the ability to increase the viability of their seed storage methods with simple and relatively inexpensive means. Since the silica gel can be redried and used over and over, it seems that we have achieved this. We intend to duplicate the experiment again using many more species of seeds. We also intend to use new seed this next time. We are concerned that some of the older seed we used may have reabsorbed some surface moisture before the test began, making it appear to dry out more quickly than it actually should. As the research progresses you will be informed of any improvements to the system which may enhance its effectiveness.

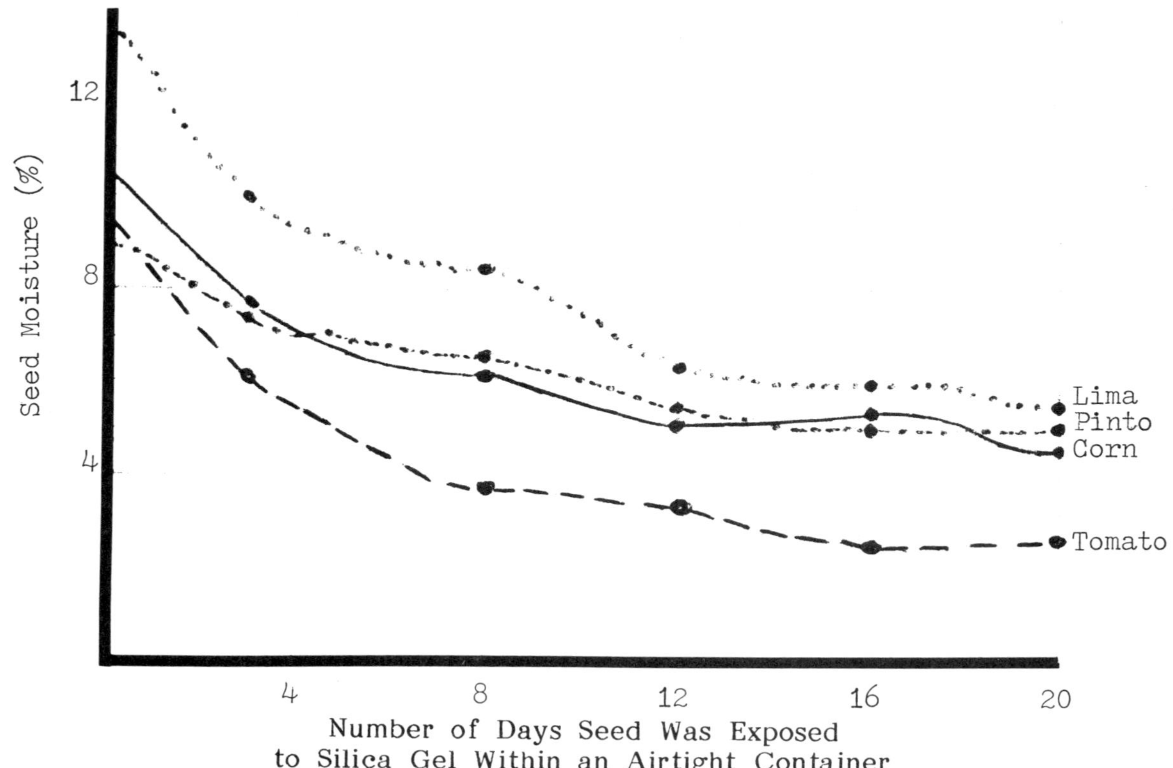

PERCENTAGE OF SEED MOISTURE

Day #	0	3	8	12	16	20
Corn	10.56	7.75	6.2	5.2	5.7	4.5
Beans (Pinto)	8.96	7.5	6.2	5.3	5.1	5.0
Beans (Lima)	13.40	9.8	8.46	6.35	6.15	5.5
Tomatoes	9.47	6.08	3.7	3.56	2.45	2.49

(Generally speaking, seed is dry enough for frozen storage when it reaches a moisture level of less than 8%. However, beans and other large-seeded legumes can go permanently dormant when dried to under 5% moisture. So a safe range for such seeds and also an excellent storage moisture is between 6% and 8%...."soft" seeds (such as peppers and tomatoes) and small seeds dry down to a much lower moisture because their mass is so small. Nobody seems to know if small seeds will be damaged when stored for long periods at a really low moisture level of, say, 2-2.5%. But it is generally agreed that they will be "safe" if we don't take them down to under 3%. So those are essentially the optimum moisture levels that we are trying to achieve. -- Kent)

(The 1982 Fall Harvest Edition, pages 20-21)

KEEPING SQUASH SEEDSTOCKS PURE

by Mahina Drees and Karen Reichhardt

To assure that each kind of squash maintains its own genetic characteristics or "purity," do not plant seedstocks of cross-compatible varieties in plots near one another. Varieties of the same species (listed below) can certainly fertilize one another via pollination by bees, which can carry squash pollen as much as a mile, though bees transport most of it only sixty to seventy yards. When cross-pollinated, certain Cucurbita mixta and C. moschata varieties apparently produce fertile offspring, so these two species should not be planted together if "seed saving" is the goal. The following chart provides a key to seed characteristics of the various species.

Species Name	Seed Description	Seed Illustration	Names of Common Types
Cucurbita pepo	oblong and white, including margin, smaller in modern squash, i.e., zucchini		pumpkin zucchini yellow crookneck patty pan acorn
Cucurbita maxima	thick and brownish, smooth, thin cellophane coating, cream margin		banana turban white giant hubbard
Cucurbita moschata	dark beige margin, not wrinkled, smaller		big cheese butternut
Cucurbita mixta	white, wrinkled with cracks in middle, pale margin, cellophane coating, sometimes brown tinge		striped cushaw silver-seeded
Cucurbita ficifolia	black or buff colored		Malabar gourd fig-leaf gourd

(The 1984 Fall Harvest Edition, page 36)

HOW TO TELL THE FOUR SQUASH SPECIES APART
by Richard Grazzini

I've scratched my head quite a bit trying to come up with an easy way. Sometimes I think referring to the list Kent prints in the Seed Saving Guide is the simplest. If you have a question about which species a variety is, hand-pollinate everything! Don't risk crossing out.

Telling the species apart by the seeds is not easy. I think that I can do it most of the time, but still cannot separate C. mixta from C. moschata. And some C. maxima seeds are difficult to tell apart from some C. pepo seeds. There are an awful lot of differences between varieties within species.

My recommendation? Grow the variety. Look at the leaves, the stem and the peduncle (or fruit stem). You should then be able to classify the variety as C. pepo, C. maxima or C. moschata/mixta.

Foliage: Touch the leaves and stems. If they are prickly, you've got Cucurbita pepo. The other three species have soft hairy stems.

Stems: If the stem is soft and round, it's Cucurbita maxima. The others (C. pepo, C. moschata, C. mixta) have hard angular stems.

Peduncle/Fruit Stem: The peduncle is the part of the stem that attaches the fruit to the vine. This is what I use as a final way to identify the species. It also works when the fruit are off the vine.

C. maxima has a soft peduncle that is swollen with a soft corky substance. The photo of a Blue Hubbard peduncle is typical of C. maxima.

C. pepo, C. moschata and C. mixta all have hard peduncles somewhat grooved and somewhat angular.

C. pepo peduncle is sharply grooved, and sharply angular. The photo is of a Connecticut Field Pumpkin. There's a cut piece of another peduncle to show the sharp angles and grooves on the pumpkin.

C. moschata and C. mixta both are smoothly five-angled at the peduncle. The C. moschata peduncle flares (widens a lot) at the fruit attachment. The peduncle of C. mixta does not flare (well, not too much). Waltham Butternut is a C. moschata; Green Striped Cushaw is a C. mixta. I have to admit, I still don't feel confident classifying a new variety as either C. mixta or C. moschata. However, very few mixta types are available....and if you guessed C. moschata each and every time, you would be right nearly every time.

Flower: Although it doesn't occur each and every time, very often the sepals (the green part right below the yellow flower petals) of C. moschata are leafy and expanded, rather than rounded. Any of the Butternuts are good to look at for an example of a leafy sepal.

(The 1984 Fall Harvest Edition, pages 33-35)

ADVICE TO PEOPLE MAINTAINING LINES OF CORN

by Mark P. Widrlechner, Ph.D.

Every time that a variety is grown out to renew the seed, it changes. Most corn varieties that were being grown before hybrid corn was developed (including most heirlooms maintained in the SSE) were genetically variable and subject to change. Many factors influence the kind of genetic changes that take place in a variety when it is grown out. Some of the factors causing change are under the control of the grower and others are not. By recognizing these factors, the grower can use them to his/her advantage in the improvement and maintenance of corn varieties.

Forces That Act to Change Varieties from One Generation to the Next

Besides selection (the grower's choice of which seed to keep), differential reproduction, population size and crossing with other varieties are major factors which can change the make-up of a variety from one generation to the next. Assuming that the grower can control cross-breeding with other varieties by following good isolation techniques (see SSE 1982 Fall Harvest Edition, pp. 18-19), let's look at the other two factors, differential reproduction and population size, in more detail.

Differential reproduction occurs when each plant in a population (field or plot of corn) contributes different numbers of offspring to the next generation. There are many factors influencing differential reproduction. Plants may produce different quantities of pollen or seed, because of their genetic make-up (genes partly control fertility, disease, stress and insect resistance) or because of environmental factors (one part of the field being better than another). Even if two plants produce the same amount of pollen or the same number of silks, they may still contribute different numbers of offspring to the next generation. One plant may silk at a time when little pollen is in the air or it may shed pollen before any silks are out. These plants will contribute less to the next generation. Over time this tendency acts to make shedding and silking times more uniform within a corn variety.

One other factor relating to differential reproduction is storage. If seed is stored for long periods of time, the seed from one plant may not keep as well as the next. By following the best storage practices (see SSE 1983 Winter Yearbook, pp. 204-212), the grower can minimize this problem.

In general, to minimize the effects of differential reproduction, it is best to plant varieties in a fairly square block of uniform ground and to keep an equal number of seeds from each ear chosen to provide the seed for the next planting.

The number of plants in a field affects the genetic stability of a variety. Over time, if too few parent plants contribute offspring to the next generation, then the crossing of related plants results in a decrease in vigor (inbreeding depression) and a loss of variation. If a population has a particular gene which occurs about one time in one hundred and only 20 plants are grown out to produce seed for the next generation, then chances are better than two in three that the rare gene will be lost. Although a grower may have limited space and effort, it is still wise to start each year with at least 30 to 50 ears per variety maintained, with every ear contributing an equal amount of seed. This may limit the number of varieties that anyone may wish to maintain, but keeps loss of genetic variation to a minimum. It is also a good idea to include the range of varietal variation in those 30 to 50 ears that are chosen, to lessen the chances that selections are related.

Notes on Selection

Although many members of the SSE are interested in keeping heirloom corn varieties as they are (the section above shows what a challenge that can be), some members may also be interested in improving or changing their varieties to suit their needs. Some characteristics of a variety are easier to change through selection than are others. Assuming that there is variation in the expression of these traits in a variety to begin with, traits that can be easily selected for include earliness/lateness, plant color, and ear height. More difficult to select are seed color (see chapter on color genes in Corn and Corn Improvement, 2nd ed., George Sprague, ed.) and disease resistance, unless the disease is present throughout the field. Most difficult (because of evaluation procedures and environmental variation) are insect resistance (unless there is uniform inoculation), flavor, nutritional quality and yield.

In general, it is best to select not just on the basis of ears, but by looking at the whole plant. It is not rare for a borer-infested, half-broken, diseased plant to produce as good an ear as a sturdy, healthy plant, but the poorer plant is likely to have offspring of inconsistent quality. Even better than selecting on the basis of one plant is the principle of basing selection on the overall performance of related plants. A simple way to do this is to plant a field such that each row comes from a different ear, making sure to reserve some seed from each ear to use in planting next year's field. At the end of the season, each row is evaluated and the seed that was set aside from the ears that produced the best rows is planted the next year. At the end of the second year, the grower chooses ears from the best plants in the field and starts the cycle over again. In choosing plants for the next cycle, the grower should avoid selecting too many plants from one row. By having related plants together in rows, it is easier to maintain genetic diversity, because the grower can make sure not to select too many plants that are closely related.

Here are a couple of other bits of advice about selection. 1) Some plant characteristics are not easy to see by the time the plants dry down (first brood corn borer damage, leaf color, time of pollen shed). With colored tags or ribbons around the main stalk, it is simple to mark plants with desired traits and find them at harvest without lots of record-keeping. 2) Selection is more effective if undesirable plants can be stopped from shedding pollen and crossing with the other plants. Some unwanted traits, such as disease susceptibility, insect feeding, poor plant form and early or late silking, can be seen before pollen is shed. Tassels can be removed from these plants before they have a chance to contribute pollen to the next generation.

The maintenance and improvement of corn varieties may not be a simple task, but it can be a fascinating and fulfilling pursuit. I have been working with corn for the last six years and the more I see, the more challenges there seem to be. The last few years I've been looking at the variation in decorative corn varieties including colored popcorns, flint and flour corns with the goal of producing higher quality, disease and insect resistant types. If you have any questions or comments about my ramblings or the subject in general, please feel free to drop me a line at: Rural Route 6, Ames, IA 50010.

(The 1984 Winter Yearbook, pages 250-252)

BACKYARD PLANT BREEDING
by Richard Grazzini

QUESTION: How do I hand-pollinate squash, so I can keep several varieties in the same garden pure?

To make them come "true," you must self-pollinate them by hand. First, in the evening, find both male and female flowers which are unopened, but firm and yellow. These will open overnight unless they are sealed. Wilted yellow flowers have already opened--don't use them. Look at the sketch of the two types of squash flowers--the female flower has a small "baby" squash below it. The flowers must be sealed so the bees don't do your pollinating for you--and bees are early risers! I usually seal with 2" masking tape placed around the top third of the flower. Other people use rubber bands, wire, or put small bags over the whole flower. I use tape because I was taught that way--and it's easy.

The next morning--wait until the dew dries if you can--pick a male flower. Take the tape off of the female flower. It should open by itself once freed of the tape--don't worry if the petals tear a little--you'll use more tape to reseal. Remove the petals from the male flower and use its stem as a handle to dab the pollen-covered anthers all over the stigma of the female flower. Reseal the now-pollinated female flower and make it bee-tight. Honeybees and squash bees (look like small bumblebees) can squeeze through tiny openings and cross-pollinate your laboriously hand-pollinated flower. Now, tag or label the flower to keep track of it during the summer.

You can make squash crosses the same way--just use a different variety as the pollen parent. Remember: to make a cross-pollinated plant "breed true," self-pollinate it by hand. Basically all cross-pollinated crops are handled like squash. However, each crop has methods and quirks specific to it. If the squash method doesn't work for the crop you are trying to save seed from, write to me and I'll see what I can do.

A warning to all would-be plant breeders: self-pollinating a naturally cross-pollinated crop can lead to a thing called "inbreeding depression" or a loss in plant vigor. Luckily for those of us who like to work with squash, squash don't seem to show any inbreeding depression.

QUESTION: How do I keep an unusual carrot (Purple-skinned) pure when it's being swamped by local Queen Anne's Lace? The flowers that I bagged and brushed set almost no seed.

This is a tough question. Cultivated carrots are cross-pollinated by insects, and will cross readily with wild carrots (Queen Anne's Lace) as well as other carrots in the area which may be flowering. Most plants which are cross-pollinated have some way of avoiding self-pollination. The carrot uses time—carrot pollen is shed hours or days before the end of the female part of the flower (the stigma) is ready to receive pollen. That's why the bagging and brushing didn't work.

Since the whole carrot flower (called an umbel) doesn't open at the same time, there should be good pollen available on some newly opened portions while the parts that opened first have receptive stigmas. All you would need to do is transfer that pollen--I'd just use the young floret as a brush and swab the older florets. You should bag the whole flower stalk as well--muslin or cheesecloth are recommended. Place the bags before the flowers open and replace them after pollination to be sure that insects aren't also transferring pollen for you!

QUESTION: What is a melon-squash and how do I use it?

The melon-squash is a winter squash that originated in the tropics, presumably Tahiti. It belongs to the same species as "Butternut" squash (Cucurbita moschata) which it resembles in skin color. It is not a cantaloupe or muskmelon.

The samples I've grown had really vigorous vines and were somewhat late in setting and maturing fruit. The fruit are quite large, and commercial seed sources seem to throw a variety of fruit shapes and sizes. Most of them have long necks and a small bulbed end enclosing the seeds. Some are ribbed, most are crook-necked.

We use it as a winter squash. The flesh is of good flavor, texture and color. It makes great "pumpkin" custards, pies and breads. Fresh, the flesh has a mild, melon-y aroma and flavor. The melon-like flavor is supposed to get stronger after the fruit is stored for two or three months. I have yet to try it that way.

(The 1981 Winter Yearbook, pages 59-60)

QUESTION: How large a sample of corn needs to be planted to maintain the various traits, given hand-pollination, but no selfing?

Corn is naturally cross-pollinated, and when self-pollinated (pollen from tassels on the same plant's silks), shows inbreeding depression (a loss of vigor, yield, etc.). If your interest is simply maintaining open-pollinated corn varieties in a limited space without losing vigor, the answer then involves a good deal of hand work. I'd plant at least 25 plants, and cover both the developing tassels and ears with small paper bags (white, preferably). The ear-bag can just be placed over the top and can be left unsealed. The tassel-bag should be tied or stapled at the bottom, and to avoid pollen falling out the bottom, bend (gently) the top of the plant over BEFORE pollen is shed (but not so early as to stunt tassel growth). Once pollen is shed, carefully cut the bagged tassels off of all the plants, then collect pollen by shaking the tassel within the bag. Combine and mix (gentle shaking) the pollen from all 25 plants, then store in a cool place until the silks are mostly out (2-3 days later). Then once a day until the silks begin to dry, remove an ear-bag, shake a small amount of the pollen mixture on the silks and rebag the ear. Once the silks dry, the ear-bag isn't needed. This should effectively mimic an isolated open-pollinated corn field. By the way, 25 plants is simply for security. You might be able to get by safely with about 15 plants.

QUESTION: What would be the method of keeping a dozen pepper varieties pure?

Peppers (and tomatoes, eggplant, lettuce, peas, beans--and okra, too, I think) are mostly naturally self-pollinated--Kim told me more than 95% self-pollinated in the cultivated types. To keep the lines pure, isolation (10-15 ft. minimum, preferably more) would be easiest. If space is at a premium, individual flowers (or flower stalks, in lettuce) can be covered with small envelopes (or bags). The covered flowers of peppers, tomatoes, eggplant and okra should be shaken daily to help pollen shedding. Shaking isn't necessary in peas, beans and lettuce, since pollen is shed (and essentially fertilization occurs) before the flower opens.

QUESTION: Does inbreeding depression ever occur in naturally self-pollinated crops? In other words, do such crops ever "run out" as some people have suggested?

If a crop is naturally self-pollinated (like peas, beans, lettuce, tomatoes, peppers, eggplant, okra), it is then also naturally inbred--and more inbreeding doesn't affect plant vigor (there's no inbreeding depression). A lot of people assume inbreeding depression and hybrid vigor are opposite sides of the same coin, but they are not. In maize (corn) it seems that the two are indeed opposite--self-pollinating corn leads to inbreeding depression while crosses between two inbred corn lines will show some hybrid vigor (how much depends on the cross). In other crops, beans for example, there's no inbreeding depression (since it is naturally self-pollinated). But an F1 hybrid bean often shows tremendous hybrid vigor (vigor then decreases in succeeding generations, and that could possibly be called inbreeding depression). Both hybrid vigor and inbreeding depression depend on the crop, the cross involved (if any) and the lines involved in the cross. The two phenomena are related--just not absolute opposites.

QUESTION: Are most of the bean varieties grown by gardeners in the U. S. day-neutral?

I was hoping someone would ask a question about day-length. The relationship between day-length and night-length can affect many things in many different plants. The most dramatic effect is on flowering (and thus on fruiting), and I assume that's what you are interested in with the beans. The common beans (snap and dry, Phaseolus vulgaris) are day-neutral. Lime beans (Phaseolus lunatus) are day-neutral but temperature-sensitive, not setting pods if the nights are too cool. The winged bean (Psophocarpus tetragonolobus) is day-sensitive; it will only flower when the day-length is LESS than 12 hours or so. That makes it a short-day plant. Basically, if a plant is day-length sensitive, it's either short-day or long-day. In the U. S., short-day plants flower in the spring or fall, and not in mid-summer, while long-day plants flower only in summer. Spinach or radishes bolting and going to seed in the summer is a long-day response. Also, strawberries produce runners only under long days (but the USDA is working on breeding day-neutral non-runnering, everbearing strawberries for you Florida and California folks). Perennial chrysanthemums flowering in the fall is a typical short-day response (why don't they flower in the spring, too? By the time the plant is large enough to be photosensitive, the days are too long)....To summarize, common beans don't show much sensitivity to day-length. The same is true of limas, although they are temperature-sensitive and will not set pods if the nights are too cool. And common beans also show some temperature sensitivity, not setting pods if the days are too hot.

QUESTION: Can watermelons and cantaloupes be hand-pollinated exactly as you described for squash? Will either of them show inbreeding depression?

Yes, watermelons can be handled like squash. The flowers are smaller and not as easy to manipulate, but it works....Cantaloupes are sometimes harder to do. Many times the self-pollinations in cantaloupes don't take well. One method used successfully is to do the selfs while the female flower (the one chosen as seed parent) is still a bud (but obviously a female flower-to-be). Some cantaloupe varieties (and some watermelons) have perfect flowers, but don't depend/assume that the perfect flowers will automatically self-pollinate. Again, in cantaloupes, you may (it is trial-and-error here) need to use bud-pollinations for the selfs. Choose a mature male flower as the pollen source and carefully open a female flower bud (same vine). I'd use tweezers and remove the sepals and petals, exposing the stigma. Put pollen on the stigma and cover the pollinated flower with a small paper bag or envelope....Inbreeding depression? I'd say not, or at worst, not much. The vine crops (squash family) seem to be able to take a good deal of inbreeding without loss of vigor.

QUESTION: By self-pollination, do you mean hand-pollination to keep the strain pure, or isolation planting?

I think that it's a good time to talk about some basic flower anatomy. First of all, flower plant species are either monoecious or dioecious. Monoecious (mono = one, ecious = house) means that both male and female parts are borne on the same plant. Dioecious (di = two) means that there are separate male and female plants. Secondly, plants of a monoecious species bear flowers which are either perfect or imperfect. A perfect flower--like a tomato flower--has both male and female parts. An imperfect flower is either male or female. (It does get even more complex--many squash, a monoecious species, bear both imperfect and perfect flowers, and often in a special sequence: male, female, perfect....) When I self-pollinate a plant with imperfect flowers (like the early squash flowers), it has to be a hand-pollination (I don't have the space for isolation even on the farm--I also don't trust the insects to not make crosses). And most planned crosses (making F1 hybrids), at least on a backyard scale, involve pollination by hand....Once again, with plants where you must transfer the pollen to cause self-pollination, self-pollination is equal to hand-pollination. With a plant bearing perfect flowers, the plant does the self-pollination for you....Isolation planting works for self-pollination, but on a small scale doing the selfs by hand is easier, more space-efficient and guarantees purity.

QUESTION: Some SSE members treat seeds in a weak bleach solution to destroy fungal and bacterial diseases. How is this correctly done? What should the solution strength be? What are the dangers to the seed?

Surface sterilization (the horticulture literature calls it "disinfestation") with bleach only kills the bacteria and fungi on the surface. Many seed-borne diseases are carried within the seed, and the bleach treatment won't help there. However, it (probably) can reduce the chances of infection in some cases, and can definitely reduce or eliminate some damping-off fungi, and therefore has a place. A 5% solution of household bleach (5 parts bleach, 95 parts water) ought to work. Cover the seeds for 60-90 seconds, rinse and sow. There should be no danger to the seeds....An effective way to eliminate certain bacterial diseases (bacterial spot and bacterial speck in tomatoes and eggplants, and black leg and black rot in cole crops) is to treat the seeds with hot water. It used to be done by all seed companies, but almost no one does it now. Why? It takes time, and thus money, and decreases germination by a few percent. But, it works! (Some seedhouses will hot-water treat seeds, if you buy pound-quantities.) Hot-water treatment: The water should be held at 122 degrees F. Soak broccoli, cauliflower, kale and Chinese cabbage for 20 minutes. Soak Brussels sprouts, cabbage, eggplant and tomatoes for 25 minutes. The 25 minutes at 122 degrees F. soak is also recommended for spinach (downy mildew control) and turnip (black rot and black leg control) seed. Control of both time and temperature is important--use a good thermometer and keep an eye on it during the 20-25 minutes recommended....

QUESTION: Why are so many of the newer commercial bean varieties white-seeded? After breeding do the white-seeded varieties become genetically stable much sooner than the colored-seeded ones?

I don't think that there is a good solid reason....the only reason that I can come up with is an esthetic one: white = purity. It may be that many consumers feel a bean with white seeds is better than one with colored. Of course, that's not true, but....Eliminating seed coat color is a fairly simple thing to do genetically. If any amount of consumer preference could be shown for white-seeded types (I'm thinking of snap-types only), it would be easy enough to justify doing.

(The 1981 Fall Harvest Edition, pages 45-48)

WARNING: SOME OF YOU ARE KILLING YOUR SEEDS!

Dr. Bruce Bugbee, who has an advanced degree in vegetable seed physiology, has warned us very clearly that we will kill our seeds if we dry them at temperatures above 95 degrees F. Some of you have gotten the idea that it's all right to dry your seeds in the oven, and it is not! Even if you set the oven at its lowest setting of 100 degrees F., the temperature will vary enough to kill your seeds.

John Withee confirms that the problem is widespread. He thinks that somewhere back down the line a garden writer or magazine has published a story about killing weevils by oven-treating seed. The problem is they didn't bother to mention that that's fine if you intend to eat the seeds, but it ruins them for growing. Two people have told me that they think they remember such an article, but they can't remember where or when they saw it. I'd really like to find out what started all of this because I really don't think that people would put their seeds in an oven unless they had been told to do so. Beans go permanently dormant if dried to below 5% moisture!

Some of the bean collectors in the SSE dry their seeds for a month or longer in the open air in a dry area of a garage or barn. Then they put them into an airtight jar and freeze them for a day to kill weevils that are under their seedcoats. After taking it out of the freezer, if they need to get back into the jar, they let it set out for a day to make sure that the contents are back to room temperature before they open it. If your seeds are cold when you take them out of the jar, atmospheric moisture will collect on them and they will absorb it and you will have to dry them again.

You can see how much damage is being done here. It is extremely frustrating when all of the time and expense involved in obtaining and distributing the seeds, plus all of your work multiplying and returning them, is ruined by a bad step in the process. And when you are dealing with something like the distribution of our Growers Network, which in some cases may be the last of that seed left alive, the problem gains its true perspective. I realize that some of you live in damp climates, but please don't put your seeds in the oven, even with just the pilot light on, unless you keep checking it the whole time with a thermometer.

(The 1981 Fall Harvest Edition, page 2)

A follow-up from Bruce Bugbee--....A correction should be made to the "Seed Killing Warning" on page 2 of The 1981 Harvest Edition. The warning should read: Seeds can be damaged, but not necessarily killed, when the temperature of the seeds is maintained above 95 degrees F. during drying....When harvested seeds are first put in a dryer, the seed moisture that evaporates cools the seed. The seed temperature can be 5-20 degrees F. higher during the first several hours of drying. When the seeds are almost completely dry (5% seed moisture), very little evaporative cooling occurs and the seed temperature is almost equal to the air temperature....As Richard Grazzini pointed out in the Fall SSE, bacteria in seeds can be killed by soaking seeds for 20 or 25 minutes in water at exactly 122 degrees F. This is a small stress on the seeds, but it does not kill them and it wipes out the bacteria....Seeds dry more thoroughly if they are dried slowly, so it's a good idea to be very careful when using high-temperature ovens....

(The 1982 Winter Yearbook, page 88)

COMPANION PLANTING

I am not publishing a Companion Planting Guide this year, because after reading the following from the Newsletter of the Henry Doubleday Research Association, I am not at all sure how much of it is true:

"....In this field everyone is quoting the book Companion Planting by Helen Philbrick and Richard Gregg, who are also quoting in many cases. We tried a number of these 'companionships' on the Trial Ground last year, and we found that: cabbage and potatoes help each other, but crowd each other out; leeks and celery are awkward to grow together and so are radishes and lettuce; we have yet to find any herbs that work as pest repellents....The traditional carrots with onions between the rows to repel carrot fly was again tried, with the usual inconclusive results....garlic is supposed to depress the yield of peas so badly that at the close range of planting, the peas should have died. In fact they grew quite well....the fundamental problem of companion planting is no clear directions, just flat statements....It is rumoured that some real work has been done in Germany, but so far there is still no hard evidence on companion planting of the kind that goes into normal gardening books on other aspects of horticulture....The companion planting craze has been running ever since 1967 and after ten years at least ten people should have tried out some of the ideas, discarded those that failed or were more trouble than they were worth, and found a few good ones....Has anyone, in any country, ever tried any of the companion planting sugggestions now quoted so widely with the two vegetables grown separately well away from the 'pairing' and the yields compared?....I appeal to anyone who has used any companion planting trick successfully to describe exactly what they did and what happened so we can try with enough experimenters to establish the case...."

Next year I will start publishing a continuing article on companion planting and cultural techniques (garden tricks). Send me any examples of these that you believe work or don't work. This will be a starting point for experiments. I'll start it off this year:

1) Alternating rows of broccoli and potatoes increases their yields.
2) A black walnut tree close to tomatoes will stunt them terribly (see also Carl Wilson's Texas listing which states that black walnut extracts may have definite insect repelling properties).
3) Onions have no noticeable stunting effect on peas.
4) Putting mineral oil on corn silks will control corn kernel worms.
5) Planting by the moon cannot be proved. (I bet I get a lot of negative feedback on this one. Several years ago I planted half of each variety in our garden in exactly the right sign, and the other half in exactly the wrong sign. Even set chickens that way. There was no difference. As my grandmother told me in the first place, "Let other people plant theirs in the moon. I'll plant mine in the dirt." If you believe in it, then you won't be afraid to test it.)

(The 1979 Seed Savers Exchange, page 38)

****

Companion Planting--Is There Any Hard Evidence?....(Continued reader input for this ongoing forum concerning companion planting evidence is solicited - Kent)....Louise Riotte (whose Secrets of Companion Planting for Successful Gardening and several of her other books are available from Garden Way Publishing Co., Charlotte, VT 05445) wrote to me: "....an experiment reported in the British New Scientist in 1970 that alternating strips of corn and sunflowers reduced the incidence of fall armyworm and gave correspondingly increased yields....Experiments by P. M. Miller and J. F. Ahrens at the Experiment Station in New Haven, CT have shown that marigolds (Tagetes--the strong-scented type) suppress meadow nematodes for up to three years and control other types of nematodes as well. A chemical produced in the roots kills them, but it is released very slowly so they must be grown all season to give lasting control. Even then results may not be readily apparent until the second season when nematode populations have been reduced....During times of grasshopper infestations, after-harvest plowing will discourage egg-laying, while spring tilling will destroy eggs still present...."

Richard Grazzini, Indiana member and a graduate student in vegetable breeding at Purdue, writes: "....There is a man here at Purdue, Dr. Al York (Entomology Department, Purdue University), who has been conducting research on companion planting. Over a number of years, his results are inconsistent--some years, under some conditions, some plants repel some insects. The biggest problem in obtaining consistent results is that the insect populations are not constant from year to year--what is a major pest one year may not even be around the next....Marigolds and nematodes are the exception. Dwarf marigold root exudate definitely repels nematodes, however there's no good evidence that the foliage likewise repels foliage-feeding insects....Dr. York is quite open-minded about the subject, feeling that the story had to start somewhere and that under the right conditions, with the right plants, it may just work....As far as practical advice goes, Dr. York did mention that a garlic/hot pepper spray has seemed to work as an insecticide in his tests....Most of the literature is based on hearsay or one person's experience, and is not based on rigorous scientific experimentation....Enclosed find one of the more rare papers on companion planting...."

(Richard sent me the results of tests conducted at the University of Illinois Dixon Springs Agriculture Center in Simpson, IL during the summer of 1973. The test attempted to repel cabbage worms (the imported cabbage worm, cabbage looper, and Diamond Back larvae) using hyssop or thyme; to repel flea beetles from eggplant using catnip or wormwood; and to repel bean leaf beetle and Mexican bean beetle from snap beans using rosemary or petunia or marigold. The experiment concludes that "the specific herbs used and their proximity to the vegetable plant did not effectively control the insect pest in question." There seemed to be virtually no difference in the damage. Richard emphasized that I should not wave this paper as a flag against companion planting, because, even though it is a scientific experiment, it is not necessarily conclusive - Kent.)

Leroy J. Schmidbauer, member from New Jersey, writes, "Marigolds again aided in reducing insect and slug damage on pepper and tomato plants.... volunteer plants without a marigold companion suffered heavy damage....After two months the marigolds were eaten totally by the slugs....I interplanted basil with pepper plants. The insects which normally chew holes in my pepper plants ate the basil leaves instead....yet basil alone in a separate part of the garden had very little damage....An article in Organic Gardening mentioned that yellow flowers (tomatoes) attracted a higher population of insects than red or orange flowers....will watch carefully this next summer and let you know....I have read that pieces of rhubarb stalk around cabbages will protect them from worms, that a rhubarb spray will protect carrots, that branches of elderberry leaves will protect cabbage, cucumbers and fruit trees, and that rubbing a cut ripe tomato on pumpkin fruits and vines will repel squash bugs....will try these methods myself and suggest that other members do the same and report back....

(The 1980 Seed Savers Exchange, pages 45 and 15)

MAKING RUBBER GASKETS FOR THE LIDS OF SEED STORAGE CONTAINERS

Automobile tire inner-tubing is an excellent rubber gasket material. Used tubes may be obtained free of charge from service stations and tire outlets....After obtaining a tube, cut it open and into large sheets which will lay out flat. The cutting can be easily performed with household scissors....Wash the cut rubber with a non-perfumed soap. Rinse thoroughly to remove all traces of detergent. Allow to air dry. Carefully remove the cardboard discs from all lids to be converted. Lay the cardboard disc on the rubber sheet and trace around it with a pen. Avoid areas of the rubber where raised ridges exist. Carefully cut out the new rubber gasket using sharp scissors....Install the rubber gasket in the lid in place of the cardboard disc. You now have a rubber gasket seal of the highest quality at no cost. Such gaskets can be produced rapidly with minimal labor.

If you have many bottles whose neck opening is the same size, you may wish to use a circular punch to stamp out gaskets. This tool easily cuts through the rubber using a mallet and a block of hardwood as a cutting surface. Such punches can be ordered through local leather shops.

(The above was submitted by the KUSA Research Foundation, PO Box 761, Ojai, CA 93023. KUSA has assembled an impressive collection of endangered grains. Financial support is needed if their work is to continue. They are also earnestly seeking a site where they can grow out their collection. A brochure and further detailed information is available. A dollar contribution accompanying literature requests would be appreciated to cover printing and postage costs.)

(The 1982 Fall Harvest Edition, page 24)

Hand-Pollination of Corn

Dr. Mark Widrlechner
(Garden Demonstration at Campout)

My name is Mark Widrlechner and I work with the U. S. Department of Agriculture at the Plant Introduction Station in Ames, Iowa. I've been interested in corn for quite a few years now. I've worked with Jack Harlan at the Crop Evolution Lab at the University of Illinois. I've got to give him credit for teaching me his techniques for pollinating corn. I think you'll find that, if you watch people pollinating corn, everybody has their own way of doing things. And so in some sense you're going to learn Jack Harlan's way as taught to me. And if you go and see somebody else, you'll learn how they go about doing it. What I'm going to try to do is to lead you through the steps of actually doing a pollination and then if we've got some time I'll try to give you a few tips on how you would use these techniques to maintain an open-pollinated corn variety.

(Photo by Mike Day)

Mark Teaching Workshop on Hand-Pollination of Corn

What we have here is a planting of a four-row block of Northwestern Red Dent. This is a good heirloom variety that's been around the Upper Midwest -- Minnesota, Wisconsin, northern Iowa -- for a long time. It was a popular dent corn back around the turn of the century. And it's too bad you can't see the ears on it, because they're really quite pretty. It's sort of a rusty color with a yellowish or a lighter cap on each kernel.

This corn is just starting to tassel. Just to get our terminology straight, the male flowers are at the top in the tassel. This brown bag is a tassel bag and it's going to be put onto the tassel at a certain stage which I'll show you. The tassel bag will be used to hold the pollen that comes out of the anthers that are held in a tassel. This tassel is just starting to shed pollen. You can see the anthers hanging there and in each anther there is a large quantity of pollen when it comes out. These have been sitting out all day, so most of the pollen has already come out of these and been blown by the wind across the field.

As you know, the ear is the female part of the plant and those flowers are located down along the stalk. At this point the husk leaves on this ear are just coming out. Eventually this ear will come to the point where there are actually silks coming out of the ear. If the silks come out on your plants before you bag the ear, it's too late for you to do anything about controlling the pollination. Around here, or pretty much anywhere in Iowa, people are growing all sorts of corn. And if the silks are exposed to the air, there's going to be pollen in the air. It's going to land on the silks, and it's too late. Whatever that pollen was, it's going to fertilize the seeds of that ear. So you've got to catch this shoot on the side of the stalk before the silks come out. And for that we have this white bag. This is a shoot bag. I don't want to advertise for any one particular firm, but I will say that the Lawson Bag Co. [P. O. Box 8577, Northfield, IL 60093] sells quite a range of sizes of tassel bags and shoot bags. If you're interested in growing out good quantities of corn, you probably should deal with a company like Lawson. Normal bags, that you would use for most other purposes like lunch bags or something of that type, if they're left out and there's a good rainstorm, will tear open or fall apart. Lawson's bags are designed to be tough and have adhesives in them that hold up to adverse weather conditions.

So, if we want to make a pollination, the first thing that we do is we look for ears where the silks have not come out yet, but will be coming out soon. Every particular variety has a time when you can see the silks just coming out. Sometimes the silks will come out right next to the stalk. Sometimes the ear will get really big and the husk leaves will come way out before you see any silks. But you get used to what your variety does and you learn how to anticipate a day or two before the silks come out and what that looks like. This isn't quite far enough along, but we will go ahead and prepare it. And the way we prepare it is, we take the leaf that is at that node where the ear is and we pull that off the plant. And it's usually a good idea to pull that subtending leaf off pretty cleanly because then this bag will fit on a lot better. Just pull that leaf downward and tear it off cleanly.

For the next step after we pull off the leaf, we take a knife or a scissors on a pocket knife and we cut back the tip of the ear. These leaves would normally cover the tip of the ear, but we cut them off. And when you cut them off, you should see the silks inside. (Question -- how do you know how far down to cut?) Well, you can cut down as far as the tip of the ear. (Questioner -- But not touch the tip, though?) If you touch the tip, there's always the chance that you're going to get quite a bit of damage at the end of the ear. At this point, the tip of the ear is down about here, so you can cut down pretty far, farther than you might think. You can cut a little bit and then a little more. This one was just ready to cut, because there's a whole mess of silks right in the middle there. (Question -- it doesn't hurt that you cut off part of the silks when you trim them like that?) No, it really doesn't hurt at all and it makes the pollinations go a heck of a lot easier than if you don't do it.

(Question -- you chose the smaller ear?) Yes, I did. (Was that intentional, because I would tend to choose the larger one that would be at the bottom.) Normally I would, too. But under these really dry conditions like this, generally only the top ear will develop. Normally this plant would have made four or five ears, but, since it's so dry out here, they're putting most of their energy into the top ear. And usually the top ear comes along first, even under good conditions. I guess there would be reasons for going for the second ear; for example, if there wasn't much pollen here. If there wasn't any pollen out here now, maybe I would decide to go with the second ear, because that one will be a few days later. But under the conditions here I don't think I'd go for any second ears unless we get some rain.

(Question -- That subtending leaf isn't involved in grain filling?) It is involved in grain filling. I won't say that it isn't. There is scientific evidence that that leaf that I pulled off would help make for a better set of seed, but it's a question of things getting in the way. And if that's really the only ear that you're concerned about and you're working with the first ear, it's going to go ahead and put resources into that first ear. I really don't think that the effect is significant enough to worry about. And it sure makes things go a lot easier.

So you've caught this before the silks came out. You cut it off with your knife and you put this white bag on it. A little earlier today, on this plant, I did that. The light is right for me to look at this white bag and see that there's some silks that have come out. You'll get to where you can see that without even taking that bag off. And every time you take the bag off, there's always a chance that pollen is going to get on the silks. So if you can tell that there's silks out of there just by looking through the bag, and there are a few on this one, then that's even better. I'm just going to take this white bag off and let you see that what we've got is about a half-inch of silks that have come out of this that I cut earlier today. And this is about the minimum length that you want these to be at the time of pollination. They can be up about an inch or an inch and a half long, but when they're this big you can go ahead and make your pollinations.

Ear Barely Visible

Tear Off Leaf

Cut Off Tip

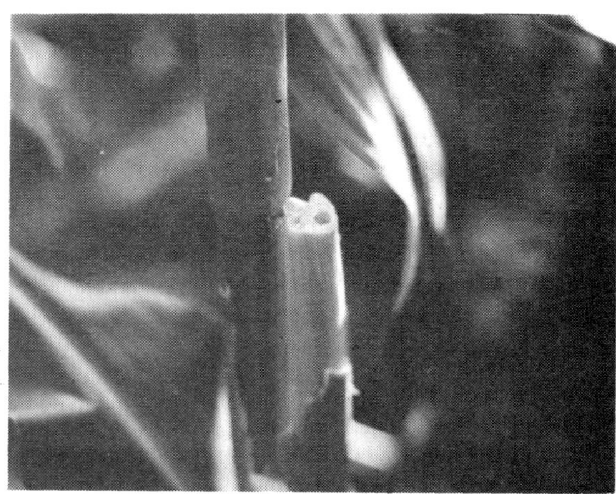
Circle of Silks

Verticle Cut Behind Ear

Wedge Bag into Cut

(Question -- So you expect normally to do the cutting off one day and the pollination the next?) If your timing is good when you do your cutting off, yes, I'd say about a day. Sometimes, we all make mistakes and cut off ears that aren't quite ready, and it may be two or three days or in bad weather even a week before something comes along. But having that white bag on there doesn't hurt it. Eventually it will come along. But if you're good at it, yes, it should be about 24 hours between when you make your cut and when you get a situation like this.

(Question -- Do you staple the ear bags?) No, I don't. I wedge them in. And they're a lot easier to wedge in if you've pulled off that leaf. If you leave that leaf on, it's sometimes very hard to pull that bag down so it's stuck solidly in there. Often I'll take my knife and make a downward cut between the ear and the stalk. Then I wedge the bag into that cut.

(Question -- How long can the silk get before it goes bad?) You really have, under normal conditions, probably four to five days to work with it. Because if they do get really long, but they don't look like they've dried out, you can go in and cut them back. And I like to keep them about the same length because it's a lot easier to see the pollen that you put on. And you can put on an even surface of pollen when the silks look like a little brush. If the silks are hanging out all over the place, it's very hard to spread the pollen evenly across that.

(Question -- You don't have to get the pollen on the tip of the silk?) No, it will go on the sides. On most plants the pollen is pretty particular about where it has to land to grow. But the whole length of the silk is receptive, so the pollen could go in anywhere. But I like to cut them into an even length just because I can see what I'm doing and I know that I can get pretty much all the silks that way.

So when the silks are at this stage, you can put pollen on them. You need some pollen. This tassel has some anthers that have already come out. And I don't know how many of you can see that very well, but most of these anthers are a yellow color. In some plants the anthers are purple right when they come out. But you can tell if you look closely at this tassel that the anthers are sort of a greenish-yellow or orange-yellow, when they come out. And then, when they get old, they turn red. But on this one, they're just starting to come out. You want to work with a tassel that's at about this stage, when only a few of the anthers have come out. And you don't really want to work with one where all the anthers have come out already. That's too late. If you look at some of those tassels over there in those next rows, all the anthers have come out and all the pollen that you were going to get out has already fallen out. So this is the stage you want, where just some of them have come out.

Give the tassel you intend to bag a good shake to get rid of any contaminating pollen that may have landed on it. Then you grab the branches

Much Too Early

Still Too Early

Just Right

Much Too Late

Put Bag Over Tassel　　　　Stalk in Corner of Bag
(Stalk Picked Only for Demonstration)

Fold Bag Back, Then Away　　　　Staple Close to Stalk

of the tassel and pull them all up together so that this brown tassel bag can slip over the tassel and get all the branches in. The sides of this bag fold up pretty nicely most of the time. And then we fold the bag up. There are different folding techniques, but whatever way that you fold it use the kind of fold that doesn't allow the pollen to just fall straight down out of the bag. You want an angle fold. And then you take your stapler, and preferably put in just one staple. No reason to waste staples or carry around an extra bag of staples with you. And this needs to stay on here until there's enough pollen in it. And, you wonder, how do you know when there's enough pollen in it? Most corn plants are on a clock. The anthers will come out in the morning and, as it heats up in the morning and the dew is burned off and it really starts getting warm, that's generally when you have peak pollen shed. It varies depending on the morning. It might be 9 or 10 o'clock. But if it's cloudy and gray, it might be 1 o'clock in the afternoon. Generally, you want to put this bag on either at this time in the evening or first thing in the morning when you get up, say at 7 o'clock in the morning. And then you come along about noon and you take these off. And hopefully then, if you put them on at the right time, there'll be a lot of pollen in this bag. If you leave it on past noon until you come home from work at 5 o'clock and it's been a real hot sunny day, it can get to be maybe 110 degrees inside one of these brown bags and the pollen will cook. And you won't get very good pollen. You'll need a lot more of it to get good seed set on your corn than you would if you caught it about noon.

So let's just pretend that I put this on now and all of a sudden it's 10 o'clock the next morning and I want my pollen. The way I get my pollen is to bend over the tassel, hopefully just enough so that there's just a little bit of a downward angle there, but not so much that I have broken the tassel or the plant. Because once I've broken this tassel, if I had put it on a day or so too early, once I've broken it, I'm kind of sunk. I can't go back the next day and start over. But if I just bend it a little bit and I don't get enough pollen, I may be able to come back tomorrow and use it again. So I bend the tassel over just a little bit and give the tassel bag a couple of good raps. Then open up the staple and I usually shake the tassel as I am pulling it out of the bag to hopefully get some more pollen into the bag. If I wanted to self-pollinate this plant, I could just take the pollen that's in here and put it on that ear shoot. But most of you probably aren't going to be involved with selfing your plants. You want to maintain the genetic variability in an open-pollinated variety. So the way you'll go about doing that is to make a pollen sample (pollen mix) of your variety.

What we would do with this variety we're standing in is to sample at least as many tassels as we had ears ready on that particular day. Hopefully even more than that, but sometimes you can't get very much to work with. To sample, we just go around and in the morning collect up all the brown bags from this variety. I take all of the tassel bags and pour their contents together and shake it up. So now all of the pollen I collected is in one tassel bag, but there's also a lot of anthers in there too. To get rid of the anthers, I turn the bag on its side with the open end slightly lower, use the thumb and forefinger of my other hand to pinch the lower corner of that opening so things won't fall out, and then gently shake the contents along the bottom edge of the bag.

Well, the anthers will come down the bottom edge of the bag first and be stopped where I have the bottom corner pinched off. I'd shake off most of the anthers because the anthers don't do much good. Then I'd take the pollen part and pour it into the shoot bag. (I usually work with a bigger shoot bag than this. They come in different sizes, but I usually work with a bigger one.)

It will be pretty obvious what the pollen is when you work with this stuff. It's a bright yellow, very fine powder and it would go into this little shoot bag. Now I'd take this little bag and just shake it around as much as I could. The pollen will accumulate down at the bottom of the bag if you hold it upright. Usually I've got scissors on my knife so I just cut one tiny corner off this bag and shake the pollen out of that hole. Or you can also try to shake it back out one of the slots. And you take that pollen and you shake the pollen onto these silks in as uniform a way as possible. Basically you cover it just enough so that you can see that you've put pollen on the silks. Try to not just dump it all in one spot on the silks, but try to spread it around. Each individual pollen grain is so small that you can't really see it. But if you've only got a little pollen, it's good to try to spread it around or else you'll get an ear where there's one seed here and one seed there and you just won't get a good ear developing. And in bad weather like this, when it's dry, an ear that just has a few seeds developing on it may abort. It may just not develop at all. It'll start, but then it's like the ear just sort of dries up.

(Question about brushing or blowing.) Oh, I guess you could. But after you practice at it awhile, you can get a pretty even distribution on it without having to brush or blow. I guess blowing would be preferable to brushing. You don't really want to be touching the pollen yourself very much because, as you're out in these cornfields and especially if you've got more than one variety to work with, there's always the possibility that your body and your hands are going to be doing the contaminating as you go from one variety to the next. So it's important to try to touch this as little as possible. If you do get a lot on you and you're worried, the quickest way to deal with that is to rub some rubbing alcohol or ethanol on whatever got the pollen. That'll usually do a pretty good job of killing it. It's not really sturdy stuff, this pollen.

(Question -- What about contamination in the air? I'm thinking about on the tassel before you put the bag on.) Remember that we shook the tassel before we bagged it. But even if we hadn't done that, since this pollen isn't such sturdy stuff and since most all of your pollen shed is between let's say 9:00 and noon, anything that came in the day before that landed there isn't going to be as viable as the stuff that just comes out. Assuming that you've got plants that shed well -- some inbred lines, for example, don't shed much pollen, but most open-pollinated varieties do shed a lot of pollen -- it'll swamp out anything that happened to lay there.

(Question -- Does pollen just have a life of that day?) Well, it depends on the weather. If it were to be foggy and 55 degrees, the pollen would probably last for two or three days. Fortunately, in Iowa when the pollen is shedding it doesn't get foggy and 55 degrees and it doesn't live very long. There is one thing to watch out for, though. When you're working to put the tassel in the tassel bag, sometimes a whole bunch of pollen will have accumulated down in the leaf whorl. It's usually a good idea to try not to get that big clump of pollen that's in the leaf whorl in your bag. That pollen has been setting down in the leaf whorl and probably isn't in as good shape as the other pollen. And the leaf whorl is often wet and if you get water in these bags when the pollen is there, the pollen gets all clumped up. So you don't really want to work with it when it's wet. Also bees and wasps will work the tassels sometimes and they could be dropping pollen into that leaf whorl. So I try to set my bag up a little above that leaf whorl. Sometimes, if it's real bad, I even go in there and rip out a couple of leaves, just to clear an area from the pollen.

You want to keep that white bag on until the absolute minute that you put your pollen on. Who knows how many pollen grains have come along and landed on it in just the few minutes that we've been talking? I don't know and I don't want to know. As soon as the pollen's put on, we take one of these brown tassel bags that we used to collect the pollen for this variety -- not a brown bag from someplace else -- and we replace the white bag with it. We take that white bag off after it's been pollinated and we put the brown bag on and we fold it around the stem. Then we take our stapler again and we staple shut the bag around the stem. Now sometimes a big-eared dent corn will produce an enormous ear. With one staple in it like that, it's possible that as that ear develops, it'll pull that staple open and the bag will just pop open later in the season. Now that's all right as long as a windstorm doesn't come along and blow your bag away, so put two or three staples in it. You won't know which one you pollinated if the bag blows away, because that bag is my marker. On big-eared lines or if the stapler is jammed, I'll just put a piece of strapping tape around there. Generally, if it's got both a staple and a strapping tape on it, nothing's going to take that thing off that plant. And if you've got the brown bag stapled and taped on there, the raccoons don't find it as easy to get into the ears. And the other thing about the brown bags marking your ear is that you can write a date or code number or variety name, whatever you want, with one of those black Sharpie permanent markers. It's pretty hard to write on these white bags. But you can put a lot of information on a brown bag and it'll be there when it's time to harvest.

(Question -- The ear has no trouble growing in the bag?) If you would staple the bag as tight as you possibly could against the stem and there were no leeway here at all, it's possible that you could stunt the development of that ear. Usually I leave a little bit of room. It doesn't seem to hurt it usually. You'll need to leave more room for big-eared types than you would for a little popcorn. For a little popcorn, a bag like this is almost a joke. But for a big dent or some of the Pueblo corns that get really long, you'd have to go with even a larger size bag than this.

Pour Pollen onto Silks
(After Removing Ear Bag)

Wedge Bag Between
Stalk and Ear

Staple Loosely

Repeat for Second Ear

(Question -- How long are the silks receptive?) I would guess at least a couple of weeks, but I keep the ears covered all the way to harvest. And the way that I tell whether it's time to harvest or not, I just lift up the bag a little bit. If it's still green, I just slip it back down and if it's not still green, then I go in and harvest. You can usually tell if you know your plants well without even lifting that up. Those bags will slide a little bit and you can check that. You don't have to leave the bags on, but if you don't leave them on then you've got to find some other way to mark that ear.

(Audience -- Mark, it just occurs to me that leaving the bag on might serve another function. It would prevent the corn earworm from depositing later in the season which could really be a problem otherwise.) Yes, that's right. But there's a flip side to that. Sometimes aphids, if there were already a few in there, won't get any predators coming into that bag. And when you open up that bag, the husks on the ear will be just full of aphids. But it does really cut down on corn earworm. If you are interested, for example, in looking for corn earworm resistance, you couldn't use a system like this, because you're keeping the earworms out of there.

Well, if there aren't any other questions about procedure, I'd just like to say a few things about maintaining varieties and about population size and sampling. Both the number of plants that you grow out in a year and what you harvest are really dependent on two things. They're dependent on what your goals are and they're dependent on the genetic variability or the variation in the population. It's easier, I guess, to talk first about variation. If you see, for example, that you have a great deal of variation in height, in plant color, in the colors of the anthers, in the colors of the seeds of the corn, then you know that you're dealing with a variable population genetically. And the more variable it is, the more plants that you're going to need to sample to keep that variability in it. Some corn varieties that have been developed by inbreeding have very little variability, and you can go down the row and every plant is going to look the same. With those you don't have to sample nearly as much to capture that type as you do when you've got something that's extremely variable. Now, at the Plant Introduction Station, for variable material we try to harvest a hundred ears from a variety. It isn't always possible to do that, but that's our goal. (Question -- how many tassels are you mixing together?) Well, that depends on the day. I would guess on an average day for a variety, there might be 25 silks that are out and they probably do 25 or 30 tassels to mix the pollen together on any particular day. That's how it would go at the Station.

The size of the sample also depends on how often these things are going to be grown out and how the seed is going to be used in the future. Because every time it's grown out, it's going to change a little bit. And if you're going to have to grow this every year and take the seed for the next year, you're going to need a bigger sample than if you grow your seed once and just use a little bit year after year and just keep going back to the original seed. The fewer times that you actually increase the seed to use in the future, the smaller a population size is necessary, let's just put it that way. For the kinds of varieties that Seed Savers Exchange members are generally working with, you should probably always grow a minimum population of one hundred plants. And you'd want to have at least 35 to 50 ears to constitute a sample. And it's probably best to make a balanced sample, too. It's better, let's say, to have a hundred seeds off of each of 50 ears than it would be to have a lot of seeds off of just a few ears. That's assuming your goal is to maintain the variety as it is.

And that gets to the second part: What's your goal? What's your purpose for doing this? If you're wanting to maintain an heirloom variety as it is, then you should also try to be unbiased in your sampling, in choosing which plants you cut the silks of and which plants you take the tassels and bag for pollen. That way, if you don't bias your sample you'll maintain that variety as it was intended to be. There are some people that, because the variety isn't adapted very well to where they grow or because they have a special disease or insect problem, want to change the variety. That opens up a whole other can of worms, which is how do you sample properly to select for the type that you're working for? That's something that I'm very interested in, but I could spend all night talking about that. And unless somebody really wants to go about doing that I'd rather not get into the details. But that's a different story, and I think you've got to be concerned about what your goals are when you start.

(Question -- If you're making a selection for your growing climate, can you take as few samples as you like and do it that way?) Yes and no. You can take a smaller sample to get the type you want, but if you do that over several generations, you run the risk of inbreeding depression. Because, eventually, the plants will be more and more closely related. And as that happens in corn, a lot of bad things happen. The plants get shorter. They yield more poorly. They're usually a little bit later. You'll find all kinds of bad things happening if you inbreed too much. So you don't want to get too small of a sample and do that year after year. But if you're selecting types out, you can get away with smaller samples than if you think you're maintaining an heirloom variety.

(Question -- Lots of times in the Seed Exchange we're having seed sent to us that isn't that pure. How can we tell the difference between that and what you've just talked about as a diverse variety?) The only way that you can deal with that is if you've got a description of what that variety is supposed to look like. In the absence of any information about what the type is, I think you have to assume that it isn't really contamination and that it is a diverse variety. I don't think you want to assume the other way. I'll give you an obvious case. Let's say you had a variety where you planted out a hundred plants and 95 of them had purple anthers and were 5' tall, but five of the plants were 9' tall and had yellow anthers. You could be pretty suspicious that somehow field corn contaminated that variety and that a small amount of the field corn had gotten mixed in with the true variety. But unless it's a pretty flagrant variation -- where you've got a whole bunch of uniforms and two or three things that are really sticking out that are different -- I wouldn't go ahead and say that's contamination. In the case that you have a description of the variety -- The Corns of New York is a good reference -- check for any description of things that you can look at before harvesting the ear. Those people were really great at describing the ear characteristics such as: at what nodes or how high up the ear is supposed to be, what color the anthers are supposed to be, what color the silk is supposed to be, how many tillers a plant is supposed to have. If you can find that kind of information, that's really useful in picking out off types. If you just have ear information and you use the kind of method that I described where you're bulking all that pollen, you could be taking pollen off of a plant that's really a contaminant and putting it into your ears.

Now there is one case where you can pick that up real quick, and that's in white varieties. If you've got a white variety of corn, you can pick up contaminants by looking at the seed of the ear when it's all done. The color of the seed to a certain extent is determined also by the pollen. If you find that you have some ears that are all white, you can be pretty sure that you have that variety. And, if you've got some other ears that are pretty much all white but have some yellow or some purple seeds in it, you can be pretty sure that you've picked up something else in your pollen that you put on there and you should just go ahead and throw that ear away. (Question -- Even the white kernels from that ear?) Even the white ones, because the pollen grain that fertilizes the embryo of that seed is different from the pollen grain that makes the color of that seed. There is double fertilization and those are independent. So any purple seed or any yellow seed on that white ear is a hint that even those white kernels could be contaminated. So you might as well just get rid of the ear and just stick with solid white ears. That only works with white corn, but it is one way to pick up contamination. (Question -- Why doesn't that same scheme work with any color when you get variations of color?) Well, for example, let's take red corn. It doesn't matter what the pollen is. The red is determined by that female, so the whole ear is going to be red -- light red, dark red -- and that's up to the female. You can pick out purples on yellow ears. But on oranges and reds or purples, you couldn't pick out other colors. It works best on white.

You don't have to go through any of this if you live in an area where people don't grow corn and you've got just one variety. Or you can stagger your varieties in such a way that you plant a real early variety and then a month later a late variety comes on and there's no overlap when they are shedding pollen. If you were to rogue out your off-type plants and you were the only game in town and there wasn't any corn for a mile, you wouldn't have to put any bags on anything. You could just let the corn do its own thing. That's a lot easier. The trouble is that in the Midwest there are very few places that you can go like that. (Question -- How far do you think pollen flies?) Well, I'd say around here it's pretty ubiquitous. It gets up there in the air -- now whether it's alive or not is another question. But it's everywhere when the field corns are at the peak of their pollination. (Question -- Do you think over a mile?) I think even over a mile. And for seed certification, depending on the situation, the distance will vary from a quarter of a mile to over a mile at the peak of pollination.

(Audience -- There is one way to avoid that in corn country, if you've only got one variety to keep pure. Most of the farmers are in a rush to get it out there, and if you wait until theirs is up and growing good and plant yours late, the pollen shed on all the field corn will be finished.) Yes, if you can get yours out of sync, or you might be able to get a crop on before the field corn has really begun shedding pollen. Now some of the really early corns from the Dakotas or from Canada are harvestable in 60 to 70 days after planting. If you're working with a corn like that you don't have much to worry about. When those things are pollinating, there isn't anything else going on. But those are just a few tricks that you can use to stay ahead of the game. It gets to be pretty time-consuming to have to put on the bags, and often we have jobs or have to be places at times when pollen shed is going on, and that makes it hard to keep up with. If you have to, you can leave these tassel bags on until you come home from work. But if you do that, like I said earlier, you're going to kill a lot of that pollen. But if you've really got a lot of that pollen, well then you just really dump it on the ears and you'll usually get something. It isn't a fatal flaw, if you can't be there at noon. But it goes a lot easier if you are.

(Question -- So even after you cut the silks off once, you can cut it back again to give you a nice, even surface there to do your pollinating?) If you work with this kind of system, the pollination works best when you've got fairly short brush coming out which makes a fairly even surface to work with. If it's all ragged and different lengths, you can cut it back again to straighten it out. (Question -- There's no problem with damaging the silks so they won't accept the pollen by cutting it again?) No, fortunately the damage on the silks is usually restricted to maybe a millimeter where you make that cut.

(Question -- Are tillers -- stalks growing from the plant's base -- characteristic of varieties or characteristic of environment?) Both. Some varieties will not tiller no matter what you do to them, and others will give you five tillers no matter what you do with them. (Audience -- I've always felt that it would be better for the development of the plant to break those off and allow it to develop more of a central stalk.) I know people think that way and there may be some merit to that. My personal "bias" -- when I'm selecting as opposed to maintaining -- is to select for plants that produce good ears on every tiller.

SEED SAVING GUIDE

We publish this "Seed Saving Guide" in each Fall Harvest Edition. It is constantly being revised and updated as we learn about new information, as we develop new techniques, and as our understanding becomes more complete. We wish to solicit input and feedback from all of the SSE's members so that this section will be as complete and correct as possible. The most extensive changes this year involve the caging of peppers, a fermentation technique for tomatoes, hand-pollination techniques for all of the vine crops, lengthy updates to the species lists for squash, and many photos taken this last summer in the Preservation Garden. I wish to express my deep appreciation to Glenn Drowns for adding his techniques and knowledge to the sections here on cucurbits and biennials. And I also wish to thank Rich Grazzini who, three years ago, gave us the original article on the hand-pollination of squash on which we have built. -- Kent Whealy

NEVER PLANT ALL OF ANY SEED OR YOU MAY LOSE IT !

Your vegetable plants and their seeds form "living chains" and you are their keeper. They will change very slowly due to either environmental factors (drought, short seasons, disease, pH changes, etc.) or genetic factors (mutations, genetic shifts, etc.) and your skill in selection will determine if these changes are for the better. You must learn how to select the right plants to save seed from and then how to prepare and store your seeds correctly. When you stop buying seeds and become plant keepers, you will have entered a whole new phase of self-sufficiency. Skills such as gardening, seed saving, and food storage provide protection for our families against the uncertainties of the world we live in.

The varieties that the seed catalogs call "hybrids" are the result of crossbreeding two varieties. For seed saving purposes they should be avoided, because seed saved from them will either be sterile or will begin reverting to one of their parent varieties. All the other seeds in the catalogs are the old "standard varieties" which run true year after year. They will come true from seed and remain pure only if you keep them from crossing. Some plants are cross-pollinating and their pollen may be carried great distances on the wind or by insects. With such plants you must either grow only one variety, widely separate different varieties, or hand-pollinate for purity. Other plants are self-pollinating, in which case several varieties can be grown with only slight separation.

Always look at your plants with seed selection in mind. Don't just look at the fruit; look at the whole plant. Select several plants to save seed from, not just the best looking or largest one. This will give your seed a greater genetic diversity and is the key to continued evolution and the ability of your plants to adapt to a variety of conditions. Select plants that have characteristics you want your next year's plants to have. Characteristics such as size, flavor, earliness, ability to survive a short season, disease-resistance, drought-resistance, insect-resistance, lateness to bolt, trueness to type, color, shape, thickness of flesh, hardiness, and storability can all be selected for.

You will need a way to mark and keep track of different varieties in the garden. This summer in the Preservation Garden we used lath from an old house that was being torn down. Use full pieces of lath because, if you break them in half, your plants will easily cover them up and you don't want to have to spend a lot of time searching for your markers. We cut 3" to 4" squares out of plastic milk jugs that we had been saving all winter. We wrote on these squares with "permanent" marking pens that you can buy at any office supply store. (Be sure that they are permanent! Write on the plastic squares and then try to wash it off.) We stapled these squares to the lath with a staple gun that you can buy at any lumberyard. We really like this system because you can write a lot of information right on the plastic tag, such as: variety name; accession number; the date that you planted it; the date it matures; and any other information you are gathering on that plant. And that tag can stay right with the seed when you harvest, process it and dry it. That way you'll have that information at your fingertips to write on paper packets or on foil packets that you are heat-sealing for frozen storage. And these plastic tags can be stapled directly to bean poles or punched with a hole so that you can tie or wire it where you want.

You must take precautions when drying and storing your seed, so that its vigor will stay as high as possible. Vigor is your seed's ability to germinate rapidly with good disease-resistance and it will be destroyed rapidly by high temperature and high moisture during storage. Seeds can be dried on a screen or on wax paper in the sun, by sealing them in an airtight container with silica gel (until they reach proper moisture levels for entry into storage), or by putting them in your oven with the pilot light on and the door cracked (WARNING: Damage to your seeds will begin to occur at temperatures of 96 degrees F. or above. Even at the very lowest setting the temperature within a stove's oven can vary enough to damage your seeds). If you have a food dehydrator, you can modify it for drying seeds. Most dehydrators are set on about 120 degrees F. But there is a screw on the rheostat that you can adjust to bring the heat down to under 95 degrees F. Be sure to check the inside temperature in several places with a thermometer to make sure that it's 95 degrees right back by the heating element. If you just measure 95 degrees up near the front cover, it may be well over 100 back by the element and you'll be damaging the seeds that are back there. We usually take seed that is already fairly well air-dried and dry it for an additional 48 hours at 90-95 degrees F.

Your seeds must be completely dry before you store them and they should break instead of bending (less than 8% moisture). This thoroughly dry seed must then be stored in a completely airtight container at as low and as constant a temperature as possible. Put each variety of small seed in an envelope and write the variety name and the year on each envelope. Put these envelopes into any glass jar that has a rubber gasketed lid that can be screwed down tight enough to make the container airtight and moisture-proof. We use everything from baby food jars (they have a lot of rubber under the lid and work great for small samples of larger sized seeds such as beans) to gallon jars full of envelopes on which we have modified the lids with home-made gaskets cut from old inner tubes. Also excellent are peanut butter jars, hot sauce jars, or any jar that has a rubber gasket under the lid instead of cardboard (the more rubber the better). Be sure to screw the lids on as tight as possible to make them airtight. Questionable lids can be taped shut with black electrical tape.

About three years ago, Dr. John Rahart from Wyoming did a fascinating seed drying experiment using a moisture absorbing agent called silica gel. Dr. Bruce Bugbee of Utah State did the moisture testing for the experiment. Silica gel looks like little plastic b.b.'s and sucks the moisture out of any material that is sealed in an airtight container with it. "Color Indicating-Silica Gel" has been treated with a dye called cobalt chloride. That dye is a deep blue when the silica gel is completely dry and gradually changes towards light pink as it picks up more and more moisture. Silica gel can be redried indefinitely in a 200 degree F. oven for 8 hours. For batches over a pound, you should stir occasionally to speed up the drying. It should always be dried slowly because at high temperatures you can scorch and ruin it. I was also told by Helmut Julinot in Toronto that he has dried silica gel in a microwave oven with excellent results. Of course, microwaves work on the moisture in a material. He was using a low-grade powdered silica gel that had no color-indicator. If it works the way we think it will on the beaded color-indicating silica gel, you should be able to set the microwave for about two minutes, watch it turn blue through the glass, and then store in an airtight container. I haven't tried this myself and would like feedback from anyone who does.

John's experiment used gallon airtight jars (although any size airtight container would do), thoroughly dry color-indicating silica gel, and small paper packets of corn or beans or tomato seeds that were at various moisture levels. A group of the packets was weighed, an equal weight of silica gel was measured out, and the silica gel and the packets were sealed in an airtight gallon jar together. At the end of a week the large hard seeds were down to about 6-8% moisture and the small softer seeds were at about 4-5%. (Those moisture ranges are just about perfect for frozen storage.) At that point the jar was opened, the seeds were separated from the silica gel, and the seeds were sealed in another container to maintain their low moisture content. These containers, whether they are jars or heat-sealed laminated bags, can be kept in a freezer with no damage to the dry seeds. The next best place would be in a refrigerator. And the next best would be in any cold area where the temperature will fluctuate as little as possible. When you take the container from the freezer, you should let it set out overnight to come to room temperature before you open it. If you don't do this, moisture will condense on the cold seeds and the effort you put into drying them will be wasted. Do not leave the container open for any length of time. Minimize your entries into the container because temperature fluctuations are not good for the seeds. Seeds stored by this method will retain their maximum vigor and will remain viable for up to five times the periods shown on viability charts.

We have also started using another type of container that we are really enthusiastic about. They are a flat bag that has laminated walls of paper / plastic / foil / plastic. They can be sealed with a Seal-A-Matic II which is available for under $20 from the Penney's catalog. They can even be sealed with an ordinary clothes iron set on "WOOL" and applied to the open end of the bag for three seconds. We've filled them with water, sealed them and stood on them without the seals leaking. You can jump on them until the bag breaks and the sealed edges are still perfect. You can cut the sealed edge off the bag with a pair of scissors, take out the seeds that you want, and then re-seal the bag. You can write on the outside of them since that layer is paper (but for safety's sake also put a label inside with the seeds as well). You can seal quite a few small packets of seeds into each laminated bag. You can put them directly into the freezer where they take up very little space compared to jars. And they are inexpensive. As I said, we're really enthusiastic about them and how well they work.

Dr. Bruce Bugbee did some follow-up tests after a year and a half on some of these bags filled with silica gel. It appears to be picking up 1/2% of moisture per year. Now silica gel would draw more moisture than a bag of seeds would, but this just shows that no container is ever really moisture-proof. And that's quite all right, because we're not trying to seal seeds away forever. I'm also not trying to suggest that everyone needs to process their seeds this way. But for those of us with huge collections, this method lets us store our seeds under optimal conditions and spread out our growouts. It is also an excellent technique that anyone can use to provide frozen back-up against loss. We set out to create a system that would let backyard gardeners set up small inexpensive seed processing centers in their homes using materials that would be available by mail-order. Anyone can purchase a Seal-A-Meal, a pound of silica gel, 100 paper/poly/foil/poly packets, and a cheap postage scale for $40-50, so it looks like we succeeded. (These heat-sealable paper/poly/foil/poly pouches can be ordered from Crystal Springs Packaging Co., PO Box 2924, Petaluma, CA 94953 ... see "Supplies for Seed Savers" later in this issue.)

****

BEANS - (Phaseolus vulgaris) - contains all of our common bush or pole beans, whether used for green snaps, green shell or dry. They are self-pollinating before the flower opens, so there is seldom any crossing. Occasionally you may notice some variation or oddities in the seed. This is more likely due to a genetically unstable variety or a difference in conditions (change of soil pH or wetness at harvest) than to any true crosses, but crossing caused by insects does sometimes occur. The seed in a crossed pod will often be a different color than seed from the parent plant. Such crosses are often highly unstable and you may have to select them for 5-10 years before they completely stabilize. (Don't offer such seeds through the SSE until they have been completely stable for three years.) Never plant two white-seeded varieties side by side, because you won't be able to tell if they cross. And seed coat color isn't always a reliable indicator for crossing in cases where one color is dominant over another. Don't plant two pole bean varieties next to each other in a row or they are liable to twine together and mix up your seed. To insure a high degree of purity, beans should be isolated by 6-12 feet.

There are several seed-borne viruses and bacterial diseases that affect beans. Some of the most common include Bean Common Mosaic Virus (BCMV), Halo Blight, Common Blight and Anthracnose. BCMV is spread by aphids, but the bacterial diseases (blights) can be spread by splashing rain, irrigation, or by people moving through wet plants. The diseases are also spread by infected seed or crop debris. BCMV can be cleaned up in a greenhouse where there is no aphid population. I think it would be a good idea to isolate new lines for the first year as they come into your collection, before you introduce them into your garden. Most of the symptoms are easily observed and include leaves that are stunted or malformed or badly curled with their tips turning downward (although slight symptoms of this type can occasionally be caused by drought or insect damage), lesions on the leaves or pods, spots or discoloration on the leaves, or foliage that becomes light-green or yellowish. Plants showing any of these symptoms should be rogued out and taken well away from your garden. It goes without saying that you should never send infected seed to anyone until you are sure that you have cleaned it up. But these are problems that you can select your way out of. Essentially, if you only save seed from healthy-looking plants, you won't have problems.

When harvesting beans, mark several of your finest plants (wooden stakes, strips of cloth, or plastic ribbons) and let the pods completely dry out on the plant, weather permitting. When most of the leaves have fallen off, pull the plants and hang them under cover to finish drying. Small amounts of seed can be shelled out by hand. For large quantities, make sure that the beans are thoroughly dry, crush or thresh the pods, separate the beans from the chaff by winnowing in the wind, label, and store. Weevil eggs, which are almost always present under the bean's seedcoat and can ruin your seed in a matter of months, can be killed by placing the thoroughly dry beans in a tightly sealed jar and freezing them for two days. It is a matter of courtesy to freeze bean seed to kill the weevils before you send seed to other gardeners. If you don't, the seeds will probably be ruined before the other gardener plants them.

RUNNER BEANS - (Phaseolus coccineus) - You can always tell which beans belong to this species because, when they are coming up, they develop with their two seed halves UNDER the ground. Many people mistakenly think that these are "fat limas," but their growth habit when emerging from the soil will always give them away. Examples would be Scarlet Runner, all of the Aztec Beans, Butterfly Runner, Painted Lady, etc. Some people grow them just for their beauty as ornamentals. They have brightly colored flowers that are very "open" in nature. Although the flowers are self-pollinating, bees and bumblebees and hummingbirds work them heavily and thus cross them. People don't seem to realize that and much of the seed being traded now is already crossed. It's common for Scarlet Runner (with its beautiful purple and black seed) to throw seed of its white counterpart. The most beautiful bean I ever saw was a Butterfly Runner with a gray background, but now, because of such crossing, the only variation I can lay my hands on has a tan background. I realize that many people like to grow different Runner Beans together and let them cross just to get beautiful color variations in the seeds. But if you intend to save seed from any of your favorites to keep them around, you should either grow only one variety of Runner Bean or separate different varieties by 1/4 mile.

PEAS - (Pisum sativum) - are normally self-pollinating, but may occasionally cross-pollinate. If you are growing more than one variety, separate them by 5-10 feet plus a tall barrier crop in between. Crossing in peas is harder to notice (and even more difficult to rogue out) than some other crops because of the similarity of the characteristics between varieties. Therefore, even though they cross slightly less than beans, you may wish to take more precautions with them.

LETTUCE - (Lactuca sativa) - is self-pollinating, but some crossing can and does occur when planted closely. Lettuce will cross easily with Wild Lettuce or Prickly Lettuce (Lactuca canadensis, or some scientific works list this plant as Lactuca serriola). When lettuce crosses with its wild relation, you will easily be able to see it when you grow plants from that seed the following year. The offspring from such a cross will have intermediate characteristics between the two and be horribly bitter. Cut down all wild lettuce within 200 feet of any lettuce plants you plan to save for seed. The USDA Plant Introduction Station recommends separating varieties by at least 10 feet. When deciding on what to save for seed, choose the plants most typical for the variety and the slowest to bolt. If you continue to save seed from the fast bolters over the years the quality of the variety will deteriorate considerably. When the seed is fully developed, cut the seed heads off, put them in a paper sack and store in a dry place until they finish drying, separate the seeds, label and store. A word of caution, if you try to wait for the entire plant to dry at once, on most varieties you will lose more seed to the wind than you will gain by waiting. The best way to get the most seed is to make several trips to harvest it.

Wild Lettuce (Lactuca canadensis)

CHINESE CABBAGE - is a cross-pollinating annual. There are three distinct groups of Chinese Cabbages:

Non-Heading Chinese Cabbage - Brassica rapa (Chinensis Group) which includes Pak Choi, Celery Mustard, Chinese Mustard, etc.;

Heading Chinese Cabbage - Brassica rapa (Pekinensis Group) which includes Pe Tsai, Chinese Cabbage, Celery Cabbage, etc.; and

Mustard Spinach - Brassica rapa (Perviridis Group) which includes Tendergreen, Spinach Mustard, Oriental Mustard, etc.

None of these Chinese Cabbages will cross with regular cabbage or any other member of the cabbage family (cabbage, broccoli, brussels sprouts, cauliflower, collards, kale, or kohlrabi), since they are all Brassica oleracea. But any Chinese Cabbage will easily cross with any other Chinese Cabbage, no matter which group it's in. Chinese Cabbages send up a seed stalk from the center of the head and the pods will turn brown when mature. Most all of the seeds will mature at the same time so the stalks can all be cut and placed in bags to finish drying. The recommended isolation distance is 1/4 mile. It is important to be very selective about the quality of the plants you save seed from. You will be far better off saving seed from the best six plants rather than 25-30 at random. Never save seed from ones which rapidly bolt to seed. Also, self-sterility can exist in this group, so always save seed from at least three plants.

MUSTARD - (Brassica juncea) - is our Western Mustard, as opposed to the Oriental Mustard Spinaches described above which are actually Chinese Cabbages. It is cross-pollinating and will easily cross with other mustards. Seed development, isolation, and method are all the same as above.

SPINACH - (Spinacia oleracea) - is cross-pollinating and its pollen is very fine and can be carried long distances by the wind. The pollen can easily drift a mile or more. The wind is the primary pollinator so, unless you can isolate different varieties in wind-protected areas, only plant one variety. Spinach plants are either male or female, so some will only produce pollen and the others only seed. You should save only your best plants, but try to leave a close to equal number of both sexes to insure good seed set, as some male plants will produce their pollen before the female plants produce flowers.

RADISHES - are insect-pollinated, so grow only one variety. Choose several of the largest and earliest roots to save seed from. Don't save seed from the early bolters. Some support may be needed when the plant reaches the flowering stage, because the seed stalk can grow five or more feet tall and have many side branches. The seed pods turn brown at maturity. Pull the plants and finish drying under cover.

BROCCOLI - is unlike the other members of the Cabbage family in that it will produce seed the first year, if sown early enough that the plants are large by the long days of summer. It will readily cross with the other members of the Cabbage family (cabbage, broccoli, brussels sprouts, cauliflower, collards, kale, or kohlrabi) if any are flowering within 1/4 of a mile. Do not cut the flower head for food if you are going to save seed from it. Treat the seed as you would Chinese Cabbage.

SUNFLOWERS - (Helianthus annuus) - will cross readily with wild sunflowers. Varieties will also cross among themselves, so separate by at least 1000 feet. Sunflowers are a good example of what some persons refer to as home-saved seed "running out." Seed doesn't actually "run out" but, unless you take proper precautions, a gradual process of undesirable crossing over several generations can make some seed practically worthless.

OKRA - [The following material was generously supplied by Ted Gibbs] - (Abelmoschus esculentus) - Okra is self-pollinating and insect cross-pollinated if different varieties are grown within one mile. Select and mark quality plants, with distinctive variety characteristics, for seed saving. Leaving pods on the plant to mature for seed will cause reduced pod production and plant aging will accelerate. Many okra variety blossoms open 41 to 48 days after planting. As this time approaches, start your daily evening seed saving plant inspection. "Bag" any blossom buds with the light green striped appearance or the slightest show of petal. Inspect your "bagged" blossom buds the next morning (no later than the evening), marking the blossomed buds with a string or ribbon loosely tied on the stem below the bag. Leave the bag on during the next full day after blossoming and then remove the bag from the marked pod the following day. Okra pollen is not viable after one day at room temperature. Your reusable bag probably has pollen on it and, if it is cool, the pollen may still be viable, so act accordingly. Cease blossom bud bagging approximately five weeks prior to your area's last average frost-free date. Dry harvested seed pods in the sun to remove dampness. The open storage life of okra seed is approximately four years.

PEPPERS - (Capsicum annuum) - are self-pollinating, but insects can and do cause considerable crossing. Steve Tanksley from New Mexico State University recently published an article in HortScience magazine entitled "High Rates of Outcrossing in Peppers." In this study they set out target plants that were genetically unique into several different fields and then traced that character's flow into the surrounding population. These target plants contained a chemical trait which could be picked up using "plant fingerprinting" (electrophoresis). The rates of outcrossing that they documented in the different fields was anywhere from 15-80% instead of the 0-5% or 5-15% that all of the standard references now give. The insects doing the crossing were mainly sweat bees. As more and more information becomes available, we are repeatedly confronted with the fact that even the self-pollinated crops are capable of high degrees of outcrossing under certain conditions.

Many of our members take no precautions to keep their peppers from crossing, but it's certainly time for us to start. We grew 100 kinds of peppers in the Preservation Garden this last summer and about 20% of them were really crossed up. Gary Nabhan told me about the Tanksley study, so we spent a day just prior to the Campout caging the plants we intended to save for seed. We used two 4-foot semicircles of wire stuck into the ground in an "X" and covered with a fabric called Reemay that was buried around the edges. Reemay is a spun polyester cloth that is thin enough to let rain and light through, but will keep insects out. Since it is spun -- like the spun glass "angel's hair" used for Christmas displays -- and not woven, I'm not sure it could be sewn, if you are thinking about applications like that. It is so thin that it tore easily when we pulled it back out of the dirt this fall. The pieces of Reemay that were over the pepper cages were about 5 foot square. We took the undamaged ones down the the river and washed them, tossed them onto weeds to dry, quartered them into 30-inch squares, picked bean pods onto them, tossed in the plastic label, tied the corners into a bundle with twine, hung them in a shed from the rafters, and our beans are drying really great. We also have friends that used it to cage just a few plants that insects really love like Chinese Cabbage. Its applications seem endless. (Reemay is available through the free catalog of Johnny's Selected Seeds, Albion, Maine 04910.)

Gary Nabhan Cutting Pieces of Reemay for Pepper Cages

If you don't use some sort of caging technique like the one just described, you must separate any two varieties by 1/8 of a mile for complete purity. The only disease that we've seen in peppers so far is Anthracnose, which are those black spots that peppers get on them. Anthracnose can be seed borne, but you can select away from it if you don't save seed from fruits that have it and rogue out those plants. Wait for the peppers that show no signs of it to reach full ripeness, pick, remove the seeds and dry. We live in such a humid climate that peppers mold when we try to dry them whole. So we have to take their seeds when they are still wet, which can be rather dangerous. We use rubber kitchen gloves, but after an hour or so our fingers still start burning and it's in our noses and we're sneezing. I urge you to use extreme caution when dealing with hot peppers. Don't rub your eyes and be careful of what you touch with your fingers.

EGGPLANT - (Solanum melongena) - pollination is like peppers, so at the very least you should separate varieties by the length of your garden or with a tall crop. Leave the best fruit on several plants for as long as possible, scrape out the seeds, separate from the pulp, dry and store. Eggplant seeds do not remain viable for long periods of time, so must be regrown every year if possible.

PUMPKINS & SQUASH - are insect-pollinated and cross very easily. All pumpkins and squash belong to one of the four species of the genus Cucurbita. For seed saving purposes, you must therefore plant only one variety from each of the four following species --

Cucurbita pepo

Typical "Pepo" Leaves

Characteristics: leaves and stems are prickly, especially when mature; fruit's stem is hard and has five sharply angular sides; and seeds are white with white margins. This species includes all summer squash, all of the true pumpkins, and varieties that are both bush and long-vining. The following varieties all belong to the species "pepo" --

ALL OF THE "ACORNS" (Des Moines, Ebony, Ebony Bush, Golden, Jersey Golden, Royal, Table King, Table King Bush, Table Queen, Table Queen Bush, Table Queen Ebony, Table Queen Mammoth, and White), Amish Field Pie pumpkin, Austrian Bush Summer, Baby Pam pumpkin, Bahce, Bela Sakaska, Bicolor Spoon, Big Red California Sugar, Black Beauty, Bloomfield pumpkin, Bulgarian Summer, Buscholkurbis Naked Seed, Casserta, Chestnut, Cheyenne Bush, Chiefini, Cinderella, Citroville de Touraine, ALL OF THE "COCOZELLES" (Green, Vining), Connecticut Field, Cornfield pumpkin, Cozini, Connecticut Sweet Pie, ALL OF THE "CROOKNECKS" (Dwarf Summer, Early Summer Golden, Early Summer Yellow, Golden, White Summer), Crystal Bell, Cupid, Delicata, Dumpling, Early Cheyenne Pie, Early Prolific, Early Prolific Straightneck, Eat All, Erken, Eskandarany, Fordhook, Fordhook Bush, Fort Berthold, French White Bush, Gem, Gills Golden

Pippin, Golden Centennial, Golden Custard, Golden Oblong, Gririt, Halloween, Howden, Huicha, Hyuga Black, Idaho Gem, Ingot, Jack O'Lantern, Japanese Pie (Abundant Life), Kahcona, King of Mammoth, Kline pumpkin, Kumi Kumi, Lady Godiva, Large Yellow Paris, Lebanon, Little Boo, Lunghissimo Bianco Di Palermo, Mammoth Gold, Mandan, Marego, Maryland Pie Pumpkin, Midwest Sweet Potato, Mihoacan, Miniature Pumpkin, Naked Seeded, New England Pie pumpkin, Oaxacan, Oaxacan Bicolor, Oaxacan White, Oland, Omaha, Omaha pumpkin, Panama, Pepinos, Perfect Gem, Pie pumpkin, Prostate, Royal Bush, ALL OF THE "SCALLOPS" (Benning's Green Tint, Early White Bush, Early Yellow Bush, Long Island White Bush, Mammoth White Bush, Patty Pan, St. Pat, Summer Bush, Yellow Golden), Showell, Small Sugar pumpkin, Spaghetti Squash, Spookie, Stickler, Straightneck, Streaker, Sugar Pie, Sweet Dumpling (Vegetable Gourd), Sweetnut, Table Gold, Tarahumara Indian pumpkin, Tatume, Thelma Sanders Sweet Potato, Thomas Halloween, Tricky Jack, Triple Treat, Tuckernuck, Uconn, Uncle Herman, ALL OF THE "VEGETABLE MARROWS" (English Vegetable, Green Bush Improved, Long White, True & Tender, Vegetable, White Bush, White Vining Vegetable), Vegetable Spaghetti, Wilbur Field pumpkin, Winter Luxury, Winter Luxury Pie, Winter Nut, Woods Prolific, Woods Earliest Prolific, Youngs Beauty, ALL OF THE "ZUCCHINIS" (Black, Black Beauty, Burpee's Fordhook, Burpee's Golden, Dark Green, Gold Rush, Gray, Green, Round, Yellow, and White Egyptian), Zikusa, and any of the small Hard-Shelled Striped and Warted Gourds.

Typical "Maxima" Leaves

Cucurbita maxima

Characteristics: very long vines; huge leaves; leaves and stems are hairy; fruit's stem is round, soft and corky; thick brownish seeds with cream colored margins and thin cellophane coatings. The following varieties all belong to the species "maxima" --

All Gold, Alligator, American Indian, Amish Pie pumpkin, Araucana, Argentine Primitive pumpkin, Argentine Summer, Arikara, Asuncion, Atlantic Giant, Atlas, Australian Butter pumpkin, Autumn Pride, ALL OF THE "BANANAS" (Blue, Chinese, Giant, Hartman, Orange, Pink, Pink Jumbo), Banquet, Bay State, Belgium pumpkin, Big Max, Big Moon, Big Red, ALL OF THE "BUTTERCUPS" (Blue, Branscomb, Branscomb White, Burgess, Bush, Russian, and Verda's), California White pumpkin, Candy Roaster, Candy Roaster (Francis Ballew's), Cherokee Indian pumpkin, Criolo, Crown Prince, Delicious (Golden), Delicious (Green), Doe, Emerald, Equadonantian, Essex, Essex Hybrid, Estampes, Flat White Boer, Forragero, German Green Pumpkin, German Sweet Potato, Gilmore, Ginny's Large, Gold Nugget, Goldpak, Golzar, Greek Small Orange pumpkin, Greengold, Guatemalan Blue, Herman's Delight, Hillbilly, Hokkaido Green, Hokkaido Orange, ALL OF THE "HUBBARDS" (Baby, Baby Blue, Black, Chicago, Chicago Warted, Golden, Green, Green Improved, Kitchenette, Large Blue, Little Gem, Minnesota, Sugar, True, Warted, Warted Green, Warted Improved), Hungarian Mammoth, Hungarian Mammoth (Cornell Strain), Iran, Ironclad, Iron Pot, Japanese, Jarrahdale, Kabule, Kara Kabak, Kentucky, Kindred, King of Giants, King of Mammoths, Kuri Blue, Kuri Red, La Calabaza (Philippine), La Kalabaza, Large Moroccan, Large Yellow Paris, Large White Manteca, Leningrad Giant, Mammoth Chile, Mammoth Gold, Mammoth King, Mammoth Orange Gold, Mammoth Whale, Mammoth (Genuine), Marblehead, Marblehead (Umatilla), Marblehead (Yakima), ALL OF THE "MARROWS" (Autumnal, Boston, Orange, Prolific, Warted), Mexican Indian, Mexigold, Mooregold, Mohawk, Ni-es-pah Long, Old English, Old Humbolt, Old Winter, Orange Giant, Orange New Guinea, Pikes Peak pumpkin, Pink Giant pumpkin, Plymouth Rock, Quality, Queensland Blue, Rainbow, Red Chestnut, Red Estampes, Red Skin, Roaster, Rouge Vif d'Etampes, Shanghai, Siva Stambolka, Show King, Sibley, Silver Bell, Smooth Green Pie, South American, Stambolka, Sweetbush, Sweetkeeper, Sweetmeat, Tokyo, Tri-Star, Triamble, ALL OF THE "TURBANS" (American, Golden, Turk's), Valenciano, Victor Watten, Vojicka, Vounicheo Blue, Vounicheo Orange, Warren, West Virginia Big, Whangaparoa, White African, White Pumpkin, Winnebago, Yugoslavian Pie pumpkin, and Zapallo Poloma.

Cucurbita moschata

Characteristics: large leaves; spreading vines; leaves and stems are hairy; fruit's stem is hard and slightly angular; fruit's stem flairs out where it attaches to fruit; sepals (little green spikes at base of flower) are arrowhead or spade-shaped in all but a very few cases; moschata leaves are slightly darker green than mixta with pointed leaf tip and slight indentions; small oblong seeds with dark-beige margin. The

"Moschata" Leaf (left) and "Mixta" Leaf (right)
Showing Slight But Distinguishable Differences

following varieties all belong to the species "moschata" --

African, African Bell, Aizu Gokwuase, Alagold, Borinquen, Butterbush, ALL OF THE "BUTTERNUTS" (Baby, Early, Eastern, Hercules, Mexican, Ponca, Puritan, Waltham, Western), Calabaza (Cuban Squash), Calhoun, Cangold, Chirimin, ALL OF THE "CHEESE" (Cutchoque Flat, Flat Warty, Large, Long Island, Tan, Wisconsin), Citroville d'Eysines, Cuban, Dickinson, East India Big Red, Florida Buff Pie, Fortuna, Futtsu Kurokawa, Golden Cushaw, Hawaiian, Hercules, Honduran, Kentucky Field, Kikuza, La Primera, Long Keeper, Longfellow, Mediterranean, Melon Squash (Tahitian), Mexican, Neck, New Jersey, Old Time Tennessee, Papaya, Patriot, Peraora, Ponca, Priester Crookneck, Sequalca, Showell Sweet Potato, Sweet Red, Tahitian (Melon Squash), Tamala, Tennessee Sweet Potato (Shumway's), Upper Ground Sweet Potato, Virginia Mammoth, White Rind Sugar, and Wisconsin Canner.

Cucurbita mixta

Characteristics: large leaves; spreading vines; leaves and stems are hairy; fruit's stem is hard and slightly angular; fruit's stem flares only slightly where attaches to fruit; sepals are generally straight even at tips; mixta leaves are slightly lighter green than moschata with rounded leaf tip and hardly any indentions; white seeds with pale margins, cracks on flat sides, and cellophane coating. The following varieties all belong to the species "mixta" --

Big White Crookneck, Black Sweet Potato, Cochita Pueblo, Chirimen, THE "CUSHAWS" [except Golden Cushaw which is C. moschata] (Albino Hopi, Australian, Gold Striped, Green Striped, Hopi, Longneck, Neckless, Old Fashioned, Solid Green, White, White Crookneck, and Winter), Gila Cliff Dweller, Hindu, Indian Vining Zucchini, Japanese Pie, Jonathan, Large White New Mexico Cliff Dweller, Mixta Gold, Mrs. Morris Potato Pumpkin, Pennsylvania Crookneck, Tennessee Sweet Potato, and Woodrey Sweet Potato.

Different varieties WITHIN the same species (one of the four groups above) will cross very easily, but there is no need to worry about any crossing BETWEEN species. Several years ago a squash expert told me that crosses between two varieties that belong to different species are so difficult to make and their progeny are so highly sterile, that he never worried about such crosses by natural means. He went on to say that what you really should be more concerned about is pollen from a neighbor's garden contaminating your efforts. These species lists will allow you to keep four varieties of squash pure (one from each species) IF your garden is isolated by at least one quarter of a mile from any gardeners who are also growing squash. If that is not your situation, PLEASE HAND-POLLINATE ANY SQUASH THAT YOU INTEND TO OFFER THROUGH THE SSE. You could wreck in one season what someone else has spent a lifetime of gardening to develop or preserve.

Seeds From One Fruit That Crossed Last Year
Produced All of These Variations This Year
What a Mess -- We Threw Them All Away

Always remember that when a cross occurs you cannot see a change in that summer's fruit, but you'll certainly see it when you grow out that fruit's seed the next year. As a hypothetical example, let's imagine that you grow a bush of Zucchini and right beside it you grow a hill of Sugar Pie pumpkin. Both are varieties that belong to the species "pepo." When you harvest the fruits that fall, they will be normal-looking zucchinis and normal-looking pie pumpkins. The cross is in those fruits' seeds. You can't see it that year! But when you grow those seeds out the next year, you'll see the mess that has been created. Every seed will produce a plant with a different size or shape of fruit. All of the fruits on a particular plant will be alike, but, even if those fruits have been "selfed" (hand-pollinated using male and female flowers on the same plant), their seeds will still be highly unstable when planted the next year. Even if you could know exactly what you were trying to select back to, it would take years of selfing and selection to erase the initial mistake and re-stabilize the strain, and in many cases it would be almost impossible to rogue out these crosses. Surely everyone can see how disastrous this is when we are working with rare varieties and then distributing their seeds. Hand-pollinating squash is really very simple and this year we have an excellent series of photographs to lead you step-by-step through this enjoyable technique. So, if you intend to offer the seeds of any vine crops through the Seed Savers Exchange, PLEASE learn the following sections and then practice them next summer in your garden.

HAND-POLLINATION OF SQUASH

Squash are insect-pollinated. In order to get pure seed for your next year's planting of a particular variety, you must keep the bees from doing their job. This is easily done, but you must become a keen enough observer to tell when its flowers are about to open. Your plant will start setting male flowers well before it produces any female flowers. This is so there will be ample pollen available when it does start setting female flowers. The female flowers have a small "baby" squash right below them, so it is easy to tell them apart from the male flowers which are just a flower on a straight stem.

Once you start studying the flowers closely, you will also be able to tell if they are "green" or are about to open or have already opened. Flowers that have already opened will be wilted, so it is too late for you to use them for your pollinations. How do you tell which flowers are about to open for the first time? They show a yellow flush of color, especially along their seams. And sometimes you will even see the sections of the flower starting to break open at the tip. When you see those signs, those flowers will open overnight unless they are sealed shut.

If you intend to hand-pollinate squash, you will be making a morning inspection and an evening inspection of your plants for quite some time.

Male Flower (left) and Female Flower (right)
(Note "Baby" Squash Below Female Flower)

In the late afternoon or evening, find both a male and a female flower on the same plant that are unopened, firm and yellow. We use 3/4 inch masking tape to seal the tips of these flowers and keep them from opening. (Warning: some cheap or generic tapes are not sticky enough and will pop open when wet with the morning dew. We used Anchor Brand purchased at a Coast to Coast Hardware Store. Or just experiment until you find one that works for you.) Tear off a 3 inch chunk of tape, wrap it around the end of the flower so that just a little bit of the tip sticks out past this band of tape, and pinch the sticky ends of the tape shut against themselves.

Tear Petals off of Male Flower & Tear Tape off of Female Flower

Wait until the dew dries the next morning, because you will be taping the female flower back shut after you make your pollination and the tape doesn't stick very well when it's wet. Find the male and female flowers you taped shut the previous evening. Pick the male flower and leave a good chunck of stem on it to use as a handle. Tear the petals off of the male flower and then hold its stem between your teeth so that you'll have both hands free. Now tear the tip off of the female flower in a straight line just at the bottom edge of the circle of tape. Then, as if in slow motion, the female flower will open wide.

Now take the petal-less male flower and hold its stem like a handle. Swab the pollen-covered anthers of the male flower on the stigma of the female flower. Try to rub pollen on each of the stigma's different sections. (Mark Widrlechner told me that at the Plant Introduction Station in Ames they routinely use three male flowers on each female flower to get a better saturation of pollen.)

Transfer Pollen

Tape Female Shut Again

Now tape the female flower shut again and the pollination is complete. Sometimes an especially brittle flower will split along one of the large seams that run up and down. If that happens, just keep taping closer and closer to the fruit until the hole is covered. You can actually cover the entire blossom with tape and the technique will still work. But be careful to not do any damage to the base of the flower where it attaches to the fruit or it will abort. The only thing left to do is to mark that fruit in some way so that you will know at harvest time which fruits contain pure seed. We use surveyor's red plastic ribbon and tie a piece in a loose knot around the fruit's stem, so it won't constrict its growth.

That's all there is to it! And that is exactly the same technique that is used on watermelons, muskmelons and cucumbers, except that their flowers are much smaller and are not so easy to manipulate. Self-pollinating a naturally cross-pollinated crop can lead to what plant breeders call "inbreeding depression." Fortunately, squash don't show any signs of inbreeding depression.

Gary Nabhan told me about a study that was done by W. P. Bemis at Arizona State University. Each week they took seed from fruits in an attempt to determine at what point there was the greatest number of fertile seeds. They found that point to be 20 days after the fruit is fully mature (skin not cut by thumbnail). During that 20-day period after the fruits were picked, the seeds continued to "plump up" and gain strength. If the squash were picked when immature, the seeds gained weight during this period, but never came near to gaining the strength displayed in the squash that were picked when mature. They tried it with a number of other cucurbits and it seems to hold for all of them. So we should all remember that there is a 20-day after-ripening period when the seeds actually improve in the fruit after you pick it.

MELONS - Most of the melons that we grow in the U. S. are "netted" and the stems "slip" from the fruits when ripe, therefore they are Muskmelons - Cucumis melo (Reticulatus Group). European Cantaloupes - Cucumis melo (Cantalupensis Group) - are the only "true" cantaloupes and they have no netting and the fruits never slip from the vines, which makes them perfect for trellis culture. Honeydew and Casaba melons make up another group which are Cucumis melo (Inodorus Group). There are several other minor groups of melons which include: Oriental Pickling Melons - Cucumis melo (Conomon Group); Vine Peach - Cucumis melo (Chito Group); Vine Pomegranate - Cucumis melo (Dudain Group) which is also called Pomegranate Melon, Plum Granny, or Queen Anne's Pocket Melon; and the Serpent Cucumbers - Cucumis melo (Flexuosus Group) which are sometimes called Armenian Cucumber or Snake Cucumber, but they are melons.

All of these different "groups" have the same genus and species -- Cucumis melo -- so any of these melons will cross with each other. You must either grow only one variety of melon or hand-pollinate different varieties or separate different varieties by 1/4 of a mile. Hand-pollinating will allow you to grow more varieties, but muskmelons are by far the most difficult vine crop to work with. Partly this is due to the small size of the flowers. But the main problem is that muskmelons just naturally abort probably 60-70% of the female flowers that bloom. You have no way of knowing which flowers the plant is going to abort, so, no matter how good your hand-pollinating technique is, you are going to have a lot of failures. You can improve your chances slightly by getting to the very first female flowers to bloom, because those are more likely to set fruit. Each time the plant sets a fruit, more of its flowers are going to abort from then on. So be sure to pinch off all fruits that weren't hand-pollinated, which will keep the plant blooming and again improve your chances.

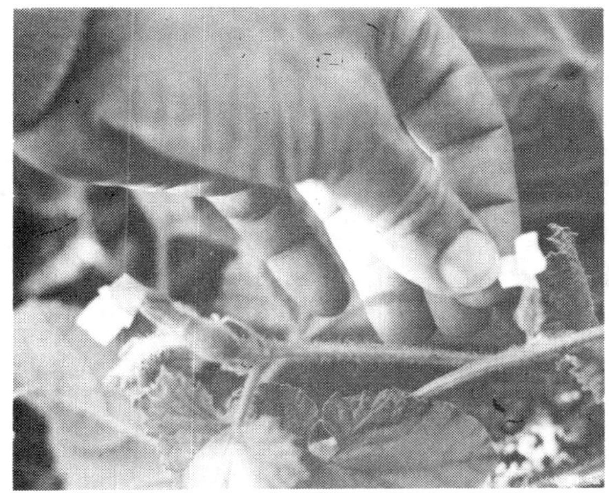

Female on Left

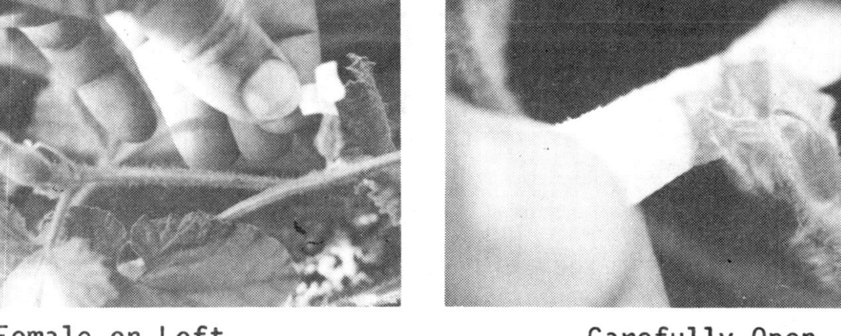

Carefully Open

To hand-pollinate muskmelons, use exactly the same technique as was described for squash. The flowers are tiny and require a much more delicate touch. It's harder to tell when a flower is ready to open, but with careful observation you will soon learn. You'll need about a 1.5" long piece of the 3/4" masking tape and then you'll have to tear that strip in half lengthwise. When sealing the female flower shut, wrap that narrow strip of tape around the tip of the flower, pinch the tape together just beside the flower, but leave the two "tails" of the tape apart. That will make it much easier to get hold of when you are opening the female flower the next morning. Be careful not to break the little stem off of the male flower when you are tearing off its leaves. That happens very easily and when it does you will have to use tape to build a stem.

Transfer Pollen

Tape and Mark

Muskmelons are so hard to hand-pollinate, that we are thinking of trying an entirely different technique next year. At the Plant Introduction Station in Ames, Iowa, they use a caging technique on cucumbers that would also work on muskmelons (the USDA muskmelon collection is in Georgia). They cover several plants with a tall cage made of aluminum tubes covered with a Reemay-like material. Then they take a miniature hive of bees and build it into that fabric wall of the cage. The hives are specially built so they can let the bees into the cage one day and into the outside environment the next day. By alternating days like that, the bees get enough food to sustain them. When the bees are in the cage, they naturally pollinate the flowers and all of the fruits that are set contain pure seed.

Next year we are going to experiment with planting a cluster of three plants and then watch them closely until they set their first female flower. At that point we would cover the plants with a cage of Reemay. My friend Lorado Adelmann is a beekeeper and I imagine he would give me a jar of bees. I'd introduce a few bees into each cage and then try to keep them alive as long as I could with a honey/water solution. When the plant had set all of the fruit that it was going to, the cage could be removed and the fruits could be marked. Now I don't know if that will work, but I'm going to try it next summer. There's got to be more than one way to skin a muskmelon.

When harvesting for seed, let the melon get overripe. Scoop out the seed and wash it in a sieve under running water. Rub the seeds and pulp with the backs of your fingers and try to force as much of the pulp as possible through the sieve, because it makes a good home for mold and bacteria and diseases. After draining the water off, place the seeds on paper plates or even a piece of glass to dry. Then put the seeds into envelopes, label and store.

<u>WATERMELON</u> - (Citrullus lanatus) - is also an insect-pollinated crop. Watermelons will cross with any other varieties of watermelon and with citron. Unless you hand-pollinate your watermelons, only one variety can be grown within 1/4 mile of any other watermelon variety.

Watermelons are relatively easy to hand-pollinate, even though the flowers are rather small. You can usually expect, when conditions are favorable and the plants are not under stress, to achieve between 50 and 75% success with your hand-pollinations. The very first female flowers that the plant produces are usually the ones that set. If you don't get those first ones, your success rate will go down each day. It's important with watermelons to use two or even three male flowers when making pollinations, because there isn't very much pollen on the male. You can tell when a male flower is ready because the anthers, the tiny circular ridged pollen-bearing structures in the center of the male flower, will look like a fuzzy little yellow ball once the pollen shows. A little trick that Glenn discovered last summer is that about 95% of the time, if you find a female flower that is going to open the next morning, the second flower back from that will be a male flower that will open the next morning also. That seems to work almost all of the time and that keeps you on the same vine.

Pollinating Watermelon

Taped and Marked

When working with tiny flowers, it's sometimes hard to tell if they are going to open the next day. Go ahead and tape any flowers that you are not sure about. As with squash, when you take the tape off the next morning, if they are ready, they'll pop open. But if the flower is still sort of greenish and the petals are closed in tight and not even starting to split, then you know it's not going to open that day. Take the tape off, remember where that flower is and come back and tape it again the next evening. Don't try to leave the tape on for two days or the flower won't develop and will just dry up under the tape. Watermelons seem to have a lot of female flowers that contain both male and female parts, but those male parts are usually sterile. So don't ever think that you can just tape such a flower shut and it will pollinate itself, because it won't.

Another tip that Glenn mentioned involved sort of a timing cycle, if you are hand-pollinating squash and watermelons and muskmelons all on the same day. You should work with the squash first, because the squash pollen will go bad by noon, unless it's a cool day. Work with your watermelons second, since their pollen is good slightly later in the day because it needs the warmth. And work with the muskmelons third, because they seem to have an even later daily cycle. As with other vine crops, you should pick off any fruits that formed which you didn't hand-pollinate. That will cause the plant to produce more female flowers instead of pouring its energy into feeding the fruits that you didn't want. And if it's getting late in the season and you're not sure if a fruit is going to mature, don't hesitate to pinch off all of the tips of the vines on that plant. That will force the plant's energy into the fruits as opposed to producing more vines.

It's always hard for me to tell when a watermelon is ripe. I've heard so many theories that it's just ridiculous. I went with the "plink - plank - plunk" method of thumping for quite awhile, but only with limited success. All of mine seem to "plunk" (maybe I'm just too anxious). Glenn came up with a method that seems to work almost every time. Look for the small tendril where the fruit's stem attaches to the vine. Watch for that tendril to turn really dry as your first sign that the watermelon is about ready. As an additional sign, watch for the light patch, where the underside of the melon touches the ground, to change color. The color of that patch will change to the next shade darker: white to cream; cream to yellow; or yellow to almost orange, on the super dark-skinned types. See if that works for you. Let watermelons that you intend to save for seed become a little overripe. Remove the seeds, wash in cool water, and then dry as you would muskmelon.

CUCUMBERS - (Cucumis sativus) - are insect-pollinated, so they will cross with other varieties of cucumbers and with West Indies Gherkin (Cucumis anguria). Remember that Armenian Cucumber or Serpent Cucumber is actually a melon (Cucumis melo, Flexuosus Group) and will therefore cross with muskmelons and not cucumbers. Unless you hand-pollinate your cucumbers, you must grow only one variety or separate any two varieties by 1/4 of a mile. And don't forget that your neighbor may be growing them also.

Cucumbers are easy to hand-pollinate, but not every pollination will be successful because the plant produces far more female flowers than it can possibly handle. Also don't be surprised if you hand-pollinate a fruit and get it to set, but when you cut open the fruit at harvest time it won't have any seed in it. This has to do with the number of insect pollinators and the fact that cucumbers tend to be parthenocarpic (capable of developing a fruit without fertilization). One study done with cucumbers showed that, if there were sufficient bees in the area, the average female flower was visited 52 times by bees while it was receptive. I know that none of us are going to be that persistent, but the more male flowers that you use, the better your chances for success. If you have the time, you should use several different male flowers on each female flower. This holds true for all vine crops, but I realize that each of us can only do so much.

When saving seed from cucumbers, let the fruits ripen past the edible stage. Depending on the variety, it will either turn orange or deep-yellow or white. The best stage for harvesting seed is when the cucumber starts to get a bit soft. Pick the fruit, cut it open, and scoop out the seeds which are each surrounded with a slimy little gelatinous sack. Put these seeds in a glass jar or a plastic container, such as a yogurt or cottage cheese carton. Let them ferment for about three days (for a more complete description of this technique, see TOMATOES). Stir every day and keep the temperature between 75-85 degrees F., if you have any control over it. When this mixture starts to separate, the bad seed and pulp will float to the top and can be poured off. The good seeds will sink to the bottom of the container. Wash them in a sieve, drain off excess moisture, spread on a paper plate, and let dry.

GOURDS - (Lagenaria siceraria) - are the hard-shelled gourds with evening-blooming white flowers. They will not cross with any of the other vine crops we have discussed. The most common edible variety is known as Cucuzzi (but is also called Italian Edible Gourd, Italian Climbing Gourd, Guinea Bean Gourd, New Guinea Butter Bean, or New Guinea Buttervine). In its immature stages it is used just like zucchini and many people like it better. All of the hard-shelled gourds used for gourdcraft belong to this group and will easily cross. (But remember that the small striped and warted gourds are Cucurbita pepo.)

Female Flower (center) and Male Flower (right)

All of these hard-shelled gourds have beautiful white flowers that open in the evening and bloom during the night. (They'll open in late afternoon if it's a hazy day). So in order to hand-pollinate them, we must reverse the morning/evening sessions that we have been using for all of the other vine crops. During the morning, find the male and female flowers that are about to open, and tape them shut. Then you have to go back in the evening, when the flowers would normally be opening, and make your pollinations. The flowers will open up quite easily and they have an awful odor that you can't miss. Except for the reversal of the sessions, the technique is exactly the same as with squash. Before frost, bring the mature fruits into the house and let them dry down slowly. By mid-winter all that will remain of the gourds will be a thin hard shell and the seeds will rattle around inside when shaken. That's the perfect container for the seeds until you are ready to break it open and plant them.

POTATOES - (Solanum tuberosum) - won't cross at all since tuber divisions are really just clones. Crossing between potato flowers affects the seed balls, not the roots. Select several of your best-looking plants, which are surrounded by other healthy plants, to save for seed. Never keep potatoes for seed that show any sign of scab or virus. You may be able to increase your production by planting small (hen egg-sized) whole potatoes, since they are less apt to be badly sprouted and often will produce a vigorous plant more quickly than cut potatoes. If you are planting small whole sprouted potatoes in the spring, don't damage the big sprout on the eye end of the potato since this will produce your most vigorous plant. Most of the other sprouts you can just break off. Some people think that yields can be improved by planting sprouted potatoes. It does appear that the plants emerge sometimes in only a few days and in some cases as long as two weeks ahead of non-sprouted potatoes. Dig your potatoes when the vines begin to dry up, because when the soil loses its shade it becomes hot which may damage your crop. How well your potatoes keep does not seem to be affected by either washing or not washing. After drying in the shade for only a few hours to toughen their skins, they are ready to store. As for storage temperature, the colder the better -- 34 to 40 degrees F. Burying them in dry sand is said to be the perfect way to store them.

Those folks who are building up collections of potatoes at some point are going to have to deal with virus problems. Potatoes that are kept for several years will gradually start loading up with virus diseases. These problems can eventually become so bad that you won't get back any more potatoes than you plant. These diseases are most commonly spread through the field by aphids. Even using a heavy chemical culture will not prevent that, since the insects usually penetrate the plant before they die.

There is one older technique that Dr. Bob Coffin, a potato breeder with Agriculture Canada, mentioned at a conference in Toronto in 1984. The technique is called "eye indexing" and is done in a greenhouse over the winter. Cut one eye out of a potato you intend to grow next summer. Make a numbered marker which will stay with the eye and write the same number on the potato with an indelible pen before you enter it back into storage. Then grow out that eye in the greenhouse and take a look at the plant. If the plant shows signs of virus (lighter or yellowish foliage, curled leaves, etc.), then throw away the plant and also the potato it came from. If it looks healthy, then mark that potato as one that you want to grow for seed the next year. Of course, it goes without saying that you should always rogue any signs of disease out of your potato plantings and get them away from your garden. It is probably also a good idea to grow any new potatoes you receive in a separate plot for a year before you add them to the garden collection you have built up.

TOMATOES - (Lycopersicon lycopersicum) - are said to be over 98% self-pollinating. But you should realize that even such a slight amount of insect pollination, over a number of seasons, may be enough to destroy the characteristics which made that variety unique. Studies have shown that if you separate varieties by just 6 feet, that 2% of crossing goes right down to nothing. So don't grow tomato varieties side by side if you intend to save seed from them. Also it is a mistake to just save seed from, for example, only your largest tomato. You will have a much greater genetic diversity in your seed samples if you save seed from the smallest, largest, earliest and latest fruits. (This would only be twice the work if you saved seed from the earliest and also large fruits at the beginning of the season, and the latest and also a small fruit near the end of the season.) Equal amounts of these seeds should be mixed.

We used what I feel is an excellent and even exemplary technique to save our tomato seeds, so I'm just going to lead you through it. This was the first summer that I have saved seed from a large number of tomatoes -- 280 varieties. I went to our local grocery store and bought a case of plastic deli containers. They were the one quart size and were a shiny white plastic with a clear plastic lid. A case of 250 of them costs $22. We tried to take seed twice during the season. We took it the first time that there were enough ripe fruits and we took it again right before frost. We'd pick a pile of tomatoes that were completely ripe and even a little soft.

I always carry a large lock-back knife, which is perfect for all the cutting you do in the garden. It's like having a butcher knife that you can fold up and put in your hip pocket. Stick the knife up into the blossom end of the tomato and then squeeze the seeds and the gel surrounding them into one of the deli containers. That works perfectly for any tomatoes like the long or paste types or any small fruited ones. For the huge flat tomatoes, cut a big "X" on their blossom side and try to squeeze them into the container. But each type of tomato is different because their cavities are different. Some of the large meaty tomatoes have just a few little cavities that run up and down along their outer edges and terminate in the little ridges on the shoulders. For those it worked best to actually stick the knife into each ridge on the shoulders and then squeeze the seeds out from the top. You'll just have to get to know each of your tomatoes. We always tried to pick enough tomatoes to fill a deli container about 3/4 full. And we used an indelible pen to write the tomato's name and number on the outside of the container. (Be sure the indelible pen you are using is really waterproof. Write on what you intend to and then briskly wash the writing.)

We put the lids on the containers so they wouldn't spill and hauled them all up to a nearby farm where we had the use of a shed. We turned an old horsetank upside down in the shed which gave us a varmint-proof "table." We took the lids off of the containers and left the containers on this table. The mixture in the containers gradually starts to separate. The good seed sinks to the bottom and the bad seed and the pulp will float to the top. After about three days, if it was warm weather, each of the containers will be completely covered with about a 1/2" spongy layer of fungus. It's nasty terrible-smelling stuff and you definitely wouldn't want it near any living areas of your house. Some people think that they need to add water to the containers, but that just slows down the fermentation. Other folks stir it daily, but that wouldn't let the layer of fungus form. Dr. Mike Courtney, tomato breeder who took over Dr. Ernest Kerr's position at the Simcoe Research Station in Ontario, told me that this fermentation process destroys every seed-borne disease that they have any problem with.

One thing you will have to watch out for is leaving the seeds in the container for too long. At the end of about the fifth day, the tomato seeds will start to sprout in the mixture. The little gelatinous sack which surrounds each seed contains a chemical growth inhibitor. That's what keeps the seeds from sprouting inside of the wet fruit. The fermentation process breaks down those sacks and also the chemical, so this procedure has to proceed on schedule. When a layer of fungus completely covers the surface of the mixture, it's time for the next step in the technique.

For this next step, you will need a sieve, paper plates, running water, and a place to get rid of a truly nasty substance. We did it in a cow yard, which was a perfect place. Take a container and tip it until the liquid is about to run out. Then take your fingers and rake out that layer of fungus and let it fall to the ground. Pick out and throw away the larger chunks of tomato that are still in the container. Pour the remaining seeds and liquid into the sieve. Hold the sieve under a small stream of running water. Now use the backs of your fingers on the hand that's not holding the sieve to rub the remaining tomato pulp through the screen. Rub a little while, then wash it, then rub a little more. It won't be long until the only thing in the sieve will be clean bright tomato seeds.

Time for Processing

Tip Container Slightly

Rake off Fungus Layer

Pick Out Large Chunks

Pour into Strainer

Wash & Rub until Clean

Let the excess water drain out of the sieve. Now reach over and hit the edge of the sieve sideways (horizontally) repeatedly against a post or something until the seeds are setting up in a ball in the bottom of the sieve. Then plop them out onto a paper plate, label the plate and put it somewhere safe to dry. We also tried to save ourselves some work by washing the deli containers and then putting the seeds back into them, since they were already labeled, but the seeds molded. You need to put the seeds on some sort of paper, so the moisture can get away. What we actually ended up doing was to roll paper plates into cones, stand the cones up in the washed containers, plop the seeds into the cones, set the container & cone & seeds back in the shed on our table, wait until the ball of seeds was dry, transfer the dry ball of seeds into the container, throw the cone away, bring the labeled containers with the dry seeds back to the house, put the seed into envelopes and dry it some more before storage.

One thing that I can't stress enough is that every step of the way there are opportunities for a few seeds from the previous tomato to be carried over on your hands or on the sieve and they can ruin all of your work. We did all of this right beside a horsetank. We would: rake off fungus; WASH OUR HANDS; rub pulp through sieve; WASH OUR HANDS; plop seeds onto paper plate; WASH THE SIEVE; etc. You've got to be so careful, because that's a summer's work right there in your hands and it's seed that you will be giving out to other gardeners.

CORNS - are wind-pollinated, so any corn (sweet corns, popcorns, ornamentals, dents, flints, etc.) will cross very easily with any other corn. So in order to keep a corn "pure," you will either have to grow it a quarter of a mile from any other corn, or you will have to hand-pollinate it. Corn is very "plastic," so by careful observation and selection you can gradually determine the characteristics that your future crops will have. Let the ears you select ripen on the stalk until the husks are dry. Then pick the ears, pull the husks back, tie several ears together by the husks, hang in a dry, well-ventilated place until thoroughly dry, shell, save only completely formed kernels, and store. (Note: an early and a late variety can be planted side by side, if the early variety stops pollinating before the silks of the late variety begin to emerge.)

I had intended to completely redo this section with an entire article on the "Hand-Pollination of Corn." I certainly had some tremendous pictures to illustrate it. But the transcript from Dr. Mark Widrlechner's demonstration at the Campout was just packed with information. So I decided to add my photos to his demonstration and the result is fantastic. So if you are interested in hand-pollinating corn, read Mark's article which follows this "Seed Saving Guide."

SORGHUMS - This group includes all milos, sorghums, broom corns, and sudan grass/sorghum hybrids. Any of these will cross with each other since they are all sorghums. All I know about keeping sorghums pure is that you have to bag the seedheads. We need to have O. J. Lougheed teach us all about the pollination of grains.

THE BIENNIALS

[Special thanks to Glenn Drowns for sharing his knowledge of and techniques for biennials, which have greatly enlarged this section.]

To save seed from biennials you must either live in an area with mild winters or have a place to store the plants over the winter.

CARROTS - (Daucus carota var. sativus) - are cross-pollinating and will cross with wild carrot, which is also known as Queen Anne's Lace. Queen Anne's Lace can be easily identified by pulling the plant and smelling the root. If it smells like a very strong carrot and is whitish in color it is Queen Anne's Lace. To save seed and ensure purity at least 1/4 mile between each variety and Queen Anne's Lace must be maintained. A careful selection process is necessary when saving carrot seed. In the autumn of the first season select 12 or more roots which are uniform and true to the variety in both shape and color. In mild winter areas (where the ground doesn't freeze) the roots can be replanted at that time. In most parts of the country the roots will need to be stored in a cellar over winter. When spring arrives, plant the roots in a block fashion, a root every 12 inches. The plants will grow to as much as five feet and become quite bushy. Six roots will supply a large amount of seed, but plant extra to ensure good pollination.

BEETS - (Beta vulgaris, Crassa Group) - are cross-pollinated. In this country there isn't a wild counterpart, but different varieties must be 1/4 mile apart. Also Beets will cross with Swiss Chard. Beets are not very winter hardy so they should be stored in a cool place over winter. Never save seed from a beet which bolts the first year, as this is a serious defect. Save your best 6 to 12 roots and replant them in the spring about 12 inches apart block fashion as you would do with carrots. Beets are sometimes self-incompatible so plant at least six roots. Self-incompatibility is when one plant needs pollen from another plant to pollinate it even though it produces pollen itself. The yield of beet seed is heavy, around a gallon jar per plant. The plants will grow 4-5 feet tall with many side branches.

PARSNIPS - (Pastinaca sativa) - Some controversy exists as to whether this crop is cross-pollinated or self-pollinated. It is best to assume it is cross-pollinated and separate varieties by 1/4 mile. Also, varieties should be 1/4 mile from wild parsnip. People call a number of things wild parsnip, but true wild parsnip can be easily identified by pulling the plant. If it is wild parsnip, you will get a very strong parsnip smell. Parsnips can be left in the ground all winter as they are very hardy. A careful selection process as to uniformity is essential to maintaining a variety.

SALSIFY - (Tragopogon porrifolius) - Use the same procedure as with parsnips. There is a wild salsify plant. It has similar foliage and has yellow flowers which produce dandelion-like blooms when ripe. Some possibilities exist for crosses between domestic and wild salsify.

TURNIPS - (Brassica rapa, Rapifera Group) & RUTABAGAS - (Brassica napus, Napobrassica Group) - are both cross-pollinating biennials. Follow the same instructions as for beets. Some crossing can occur between turnips and rutabagas and also turnips will cross with other turnip varieties and rape. Some crossing can occur but isn't likely with radish, mustard and Chinese cabbage. To prevent crosses between the last three vegetables and turnips, separate by length of your garden. Rutabagas will cross with rutabagas and rape. Rutabaga varieties will also cross with the same vegetables as turnips with slight chances of producing a cross between the related species.

CELERY - (Apium graveolens var. dulce) - & CELERIAC - (Apium graveolens var. rapaceum) - Both will cross with one another and each will cross with other varieties of itself. Again you should save 6 to 12 plants over winter and then select the best 4 to 6 the next spring. The plants will grow quite large, to five feet in some cases, and each healthy plant will produce an abundance of seed. All of the seed will not ripen at once.

ONIONS - (Allium cepa, Cepa Group) - & LEEKS (allium ampeloprasum, Porrum Group) - are usually very hardy. Onions are cross-pollinating, so grow only one variety and though not confirmed, leeks will probably cross with onions. This means you can only grow either one onion variety or one leek variety, not one of each. To get a good supply of seed you will probably want to plant between 12 and 15 plants as the seed production is not as high as with beets or carrots. The seed will be produced in a ball-like structure at the top of the plant. It needs to be harvested when the ripe black seed becomes visible or your crop will shatter onto the ground. Under no condition save seed from plants which bolt to seed the first season, because this is highly undesirable.

CABBAGE FAMILY - (Brassica oleracea) - (cabbage, broccoli, brussels sprouts, kale, collards, cauliflower and kohlrabi) - All of these will cross with each other so you must choose one variety from the entire group. You cannot save seed from one brussels sprout and one cabbage, for example, because the two will cross. Be selective for what you save seed from and in most cases support will have to be given to the plants during flowering as they will grow quite large. A good healthy cabbage plant will yield as much as 1/2 pound of seed. Remember to save at least three plants for seed, as many members of this group are self-incompatible, so if you only have one plant you will not get any seed.

PARSLEY - (Petroselinum crispum) - Let the plant overwinter in the garden and it will produce an abundance of seed the second year. Different varieties will cross, so grow only one.

ENDIVE - (Chicorium endiva) - is really a biennial, but, if you have a long season and can plant it very early, it will seed the first year.

And seed saved from this will not be genetically inferior. If winters are mild, warmer than 0 degrees F., the plants will overwinter and seed the next year. The plants are self-pollinating. Follow the same procedures as with lettuce.

CHICORY - (Chicorium intybus) - Both leaf and root types exist and both are biennials. Plants will cross with wild chicory, a common roadside weed. The plants have blue flowers. When saving seed the plants of root chicory can get to seven feet tall, so some support may have to be given. It is cross-pollinating so grow only one variety.

SWISS CHARD - (Beta vulgaris, Cicla Group) - should be treated the same way as beets and it will cross with beets. So grow only one variety of either Beet or Swiss Chard, not one of each.

PERENNIALS - Horseradish is propagated by planting sections of either the main root or the smaller lateral roots of the parent plant. We do this in the spring, but I am sure that in milder areas it could be done in the fall. This is very easily done, so don't plant horseradish where you don't want it to always be. Jerusalem artichoke is multiplied like potatoes and is also very hard to kill out once started. Rhubarb is propagated by division of the crown, since the seeds don't often come true. Asparagus can be multiplied either by dividing the root crowns (make sure each piece contains at least one growing point on the top), or by saving the pea-sized berries of the female plants in the fall before frost.

HOT WATER TREATMENT OF SEEDS is a method of controlling the seed-borne phase of diseases such as black rot and black leg in the cabbage family, bacterial canker and target spot in tomatoes, and Septoria spot in celery. You will need an accurate thermometer, electric frypan, large sauce pan, kitchen sieve, and paper towels. Try a practice run without the seeds. Heat some water to 50 degrees C. Pour a little of this water into the warm electric fry pan, fill the sauce pan 2/3 full and set it in the fry pan. Regulate the temperature by either turning up the fry pan or taking the sauce pan out of the fry pan. When you can maintain 50 degrees C., pour in the seeds, stir until they are all wetted and not floating, then stir gently through the whole process. Treat broccoli and brussels sprout seed for 20 minutes at 50 degrees C., cabbage for 30 minutes at 52 degrees C., cauliflower for 25 minutes at 52 degrees C., celery and pepper seeds for 30 minutes at 50 degrees C., and tomato for 25 minutes at 55 degrees C. Then sieve the seed and spread it on paper towels away from direct sunlight, dry and store.

GERMINATION TESTING - In fairness to our fellow members, we should all test the seed we send out. Take 10, 25, 50, or 100 seeds for each variety, roll in damp paper towel, put in a plastic bag, and place in a warm location. Count the sprouts after 7-10 days. Seven sprouts out of 10 seeds would be 70% germination, etc. Basically the idea is to be sure that at least some of your seed is going to sprout. It's much better to discover now that you have more to learn about saving seed, than for another Member to discover the problem next spring when they are waiting for your seed to come up.

ISOLATION - In most sections of this "Seed Saving Guide," I have used isolation and planting only one variety of a cross-pollinating crop to maintain purity. It should also be noted that even self-pollinated plants may cross if conditions are just right. So if your aim is absolute purity and you are saving seed from more than one variety of a self-pollinating crop, separate the varieties by at least a row or two of another crop.

Additional Reading

Growing Garden Seeds by Rob Johnston, Jr. (available from Johnny's Selected Seeds, Albion, ME 04910, for $2.30 postpaid, or $1.95 with seed orders). Covers each species thoroughly with information on flowering, pollination, culture for seed, distances to separate varieties for purity, seed production, harvest, threshing, and cleaning.

Growing and Saving Vegetable Seeds by Marc Rogers (available from Garden Way Publishing, Charlotte, VT 05445). Contains a very clear and illustrated description of hand-pollination techniques, for both corn and vine crops. Discusses saving seed from biennials at length.

Agricultural & Horticultural Seeds, FAO, Rome, 1961.

The Seed Grower by Charles Johnson, 1906.

Seed Production & Marketing by L. R. Hawthorn and L. H. Pollard, Blakiston Co., 1954.

Cucurbits by T. W. Whitaker and Glen H. Davis, Interscience Publ., New York, 1962.

Species & Varietal Crosses in Cucurbits by A. T. Erwin and E. S. Haber, Agricultural Experiment Station, Iowa State College, Ames, IA, 1929.

Principles of Plant Breeding by R. W. Allard, John Wiley and Sons, NY, 1960 (available through Rob Johnston's catalog, address above).

Principles and Practices of Seed Storage by Oren L. Justice and Louis N. Bass, USDA, Agricultural Handbook No. 506, 1978.

Viability of Seeds, edited by E. H. Roberts, Chapman and Hall Ltd., 11 New Fetter Lane, London, 1972.

Hybridization of Crop Plants by Walter R. Fehr and Henry H. Hadley, 1980, 765 pages, from the American Society of Agronomy, Inc., 677 S. Segoe Road, Madison, WI 53711.

Glossary of Plant Genetics

By Robert L. Johnston, Jr., President
Johnny's Selected Seeds

ADAPTION. The relationship of a plant or variety to its environment or growing method.

ALLELE. One of two possible forms of a single gene which can express itself as either recessive or dominant.

ANTHERS. The pollen-bearing structures which, supported by the filaments, form the stamens within a flower.

ASEXUAL REPRODUCTION. Reproduction of the next generation by vegetative tissue (i.e. tubers, cuttings, cell culture, etc.) instead of by seed.

BREEDING. The art and science of changing an organism genetically.

CHROMOSOMES. Strands within each cell's nucleus (being characteristic in number for each species) made of the complex chemical DNA. See "Gene."

CLONE. An organism or group of organisms derived by vegetative (asexual) rather than sexual means.

CROSS-POLLINATION. Transfer of pollen from the anther of a flower to the stigma of a flower of another plant.

CULTIVAR. A plant variety originating under cultivation. Generally, a named variety.

DOMINANCE. The interaction between two alleles such that one allele (the "dominant" one) is the only one to express itself when paired with the other allele (the "recessive" one).

ENVIRONMENT. The sum of external conditions which affect growth and development of an organism.

F1. The first filial generation or first generation after a cross, usually referred to as F1 hybrid or simply a hybrid.

F2. The second filial or second generation after a cross grown from seeds obtained by self-pollinating F1 individuals.

F3, F4, ETC. F3 is the third filial generation, F4 the fourth, etc.

FERTILITY. The ability to produce functional seeds and, in turn, viable offspring.

FERTILIZATION. The fusion (joining into one) of the nuclei of male and female gametes.

GAMETE. An egg or a pollen grain formed after meiotic division. A sex cell.

GENE. The unit that carries hereditary traits. It is a section of a chromosome just long enough to spell out one "instruction" in a plant's "recipe."

GENOTYPE. The genetic constitution of an individual.

GERMPLASM. The sum total of hereditary materials in a species.

HETEROSIS. Extra growth rate, size, or productivity generally shown by F1 hybrids.

HETEROZYGOUS. Having unlike alleles at one or more loci. Indicates some degree of diversity within a variety or plant population. Generally speaking, a necessary condition to maintain vigor in cross-pollinated crops. Opposite of homozygous.

HOMOZYGOUS. Having like alleles at one or more loci. A self-pollinated variety or an inbred is homozygous at all loci.

HYBRID. The product of a cross between genetically unlike parents.

IMPERFECT FLOWER. A flower having stamens (male) or pistil (female) but not both.

INBRED. A plant resulting from continued inbreeding. An "inbred line" is a nearly homozygous line usually originating by continuous self-fertilization, accompanied by selection.

ISOLATION. The separation of one plant (or group of plants) from another so that mating between them is prevented.

LINE. A group of closely related plants derived from pollination of a single plant or a small number of plants.

LINKAGE. An association of two or more genetic characters due to the close proximity of their controlling genes on the same chromosome.

MALE STERILITY. The absence or nonfunction of pollen in plants.

MEIOSIS. The process of cell division in the anthers and ovary during which there is an exchange of genetic material, and after which the pollen grains and ovules (sex cells) form, bearing half the number of chromosomes as the mother plant.

STIGMA. The surface of the pistil on which pollen grains are deposited during fertilization.

MONOECIOUS. Male and female flowers formed separately on the same plant.

MUTATION. An unexpected, heritable change in a gene.

PARTHENOGENESIS. Development of a fruit without fertilization of the ovary. Plants or varieties that are capable of this are called "parthenocarpic."

PEDIGREE. A record of the ancestry of an individual, variety, or family of plants.

PERFECT FLOWER. A flower which has both stamens (male) and pistil (female).

PHENOTYPE. The appearance of an individual, opposed to its genetic make-up (genotype).

PISTIL. The female reproductive organ which produces female sex cells and bears the seed. Usually consists of an ovary, a style, and a stigma.

PROGENY. Offspring of (usually) a single plant.

RECESSIVE. The allele which expresses itself only when homozygous. It is masked by the dominant allele in the heterozygote.

RECIPROCAL CROSS. A cross-pollination made in the opposite direction from the initial cross, in which the male and female are reversed.

ROGUE. An unwanted or atypical plant in a finished breeding line or variety. "Roguing" is the removal of these individuals to purify the stock.

SELECTION. Retention of desirable individuals in a plant population during a breeding program.

SELF-POLLINATION. Transfer of pollen from the anthers of a flower to the stigma of the same flower or another flower on the same plant.

SIBS. Individual plants derived from the same parents.

SPECIES. The unit of taxonomic classification into which genera are divided. The maximum interbreeding group, isolated from other species because of reproductive incapacity between them.

STAMEN. The male reproductive organ which produces the male sex cells carried in the pollen grains. Usually consists of a filament and anther.

STYLE. The usually elongated portion of the pistil connecting the stigma and ovary.

VARIATION. The occurrence of differences among individuals due to differences in their genetic make-up and/or the environment in which they were raised.

VARIETY. A subdivision of a species. A distinct group of individuals.

ZYGOTE. Cell formed by the union of two gametes.

****

Supplies for Seed Savers

JOHNNY'S SELECTED SEEDS (Albion, Maine 04910 207/437-9294) Rob Johnston is carrying a material in his catalog called "Reemay", which is a white spun polyester cloth that is 67" wide. It is extremely lightweight, lets the light and rain in, holds heat in, but keeps insects out. It's perfect for many caging techniques, when stretched over some sort of support and it's edges are buried. Some people use small pieces of it to tie around individual blossoms. Seeds or pods or even whole plants can be placed on pieces of it, tied into bundles, and hung under cover to dry. Even though it is thin and spun, it might be possible to sew it for bags, but we haven't tried that yet. Reemay's applications seem endless. A brochure is available, or just request Johnny's free catalog.

CRYSTAL SPRINGS PACKAGING COMPANY (PO Box 2924, Petaluma, CA 94953 707/778-0567) John R. Quinn, one of our "Listed Members", handles all types of packaging, including the following items: 2 or 4 mil. zip-lock plastic bags in sizes from 2" X 3" on up (minimum of 100, samples of beans & corns look great in them); all sizes of tuck-top cardboard boxes from 3" X 2" X 1" on up; heat-sealable paper/poly/foil/poly bags for air-tight storage of seeds (can be sealed with a Seal-A-Meal II from Penny's, see "Seed Saving Guide"); many sizes of inexpensive glassine envelopes which some people use to cover hand-pollinated blossoms; many sizes of muslin draw-string bags which are excellent for bagging blossoms and can also be used to hold seeds that are being dried in an airtight container with silica gel (see "Seed Saving Guide" for a description of this excellent drying technique); beaded color-indicating silica gel in 8 oz. & 1 lb. & 5 lb. sizes (prices as of last issue were $3.25 & $5.25 & $23.00 postpaid); two types of large heavy-duty heat sealers for under $100; double-faced aluminum tags with brass grommet and wire ties which can be permanently written on with a ballpoint pen and used to mark plants or pollinations; plastic stakes in 3" or 4" or 5" sizes for the same purpose; bubble pouch mailers and styrofoam-lined manila envelopes for mailing a large number of packets of seeds; and coin envelopes. John also has all sizes of cardboard boxes, but he says you can probably find them cheaper locally because of shipping costs. Just drop John a card and he will send you prices on what he has available.

Hodgepodge

Winter of 1984 near Princeton, Missouri — Our Home Is Finally Completed

HODGEPODGE

Tomato Technique - Dr. Clarence Ryan, Chairman of the Agricultural Chemistry Dept. of Washington State University, has discovered a way to make tomato plants resistant to insects. He activates a tomato plant hormone by simply damaging or wounding the tomato plant leaf. Once the hormone has been activated, insects that attack the plants die of severe digestive disorders. The hormone is completely harmless to humans. Dr. Ryan has been awarded large government grants so he can try to discover if other vegetables and fruits have a similar hormone. The discovery has staggering potential as it may soon make it possible to raise plants that are completely resistant to insects.

Ernest Young, Winfield, Kansas, used this technique on 15 tomato plants last spring. He crushed the bottom leaves of 14 plants and one he left alone. The 14 were completely undamaged by insects, while the one was severely attacked.

Curt Baucom, Wellington, Kansas, tried the technique when tomato plants he was growing for wholesale became infested with white fly. He mashed the bottom leaves on half the plants and sprayed the rest with insecticide. The white flies disappeared and stayed gone from the damaged plants. White fly numbers on the sprayed group were greatly reduced, but only temporarily. Curt says it works on aphids, too.

These are the only applications of the technique that I know of. I plan to try it next year and hope you will, too. Try it on plants other than tomatoes (any plant you've had insect problems with). Write me about your experiments and any conclusive results and next year I'll write an article about our findings.

(The 1977 Seed Savers Exchange, page 14)

Sub-Acid Tomatoes — There are reports that many tomato varieties are becoming so sub-acid that it is now dangerous to can them in only a water bath. I've also read that such tomatoes really aren't sub-acid, but simply have a higher sugar content which mashes their acid. If there really is a danger, then what is the safe level of acid for tomatoes? Is there any kit on the market so you could test them like you would test the acidity of soils? Answers hopefully next year.

(The 1977 Seed Savers Exchange, p. 9)

Last year I asked about the truth of reports that many tomato varieties are becoming so sub-acid that it is now dangerous to can them in only a water bath. Here's the feedback:

"Normally, the acid in tomatoes and other fruit prevents the growth of botulism bacteria. However, molds growing on the surface of home-canned high acid foods can reduce the acidity to the level where botulism can grow and produce toxin. Never eat—or taste—home-canned foods which show ANY signs of damage. A minute amount of botulism poison can make a person fatally ill."—Mrs. Judy Troftgruben, University of Illilnois Extension food and nutrition specialist.

The USDA, after a two-year study of more than 350 tomato varieties, has released a report on "Low Acid" Tomatoes, Canning, and Botulism....It shows: 1) There have been no reports of botulism in home-canned tomatoes since 1974 and only five cases since 1950. 2) Botulism bacteria will not grow below pH 4.8, a pH value rarely, if ever, found in tomatoes. 3) Only five varieties of tomatoes come close to pH 4.8--Garden State (a 1940s tomato no longer commercially available), Ace, Ace 55 VF and Cal Ace (commercial varieties produced for the fresh market, which should not be canned) and Big Girl (pH 4.6). There has been little or no change in the mean pH of tomato varieties introduced over the last two decades. 5) Yellow, orange, light-colored and small tomatoes, which the public considers "low acid," are actually of relatively high acidity (low pH). Their flavor is due to a high sugar content which masks the acidity. 6) Tomato acidity decreases when the fruits become over-ripe. The pH of very over-ripe tomatoes can, in some cases, reach the low acid range compatible with botulism growth. The report summarizes: Don't be concerned about the acidity of specific tomato varieties. Just don't can over-ripe tomatoes and follow a reliable canning guide explicitly.

(The 1978 Seed Savers Exchange, page 19)

S. B. Powell, Federal Information Center, General Services Administration, Room 5716, 7th and D Streets, SW, Washington, DC 20407 - In reply to your request for information concerning sending seeds to and from Canada or other countries, this is to advise you that small amounts of seeds are not inspected under the Federal Seed Act. Therefore there should not be any problems or restrictions in mailing the quantity of seeds you inquired about.

(The 1978 Seed Savers Exchange, page 19)

Paul Jackson, Executive Director of the International Association for Education, Development and Distribution of Lesser Known Food Plants and Trees - It is very important to find all of the old varieties that can be found. Some of these are very superior in flavor and food value to the new ones that have been developed with purely the commercial in mind. Too many good things have gone by the wayside simply because it was impossible to make enough money from them. They need to be saved for their own individual value and for the value of their particular genetic strains in future development.

(Paul's organization plants and maintains lesser known fruit and nut trees as street trees in extreme poverty areas of the tropics. He has established a nursery and seed bank (non-profit) to provide a source for rare and commercially uneconomical species of edible plants and trees. He also puts out an excellent small magazine called "Good & Wild." A subscription and membership in his organization are $7.50 per year. Address: Box 599, Lynwood, CA 90262.)

(The 1978 Seed Savers Exchange, page 27)

Bernard M. Lesse, Chief Examiner, Plant Variety Protection Office, USDA

Grain and Seed Division, National Agricultural Library Building, Beltsville, MD 20705 - I have been a plant explorer for USDA and Rockefeller Foundation. If you have specific questions concerning the work that private industry and the Government are doing to preserve germplasm, I will be glad to offer my assistance....For additional information concerning the maintaining of germplasm please contact Dr. Louis Bass, National Seed Storage Laboratory, University of Colorado, Fort Collins, CO 80521.

Dr. Louis N. Bass, Laboratory Director, National Seed Storage Laboratory, Colorado State University, Fort Collins, CO 80532 - Thank you for your letter requesting information about the National Seed Storage Laboratory and plant exploration trips. In regard to seed hunts, or plant collections trips, the National Seed Storage Laboratory does not send out collecting teams in the manner normally thought of as plant exploration. These teams are sponsored, in general, by the Department of Agriculture....The plant collecting team goes to the specific area, collecting in the wild and from market places the kinds of seeds desired. The collected seed samples are returned to the Plant Introduction Stations, where they are increased and made available to seed research people throughout the United States. A portion of that seed increase is then placed in the National Seed Storage Laboratory for long-term preservation. We also obtain seed directly from individuals, seed companies, state experiment stations, other federal agencies and so forth. We are continually looking for sources of old varieties that are no longer carried in seed catalogs, but are being maintained by individuals for their own use.

(The 1978 Seed Savers Exchange, page 28)

North American Fruit Explorers (NAFEX) is a not-for-profit organization established a few decades ago for the purpose of preservation of old or antique fruit varieties, devising and employing unique cultural techniques, among other aspects of fruit culture. Its membership, which is broad-based and covers nearly all the North American continent as well as many foreign countries, is skilled in plant propagation and united by a tremendous enthusiasm for good fruit. Each year they search out long-abandoned homesteads and collect thousands of fruit and nut plants and propagate them in their gardens and back yards. Many of them search National Park and Forest areas for the long-abandoned homesteads of hill farmers who were moved out of these areas during the 1930s. They are always on the lookout for long-lost fruit varieties, fruit of unusual eating quality or chance seedlings that are immune to disease or insects. They are in no way set up to handle questions from letter writers.

Although it is primarily geared to the amateur, professionals, both active and retired, contribute to their quarterly publication Pomona, which consists of excellent articles written by members such as: Fruit Choices for Southern Ontario, Grafting Dormant Scions, Chip Budding, Grape Problems in the Southeast, Scion Collection and Storage, New Hybrid Chestnuts, Improve Raspberry Yields, No Need for Grafting Wax, Persian Mulberries, Fruit at 8,600 Ft. in Dry Climate, Luther Burbank Plant Creations, The Great Find--A Rugged Peach, Iranian Mulberries Will Take on Russian Stock, Pawpaw Cure, Li and Lang and Other Ju-jubes, All About Tree Fruit Viruses, Stop Peach

Leaf-Curl, A Valuable Gooseberry, Cherryplum Breeding Project, Testing Oriental Persimmons for Maryland, Trickle Irrigation for Apples and Peaches—and this is less than a quarter of the articles in only two issues.

This year (1979) their Milo Gibson Award was presented to W. E. Dancy for his work in locating and propagating grape varieties bred by T. V. Munson. This valuable collection of grapes, many particularly suited to conditions of the Southwest, was almost lost. But Mr. Dancy's search recovered 344 named varieties, which are now preserved in the T. V. Munson Memorial Vineyard at Grayson County College in Denison, Texas.

In 1980 NAFEX started a scion and plant exchange program patterned after the Seed Savers Exchange. Robert Kurle and Margaret Hudspith, who head the project, had expressed their enthusiasm about the SSE and how well it is working, and wanted to do something similar with fruit. Now their scion exchange is a reality. In each of their quarterly issues of Pomona there is now a section listing varieties available and varieties sought.

Recently, NAFEX officers decided to form an Organic Culture Committee to actively seek out viable methods/materials and make such information available through the Pomona publication. Areas of investigation include variety evaluation, soil fertility-related materials, cultural techniques, bio-foliar feeding materials, botanical insecticides and biological pest control....An extensive lending library of books about fruits is maintained for NAFEX members. Seriously interested fruit (and nut) hobbyists can apply for membership ($6 annual fee) to Mary Kurle, NAFEX Secretary, 10 S. 55 Madison St., Hinsdale, IL 60521.

(The 1978 Seed Savers Exchange, page 28
The 1979 Seed Savers Exchange, pages 36-37
The 1980 Seed Savers Exchange, page 43
The 1984 Fall Harvest Edition, page 103)

J. L. Hudson, Seedsman (A World Seed Service, P.O. Box 1058, Redwood City, CA 94064) - For at least 3,000 years teosinte, the closest relative and wild ancestor of corn, has naturally crossed with cultivated maize varieties in Mexico. This results in a great genetic diversity that is drawn upon to produce improved varieties. The genetic diversity in these areas is greater than could be produced by 100 years of artificially induced mutations. Ironically, the introduction of modern agricultural methods and hybrid corn varieties to these last areas of teosinte's occurence is causing teosinte to rapidly become extinct. If this happens, we will lose the ancestral genetic base of corn and we will no longer be assured of its continued development. The cultivation of genetically uniform crops can have disastrous results.

I barely got a chance to glance through the 1979 SSE before it was borrowed (borrowed twice now). I'll have to tie it to my desk next time....You asked about a computerized list of all vegetable varieties still available. How I wish such a thing existed!....I imagine those madmen in Europe trying to legislate which vegetable varieties can be sold have some sort of a list, but I don't know how to get hold of it....You might try the American Seed Trade Association or the publishers of "Seed World." But these organizations seem to be at about the extreme opposite end of the spectrum in relation to our concerns for old varieties....Two books that I usually use for variety names and descriptions are The Vegetable Garden by Vilmorin and the now hard-to-find The Seed Grower by Charles Johnson, 1906, but these are only for old varieties....I recommend you read Vegetable and Flower Seed Production by Hawthorne and Pollard.... One other book with some information is Early Garden Catalogs by John Harvey, published in London, 1972, by Fillimore. It is about varieties grown from the 1400s to the 1800s. Most of the book is commentary and information on how they determined what the varieties were. As there were no catalogs in the early days, they studied old bills of lading, etc., but there is a list of varieties in the book....If a plant is rare or extraordinary, it may still be worthwhile to collect even if it may have crosses....there are many gardeners who would rather put up with roguing out the crosses rather than having no chance to grow the plant at all....I must compliment you again on the 1979 SSE. I am so glad that this work is being done. You seem to have found the best method of distribution and perpetuation of these vegetables. So much of this work simply cannot be done by seed companies due to the economics of it....

I understand that supporters of the (plant patenting) legislation are holding up the "Sugar Snap" pea as an example. I have heard them claim that this "totally new" vegetable couldn't be patented, and this is why we need the law changed to allow it to be patented....I have also heard it claimed that it was developed by a small breeder with the cooperation of the patent law. This type of pea was cultivated in Europe prior to 1880, and though "Sugar Snap" may be a "new" variety that may or may not differ from those grown 100 years ago (we will never know), it is far from a totally new vegetable. It is a fine variety and in no way do I mean to belittle the years of effort that went into it; it's just that this is an example of how an old variety could be resurrected and patented. One of the aspects of the law as it stands that I find disturbing, is that a variety need not be developed to be patented, but "discovered" varieties can also be patented as long as they differ from those in cultivation, and the varieties whose descriptions have been published. A variety of corn which has been in existence for a thousand years in some remote valley, upon its "discovery" by an American, could conceivably be patented because its description had not been published....Unfortunately, the closer I look at the law, the more questions I have. I thought it would answer my questions, not confound them....

If you have a large turnover of members, you might suggest that anyone who decides they no longer want to maintain a variety should send you whatever stocks they have left, or perhaps they would consent to be listed a final time in a "Closeouts" section or something, so that what they have is not lost. If people find it too much trouble to do the packing of the seed to exchange, perhaps certain people could do the distributing for others. Maybe there are people without the garden space, or shut-ins who would like to help but can't grow seed themselves. I imagine one person could distribute for ten....

The money must be out there, it's probably just a matter of finding it. I remember I met a man about ten years ago whose job was to tell rich people who to give their money to. They actually paid him as a consultant to find good causes to donate to. Your work is so important that I feel it will make it because it has to....

You might consider encouraging people to describe, photograph or make herbarium pressings of the old varieties, as this might be useful in the future to prevent them from being patented after someone "discovers" or resurrects them....

(The 1978 Seed Savers Exchange, page 30,
The 1980 Seed Savers Exchange, page 46,
The 1981 Winter Yearbook, pages 61-62,
The 1981 Fall Harvest Edition, page 63)

The potato was first cultivated in about 200 B. C. by the Inca Indians of Peru. Today at the International Potato Center in La Molina, Peru, a team of 86 scientists are working to create a potato that will grow fast in the humid tropics where millions of the hungriest poor live. These scientists have now developed a potato which produces edible tubers in 31 days and matures in 60 days (compared to 120 days for an Idaho or Maine potato). This rapid growth should prevent all storage problems and make it an important food source in areas where it is presently unknown. The Center's scientists are now working to make their potato completely disease-resistant, so that it can be grown entirely without costly pesticides (often out of reach for poor farmers).

(The 1978 Seed Savers Exchange, page 30)

Paul Rogers, horticultural consultant to Old Sturbridge Village, oversees Preservation Orchard - four acres of 19th century and older apple varieties, many of which were thought to have become extinct. In the late 1800s, the Royal Horticultural Society of London published a list of 1000 apple varieties. Hundreds of these were once grown in the U. S. Slowly these old varieties are being rediscovered--growing in long-abandoned orchards, on old homesteads, wild along roadsides and deep in the woods. Preservation Orchard now contains 119 varieties and the search continues for about 30 more excellent varieties of a century ago. When a variety is found, a panel of experts classifies it by examining fruit, leaves, branching and other factors. An order blank and list of apple varieties and scions available for grafting from Preservation Orchard can be obtained by sending a SASE to Worcester County Horticultural Society, 30 Tower Hill Rd., Boylston, MA 01505, phone 617/869-6111. Shipments are made in March. (Submitted by Helen Knox, who also highly recommends All About Vegetables, edited by Walter L. Doty.)

(The 1978 Seed Savers Exchange, page 31)

Dr. Garrison Wilkes, Prof. of Biology, Univ. of Mass. - Up to the present time we have been able to return to areas of genetic diversity to collect germplasm for further breeding programs. Suddenly in the 1970s we are discovering that Mexican farmers are planting hybrid seed corn from a Midwestern seed firm, that Tibetan farmers are planting barley from a Scandinavian plant breeding station, and that Turkish farmers are planting wheat from the Mexican Wheat Program. Each of these classic areas of genetic diversity is rapidly becoming an area of seed uniformity.

The reason for alarm and concern about the loss of native strains is the irreplaceable nature of the genetic wealth. The only place genes can be stored is in living systems, either living branches, such as the bud wood of apple trees, or in the living embryos of grain or vegetable seed. The native varieties become extinct once they are dropped in favor of introduced seed. The extinction can take place in a single year if the seeds are cooked and eaten instead of saved as seed stock. Quite literally, the genetic heritage of a millennium in a particular valley can disappear in a single bowl of porridge.

The genetic diversity of our food crops is a national wealth which the varied racial and ethnic members of our nation have brought from their homelands. It is unthinkable not to preserve and maintain these and other reserves of genetic diversity that still exist for future plant breeding needs.

(The 1979 Seed Savers Exchange, page 32)

Searching for Seeds of the Southwest Indians - Gary Nabhan is a faculty member of the Plant Sciences Department of the University of Arizona and is also associated with the Arizona-Sonora Desert Museum. Gary spends considerable time searching for and learning, writing and teaching about crops Southwestern peoples have depended upon historically. He is a valuable addition to our membership. He has written me several fascinating letters over the last year, which include:

Searching for rare and endangered seeds sometimes approaches the proverbial wild goose chase. I have been involved in such searches for Southwestern beans, maize, gourds, devil's claws and sunflowers, often following clues from one remote Indian village to another. When such seed is found, the elderly Indians often present part of it as a gift, so that it will be available to future generations. Also many times we can provide varieties of seed that Indian tribes may have lost, which they, in turn, increase and return to us....Indian sunflowers, grain amaranths, teparies, tobacco, Sonoran panic grass, devil's claw and tan-kernelled chapalote maize (the prehistoric popcorn of the Southwest) have become acutely diminished and are endangered. Hopi sunflowers, already rare, are "losing their genes" via inadvertent hybridization with modern mammoth Russian sunflowers....Many native varieties contain valuable characteristics in their genes: Teparies are not susceptible to common bean blight; Hopi blue maize will grow in the most arid conditions; and Hopi lima beans are resistant to root knot nematodes. If these varieties are lost, future farmers will be left with much more vulnerable, less adaptable crops....A young Papago woman told me about how her grandparents' heirloom seeds were accidentally lost: "They were talking about the old foods one day and decided to go look for their seeds....They looked for a long time, but never did find them....They just sat down and couldn't believe it was all gone"....

Dr. Garrison Wilkes
Professor of Biology
University of Massachusetts

Dr. Gary Paul Nabhan
Native Seeds/SEARCH
(Photo by Jerry Johnson)

Dr. Erna Bennett Discussing Local Wheat Varieties
With Women In Karpension, Greece
(Photo Courtesy of Florita Botts, FAO)

The U. S. government does not see the urgency of supporting collection programs, and the National Seed Storage Laboratory did not even receive a budget increase during its first 15 years of existence. The USDA has not even had one full-time plant explorer since 1970. Funds allocated for collection have never been 1/1,000th of what is spent on producing artificial hybrid varieties, making these collections scanty at best. As agricultural botanist Jack Harlan warns, "If you are willing to entrust the fate of mankind to these collections, you are living in a fool's paradise"....Southwest Indian farmers constantly expose old varieties with uncommon gene combination to rigorous environments in which new traits are selected. Thus the traditional agricultural ecosystems are the real source of gene pool renewal. Unless others quickly become concerned with the fate of traditional agriculture, we may soon enter the nightmare which Jack Harlan describes: A future where all native seed stocks will be in seed banks and experiment stations, no longer part of any tradition....

There are many stories of viable seeds discovered in prehistoric cultural deposits in the Southwest. These stories include a jack bean, a white runner bean, Jacob's cattle bean and pod corn. I have done extensive research and published a paper on this subject. There is no solid documentation to substantiate any of these claims....Your Seed Savers Exchange is invaluable to me, and to friends. By the end of next summer, we'll hopefully have enough of the really rare Indian seed stocks to offer those through the Exchange....

(<u>The 1979 Seed Savers Exchange</u>, pages 32-33)

<u>Dr. Erna Bennett</u>, Genetic Resources Expert for the United Nations - When collecting native seed specimens, the less developed an area is, the less a zone has been touched by modern conditions, the more important it is. Clearly, therefore, the explorer must aim to get away from every sign of progress. When we travel in primitive parts of the world where modern varieties don't occur at all, or didn't until quite recently, we notice that there is no crop ever completely free of disease, but there is no crop ever completely devastated by disease. Disease and variety live in a kind of genetic balance with each other....If we start to rely internationally on single varieties and if genetic erosion continues unchecked, continent-wide famines may be the result within a few years.

(<u>The 1979 Seed Savers Exchange</u>, page 33)

<u>Dr. Paul Ehrlich</u> - There's just no such thing as a perfect strain of plant that is going to grow forever in a stationary climate and be attacked by only one pest which never gets any better at its job. The bugs are always evolving and the weather is always changing. The variety of grain or vegetable that has a slight edge over all other strains right now can easily be inferior to another variety ten years from now. In a case of an unexpected attack by something like the corn blight which swept the Midwest a few years ago, today's "best" strain might even be wiped out completely. This co-evolutionary race is a very close and a very chancy one and we need all the variability on our side that we can get.

So here we are doing just the opposite of what we should. We're refining our food crops down to just a few monocultures. We're losing the genetic variability of our food supply at the very time our plants' enemies are developing increasing resistance to pesticides and we're entering an age of rapidly changing climate. Man, that is dangerous.

(The 1979 Seed Savers Exchange, pages 33-34)

American Chestnut - About a year ago I read that approximately 30 American chestnut trees had been found surviving in West Virginia. They had all been infected with the fungus disease for years, but had survived, implying some disease resistance. WVDA personnel were attempting further selection to try and develop a truly resistant tree.

I did some letter writing, and the following is from an exchange of letters with J. Bruce Given who is Chestnut Project Leader for the West Virginia Department of Agriculture:

Our work with the American chestnut is entering its seventh year. We have now seen enough to be convinced there is nothing whatever to prevent the restoration of the species in original purity....We don't usually consider using a tree until we can see it and determine if it shows satisfactory disease resistance. If no disease is apparent we do not use the tree, because we believe there never has been an American chestnut immune to blight, so the disease-free tree has just been lucky (isolated)....If a tree is capable of making perennial cankers and holding on to produce annual crops of burrs, we consider there is some resistance. Rarely are there any seeds because you can't find two such trees growing close together. Occasionally if we harvest lots of burrs, we get a few nuts. We don't know if pollen has been carried in a great distance or perhaps an occasional bloom may be self-pollinated....Normally Chinese chestnuts are much larger than American, but they are quite variable and sometimes the nuts are quite small and sweet. Older people, having such a tree and remembering the American nut, assume their tree is native. However, if it were native there would be only burrs, no nuts, because of lack of pollen. Therefore, when they offer to send us nuts, we can be sure it is Chinese. If they report burrs and no nuts, then we are interested. I believe I am correct in saying all seed shipped to us has proven to be of species other than American....

All chestnuts have hairy leaves, new growth stems and buds in early summer. The American sheds the hairs about midsummer and becomes bald (glabrous). Others retain the hairs....On the Horticultural Farm at West Virginia University, there is a plot of American chestnut trees which were planted about 40 years ago by the late Professor Roger Pease. They have all blighted back to the ground one or more times and no one there now cares about them. I obtained permission to maintain the plot and this year collected about 1,000 nuts. Obviously there is some resistance or we could not have gotten them....We may try to speed up the generations in our project by planting large quantities of seed from the disease-resistant sources we now have and produce thousands of seedlings. When they are about 1" diameter, we will inoculate them with the fungus and then use the best surviving scions to top work bearing-size Chinese trees....<u>We do not sell trees and are not making general distribution</u>. It will be years before our project has reached to point where that is possible. However we can make a limited number of seedlings available to persons who are also working on American chestnut projects....I am simply amazed at the scope and extent of your Seed Savers Exchange. How you are able to do so much without apparent finance is simply beyond me. It must be truly a work of love....

(The 1979 Seed Savers Exchange, pages 35-36)

Carpathian Walnuts - In the 1930s Rev. Paul C. Crath was engaged in missionary work in the Carpathian Mountain highlands of the Ukraine district of northern Europe where he often noticed the hardiness of the English walnuts growing in that severe climate. His investigations showed that these trees represented the survival of the hardiest of stock brought there centuries earlier from southern Europe and southern Asia. These trees were all seedlings, since grafting was unknown there, and the nuts from different trees differed widely in size and quality.

He became so impressed with their hardiness and the excellence of the nuts that in 1934 he began a systematic search for the best trees of the region. He located about seventy trees that he considered worthy of propagation. (One tree produced over thirteen hundred pounds of nuts in a single year.) No tree was considered that showed even slight injury from the severe winter of 1928-9 when temperatures remained from 40-45 degrees F. below zero for weeks at a time. From these very hardy trees, selection was based on production of good crops of thin shelled, easily cracked nuts with large kernels of good quality. Seed was saved only from trees that were growing in isolated locations, to avoid cross-pollination, so that they would closely resemble their parents. Seed from these best trees was planted in a nursery that Rev. Crath established near Toronto.

Carpathian Walnut seedlings grow rather slowly above the ground for the first two or three years during which time they are establishing a large, strong tap root. After that they grow rapidly, outgrowing even our native black walnut. The nut and kernel are practically the same as the English walnut, which will only grow on our Pacific coast, and in some cases even larger and better-flavored. They have a light-colored bark and dark green, glossy, almost tropical-looking foliage, making them a beautiful ornamental landscape tree.

Louis Gerardi Nursery (RR 1, Box 146, O'Fallon, IL 62269) normally has for sale trees of grafted Carpathian walnuts of the following varieties: Hansen, Helmle, Merkel, Colby, Henry, James, Lake, McKinster, Fateley and Royal. However, last spring they had an extremely poor take on grafts and this spring will have only scion wood available for 75 cents a foot. They also have a very large selection of varieties of each of the following: northern grafted pecans, grafted hicans, grafted hickories, grafted black walnuts, grafted heartnuts, grafted pawpaws, grafted Chinese chestnuts, grafted filberts and hazels, grafted persimmons, and chinquapins. Write for their prices. Colby and Lake varieties of grafted Carpathians are available from Bountiful Ridge Nurseries, Princess Ann, MD 21835. (I would appreciate any information on other sources of grafted Carpathian walnuts - Kent.)

(The 1979 Seed Savers Exchange, page 37)

An update to the above:

Joe Gady writes, "Pataky's Nut Grove Nursery (1116 Hickory Lane, Mansfield, OH 44901) has Fatley and McKinster types....Miller's Nursery (909 W. Lake St., Canandaigue, NY 14424) has a seedling Carpathian strain that Paul Crath found...."

Audrey O'Connor of The Cornell Plantations, Cornell University, wrote that she talked to Dr. MacDaniels, the famous nut breeder and grower, for me. She reports he said, "The picture changes from year to year." He recommends writing to Gerardi Nursery, O'Fallon, IL 62269. (This is the company I wrote about last year. If their grafts took this spring, they will have ten different grafted varieties - Kent.) Other possibilities: Decorah Nursery, Box 125, Harpers Ferry, IA 52146; Earl Ferris Nursery, 276 Bridge St., Hampton, IA 50441; Emlong's at Stevensville, MI 49328; Talbott Nursery, Route 3, Linton, IN 47441; and Miller's Nursery (address above)....

Several years ago Rodale wrote: "The best of the Carpathians so far developed are Metcalfe, McKinster, Colby, Weng, Orth, Morris and Deming." (Sources, please - Kent.)

(The 1980 Seed Savers Exchange, page 47)

Avant Gardener - Thomas and Betty Powell are editors of Avant Gardener (Box 489, New York, NY 10028). Their bi-monthly, prize-winning newsletter is filled with new and unusual news on ALL aspects of the world of horticulture. Subscriptions are $15.00/year and well worth it. Some quotes: "....gather a batch of the insect that is damaging your plants, grind them up in a blender with some water, then strain the mixture through cheesecloth, dilute it and spray it on the affected plants. Gardeners report these 'insect homogenates' have eliminated infestations of slugs, grasshoppers, fall armyworm, various beetles, aphids, cutworms, scale, Western grape skeletonizer, mountain ash sawfly, fungus gnats in the greenhouse and others....more than 20 years of tomato trials at Cornell have shown that small tomato plants with 5 to 7 leaves give higher total weight and bigger, more attractive fruits than do larger older plants...."

(The 1979 Seed Savers Exchange, page 39)

Turkey Red Wheat - For three years I've been asking for a source of Turkey Red wheat. The last field I saw of it was in the mid-1950s near my home in the middle of the southern Kansas wheat belt. I figured the Mennonites around Nickerson, Kansas would still have it, but was never able to contact anyone there.

David Crockett, Ohio member, wrote that Turkey Red wheat was available from a man at Kansas State University. David told him about the SSE, and one Saturday afternoon Carl Overly, Agronomist at Kansas State, called me. He said they had some Turkey Hard Red winter wheat which had been brought from the Crimean area to Halstead, Kansas around 1870. Until last year they had always kept it as a standard by which they judged other wheats. But now there are thousands of other wheats and not all of them can be kept, so Turkey Red was about to go into the Archives.

But they are going to grow it again this winter at Kansas State. It will be available through me to a few selected growers. Write me if you are interested. I want a person(s) who is adept at growing wheat who will multiply it and make it available by the pound through the 1981 SSE. Mr. Overly is in no way able to handle any orders or correspondence. This is through me - Kent.

(The 1980 Seed Savers Exchange, page 44)

Comfrey may be hazardous to your health - In her magazine Landward Ho Velma Wilkerson writes: "It has been discovered that comfrey contains pyrrolozidine alkaloids. These are accumulative, carcinogenic chemicals which can cause irreversible liver damage in humans and a number of animals, often resulting in death....A number of other plants also contain pyrrolozidine alkaloids. One of these, tansy ragwort, is responsible for many fatalities among cattle and horses grazing on tansy-infested pastures....however....comfrey contains only about a third of the pyrrolozidine alkaloids as tansy...."

Organic Gardening and Farming reports that a world authority on alkaloids, Dr. D. J. C. Culvenor of the CSIRO Animal Health Division at Melbourne, Australia, has warned against eating or drinking comfrey by human beings and animals. Young leaves contain the most alkaloid, mature leaves and teas the least. Comfrey ointments and poultices contain only very small amounts of this alkaloid, and appear safe for continued use.

(The 1980 Seed Savers Exchange, page 45)

Cucurbita Crossing - Prof. J. R. Baggett (Dept. of Horticulture, Oregon State University, Corvallis, OR 97331) writes: "The question of crossing between Cucurbita species is interesting and maybe comes down to deciding how serious a few crosses might be....Actually, the few crosses that could occur would probably not be perpetuated beyond the F1 generation because of sterility. As you have clarified in your 1979 Yearbook, the important part is knowing the species of each squash and pumpkin grown. Identifying species by peduncle and plant characteristics, using the information provided by Whitaker and Davis in Cucurbits, and Whitaker and Bohn ("The Taxonomy, Genetics, Production and Use of the Cultivated Species of Cucurbita," Economic Botany, Vol. 4, No. 1, 1950) is not difficult in most cases. Sometimes peduncle separation of C. pepo and C. moschata is confusing, but if plants are also available there should be no great problems. Seeds are also fairly good. Seed of C. pepo and C. maxima are easy to recognize, and each is quite different from C. moschata and

C. mixta. These last two look alike, generally, but Whitaker says C. moschata is scalloped and obtuse on the margin while C. mixta is barely scalloped and acute on the margin (use a hand lens)....The only good reference source on old varieties I know is Vegetables of New York, Vol. 1, Part IV: The Cucurbits, by W. T. Tapley, W. D. Enzie and G. P. Van Eseltine, 1937, J. B. Lyon Co., Albany, NY."

David B. Gerguson (Agricultural Research Manager, David & Sons, 5626 E. Shields, Fresno, CA 93727) writes: "The interspecific crosses in cucurbita are so difficult to make and the progeny of the crosses are so highly sterile that I do not worry about species intercrosses by natural means. Example: I have a C. mixta X C. pepo cross. I have worked very hard to select for fertility in the progeny, and at the end of five generations it is now up to the 20% fertility level. I hardly believe the ordinary person will make the effort to obtain the crosses, let alone perpetuate them. The risk of outcross in their gardens between species is not as great as their risk of being contaminated by pollen from a neighbor's garden....Good general publications on cucurbits are indeed rare. I think you quoted the two great bibles (Cucurbits by T. W. Whitaker and Glen H. Davis, Interscience Publ., New York, 1962 and Species and Varietal Crosses in Cucurbits by A. T. Erwin and E. S. Haber, Agricultural Experiment Station, Iowa State College, Ames, IA, 1929) last year, and Whitaker's book is no longer available and has become a rare book...."

(The 1980 Seed Savers Exchange, page 46)

Robert Kurle, former Secretary of NAFEX - Miles T. Roberts, Rt. 2, Villisca, IA 50864, has over 400 varieties of corn and perhaps several hundred varieties of beans. He claims to have a non-hybrid corn that yields as well as any of the hybrids and has a higher protein content. He tried to sell it, and the seed companies were not interested. I haven't heard from Miles for some time. I hope he is okay. Age is catching up with him....(I tried to call Mr. Roberts. No phone. I've written twice. No reply. I hope at least some of you realize how little time we have! - Kent)

(The 1980 Seed Savers Exchange, page 46)

Hulless Oats - There is growing interest in hulless oats in the Northwest. They produce about 70% the yield of regular oats, but since oats are 30% bran, it all evens out.

SOURCES:

David Sealander (Route 4, Box 371, Idaho Falls, ID 83401) has two varieties of hulless barley (Godiva and Hyproly) and eight varieties of hulless oats including a sizeable amount of Terra, a Canadian variety. Send a self-addressed stamped envelope for prices. (Editor's note: Currently, David Sealander still has these varieties on hand but not much quantity to sell except for Godiva. He suggests that folks interested in information and small quantities of seed should contact their USDA Experiment Stations.)

Kester's Wild Game Food Nurseries (P. O. Box V, Omro, WI 54963) offers a Japanese hulless barley called Lukishinrikii which "is newly introduced in the U. S....loses its hulls when it is threshed like our wheat....high protein content." Catalog $1, refundable with order.

Great Divide Exchange (P. O. Box 1108, Whitefish, MT 59937) has hulless oats available. Send stamped envelope for prices.

O. J. Lougheed of Wauconda, WA 98859 writes: "....Self-Reliance Seeds has found a home in Ferry County, a couple miles from the Canadian Border.... Even with our late start, after double rotavating an acre of established alfalfa, we have managed to seed trial and multiplication plots for over 80 varieties of wheats, 20 of fiber and oil flax, 30 of dry beans (including 5 of soy) and 5-10 varieties each of hulless oats and barleys, proso and pearl millets, beardless barleys, field and sweet corn, snap beans, shell peas and spring rye....We hope to have a catalog and price sheet available by January. Interested people should drop us a postcard to be placed on the mailing list....To ask specific questions, please send a letter with a self-addressed stamped envelope....Self-Reliance Seeds is looking for committed gardeners and small-scale farmers to grow and test small quantities of seeds...." (Editor's note: O. J. is currently at Ghost Ranch in New Mexico where he is maintaining a sizable collection of garden grains. He offers a large portion of his collection as a Listed Member of the SSE in each year's Winter Yearbook.)

(The 1980 Seed Savers Exchange, page 47)

The Cheyenne Community Solar Greenhouse is a prototype, the first of its kind in the U. S. The greenhouse is three-sectioned, 5,000 square feet and utilizes 55-gallon drums containing water as passive solar collectors. This was enough to keep the greenhouse always above 40 degrees F. even during a recent record cold Wyoming winter.

Indeed a community greenhouse, its work force includes senior citizens, high school work experience programs, youngsters in the Summer Program for Economically Disadvantaged Youth, and juveniles in trouble who are offered greenhouse work instead of court fines. There is much interaction between senior citizens and youth, an interaction that is becoming increasingly rare. They are now also involving the mentally and physically handicapped.

The food they grow is all organic and they offer training sessions in solar greenhouse operation and food production. One-third of the greenhouse is a commercial greenhouse and store which helps make the project more economically self-sufficient. The greenhouse also maintains outside community gardens where people without land can grow gardens.

Slide shows and audio tape presentations on the history, design, maintenance and management of the greenhouse project are available to be used as an organizing tool for groups interested in similar projects. Contact Shane Smith, Greenhouse Director, Community Solar Greenhouse, 1603 Central Ave., Suite 400, Cheyenne, WY 82001.

(<u>The 1980 Seed Savers Exchange</u>, page 47)

<u>Canning Centers</u> - Anita Thomas (Route 2, Box 199, Aurora, MO 65065) writes: "We have a Canning Center in Aurora where people can take their produce and can it in large quantities. I have used it for two years now and love it...."

I also read that three Community Canning Centers were made available to the citizens of Benton, Carroll and Madison counties in Arkansas in 1975 through a grant from the Office of Human Concern. These canning centers use large professional canning equipment so that large amounts of home-grown produce can be processed quickly and efficiently.

I suggest that persons interested in starting a Community Canning Center contact Anita Thomas for information about the one in Aurora (always include a stamped envelope when you request a reply - Kent). Also write to the Ozark Producer-Consumer Alliance (Box 301, Berryville, AR 72616) which helps run the three Canning Centers in Arkansas. (Editor's note: A recent letter addressed to the Ozark Producer Consumer Alliance was returned with the notation "Attempted--Not Known.")

(<u>The 1980 Seed Savers Exchange</u>, page 48)

<u>Alma Gem Melon</u> - I recently read a fascinating article written by Judith Joy for the Centralia (Illinois) <u>Sentinel</u> about W. S. "Uncle Billy" Ross. It seems that in 1885 when Uncle Billy moved back to Alma, IL he brought with him the seeds of a green-meated muskmelon which became known as the "Alma Gem."

Soon melon patches sprang up all over Alma. Before long the demand for manure for fertilizer was so great that it was shipped in by the carloads from the Chicago stockyards. Guano was even shipped in from Chile. Eventually over 2,000 acres of melons were in cultivation. During harvest strings of wagons came from all directions to the railroad station. By noon each day over 1,000 wagons had transferred their melons to the waiting fruit cars. On the peak day in history, 26 carloads of Alma Gems were shipped. All of Alma prospered.

There are many versions as to how the industry crumbled. One is that the growers got greedy and started harvesting their melons too green and before the flavor fully developed. Another story says that a rust disease killed the vines. Whatever the case, no known seed of Alma Gem survives, although some persons say that it is a parent of Rocky Ford.

I wrote to Judith Joy and she ran a short article on the farm page of the paper asking anyone who might be keeping the seed of Alma Gem to contact me. I have received no replies to date. How can something that good be lost and gone forever?

(<u>The 1980 Seed Savers Exchange</u>, page 48)

Edward Lowden, 91-year-old plant breeder and developer of the Lowden black raspberry, writes from Ontario - I am writing this letter to let you know I am very appreciative of what you are doing....Mr. Topp tomato, a very valuable tomato with frost-resistant genes, has been lost....around 1917-1921 I used to buy its seed from Gurneys in South Dakota....it had valuable characteristics that surpassed anything I have ever seen in other tomatoes. It was a terrific cropper of small 1.5-2" fruits in clusters of 30 to 40....but its most valuable characteristic was its ability to withstand light frost in the fall that ruined other tomatoes.... the Imur Prior Beta tomato had the earliness of many of the Sub-Arctics from Beaverlodge, but I soon saw that it did not have the great hardiness of Mr. Topp that I had known years ago....I would give anything to secure that one tomato gene again....maybe someone somewhere knows of it....

(The 1981 Winter Yearbook, page 60)

Leland W. Hudson, Horticulturist (Regional Plant Introduction Station, Washington State University, Pullman, WA 99164) -I am the horticulturist at the Western Regional Plant Introduction Station here in Pullman. As part of my work here, I am in charge of collections of lettuce, cabbage, lentils, chickpeas, horsebeans (*Vicia faba*), various *Allium* species (including garlic, but not including *A. cepa*, or the common onion) and beans. All told, there are about 15,000 items in this collection, of which 10,000 are beans or *Phaseolus* species....We grow this material out and increase it, conduct preliminary evaluations and store it in controlled conditions (about 40 degrees F. and 30% relative humidity). Under these conditions, bean seeds should last 20-30 years. Some, no doubt, will last longer, and some not as long. It appears that longevity of bean seed in storage may be genetically controlled, hence some lines will retain their viability for quite an extended period of time and others will not. I suspect that there are genetic/environmental interactions here, too, in that the longevity of seeds in storage may be related somewhat to the environment while they are growing as well as the storage environment....

The Great Northern bean originated with Son of Star, a Hidatsa Indian, and it has come down to us in a number of selections made by various states, notably Idaho, Montana, and especially Nebraska. It is present in most supermarkets and sells as the Great Northern bean. However, these present Great Northerns are selections out of the original and by no means contain all of the germplasm originally present. Any time you select for a character, you automatically reject its counterpart, and there may be unnoticed advantageous genetic linkages with the rejected materials providing for a narrower gene base. If this process of refinement and selection is carried on for a number of generations, the result is that the plants lose their ability to compete in the environment, and as the saying goes, Grandma's beans just simply ran out. So select and choose if you must, but remember the risks that you take....

We will grow about 900 lines of beans per year, of which about 400 are in the greenhouse; the rest are in the field. We grow two crops per year in the greenhouse, seeding one the first of July. As the short days of fall come on, those short-day plants then are forced to bloom, and we harvest a crop in December and start another one the first of January. This material gets up and gets exposed to short days before the longer days of spring come along in March. We then are able to harvest a second crop in May and June....There are bees that will produce a great deal of outcrossing in beans, not only Phaseolus coccineus, but also P. lunatus and P. vulgaris. Experiments in various parts of the world have shown that in common beans, you would not ordinarily expect more than 5 or 6% outcrossing in most places. However, in some parts of Puerto Rico and Hawaii, there are bees which may up this to around 30 or 40% of outcrossing. Some types of lima beans, runner beans and common beans are more susceptible to outcrossing than others....I have not experienced any particular problems with outcrossing in fields in this area....

Freezing of seed in a controlled atmosphere in a sealed can has possibilities. Some crop seeds respond better to this method than others. Dr. Louis Bass' group at the National Seed Storage Laboratory on the Colorado State University campus (Fort Collins, CO) has been working with cryogenic storage of seed for several years. I'm sure they are in a better position than I am to advise you in this regard....I converted a reach-in freezer with about 100 cubic feet capacity to a seed storage unit which kept conditions at about 40 degrees F. and 35% relative humidity. The conversion cost only a few dollars and served us well over a period of quite a number of years. However, as with all such types of units, one should provide a limit thermostat which prevents the unit from overheating or getting too cold. These thermostats merely cut off all power to the unit. The insulation plus the cool temperature of the contents will keep the interior at a reasonable condition for a day or more if the doors are kept shut. This gives you a chance to troubleshoot the unit and correct the problem before conditions inside reach the point they affect seed viability.

(The 1981 Winter Yearbook, pages 60-61)

George L. Herter (founder and former president of Herter's Inc), writes from Minnesota - Dear Mr. Whealy: I do not remember when I have enjoyed a letter as much as yours. Seemed almost like a visit over the fence with you....The seed companies now are all buying their seeds from Japan, South Korea and The People's Republic of China, by the pound, and packaging them here so they do not have to label them grown in Japan, Korea or China....No one but you is trying to save the very valuable non-hybrid vegetables that are all rapidly going to complete extinction....I hate to say this, but if you fail with your seed storage program, within three years most of the good non-hybrid varieties will be extinct....

Ben Quisenberry, 93-year-old tomato seed grower from Ohio - (June 10, 1980)—I'm sorry to disappoint you, as we are all sold out of our 1979 tomato seed; however, our 1980 (500 plants) crop looks very good, so we plan to furnish seed to you this fall, if all goes well....(October 3, 1980)—This summer when my tomato garden of 500 plants needed me most, I was in the hospital. When I returned to my garden, I managed to salvage from the large weeds about half of the seed I needed to take care of my trade. Before I went to the hospital I had 32 varieties, now I have only 9 varieties. However I am giving you, and your friend Starr, a few seeds of each of the nine varieties.... (December 5, 1980)—Didn't get any seeds from the following: Tasty Evergreen, Long Keeper, Stump of the World, Square (very unique, late, slow growth), Tomango, Tiger Tom, San Marzano, Large Red Cherry, Goldie (a large, light yellow), White Tomato (source: Mrs. John Schutter, Algona, IA), Beefmaster, Indian River and Czechoslovakia mailed to me 12 varieties, small, varieties....I'll confess I was and am discouraged--have wanted so much to perpetuate good tomato varieties, and no doubt you too. If you can think of any way I can help you, please let me know....

Dr. Louis N. Bass, Director of the National Seed Storage Laboratory - The early history of agriculture in this country reveals a continual yet virtually ignored loss of valuable and often irreplaceable plant germplasm--the substance by which plants transmit their hereditary characteristics....no adequate provisions were made for the preservation of original stocks....

(The 1981 Winter Yearbook, page 62)

Rick Manning, 710 W. 24 1/2 St., Austin, TX 78705 - I represent Austin Community Gardens, an urban gardening group with over 600 active members. We are looking for sources of locally (Texas) grown, bulk seed to sell to our gardeners. We represent a growing market for these seeds and would like to support a small, local seed supplier.

Burpee's 1888 Catalog - a beautiful and faithful reproduction illustrated with the original engravings. Provides a rare, accurate picture of the 19th century garden. 128 pages. $2.00 + $1 handling from whichever outlet is closest: Burpee Seed Co., Warminster, PA 18991 or Clinton, IA 52732 or Riverside, CA 92502.

(The 1981 Winter Yearbook, page 63)

Peter T. Reich, South Carolina - I am currently doing research on people and events that shaped United States Agriculture. Some of these people and events have had tremendous impact on this country, yet are little known. Word of the Seed Savers Exchange reached me through Rodale Press (Organic Gardening). To say the least, I think your organization may play an important role in U. S. agriculture....

Carol Mohawk, "Akwesasne Notes," Mohawk Nation) - I read about your Seed Savers Exchange in the February 1981 issue of the "Cherokee Advocate." Would it be possible to receive a complimentary review copy of your yearbook? I would like to review it for our newspaper....and I feel it would be an aid in my writing as well as in my teaching. The class I have been teaching is entitled Native People, Colonialism and Food. One of the areas we discuss in class is the flooding of the market with hybrid seeds versus the preservation of our native seeds. The students are beginning to see the relationship between continuing the "purebred" seeds of our ancestors and the health of our ancestors versus hybrid seeds, chemical fertilizers and disease today....

(The 1981 Fall Harvest Edition, page 56)

Dr. Jack Harlan, Professor of Plant Genetics, University of Illinois - There is a heroic tale about the seige of Leningrad during World War II. People were dying of cold and starvation, reduced to eating rats, cats, dogs, dried glue from furniture joints and wallpaper, or anything else that might prolong life. All this time, truckloads of edible seeds were in storage at the All-Union Institute of Plant Industry. The seeds were too precious to be sacrificed even at the cost of human life, and the collections survived for future use. We may pray that such a threat will never occur again, but prayer may not be adequate to save priceless genetic resources....

(The 1981 Fall Harvest Edition, page 57)

Leonard L. Cope, Kentucky - I am an 82-year-old man, retired from my railroad job....I have been married over 60 years, gardened all my adult life and gardened organically for 30 years. I think it's the only way to garden, healthwise and other-wise. Also I am a strong believer in only planting open-pollinated varieties that a man can raise year after year. I don't want any hybrid, regardless of how good it's claimed to be....

Many years ago we were growing a special sweet corn that was a new open-pollinated sweet corn called Aunt Mary's. We ordered our seed from a seed company in Michigan. The wife and daughter and I were all very fond of it...One spring the company had a note in my order: "We cannot supply open-pollinated Aunt Mary's corn. We are sending you this hybrid of it"....When we had our first to eat, boiled on the cob, we all three agreed it didn't taste nearly as good as all the Aunt Mary's corn we had grown for several years. It just happened that we did have some seed ears we had saved. So since that day we have grown our own open-pollinated Aunt Mary's sweet corn. I don't order anything from that company now and I don't order any hybrid seed from any seed company....

(The 1981 Fall Harvest Edition, page 60)

Alex Caron, King City, Ontario - I am the Vice President of the Canadian Organic Growers (C.O.G.) and also a member of the much smaller Natural Farmers of Ontario....C.O.G. is moving forward quite rapidly and is almost incorporated as a charitable organization. I feel that we have an untapped resource in our 340 members....

We are beginning a fundraising effort for a number of projects we want to undertake, including a Heritage Garden and a Seed Exchange. The Heritage Garden is at my farm....Maybe since we are a Canadian organization, we could take on the preservation of Edward Lowden's collection as a special project.... Lowden is only 40 miles from me....One of the things I have done on my own is to obtain some of Lowden's raspberry and blackberry stock. Shipping rootstocks to the U. S. is very difficult, so perhaps that part of Lowden's collection can be preserved up here....Kent, I would be happy to work with you and to help out in any way that I can.

(The 1981 Fall Harvest Edition, page 62)

Rachael Sumption, Florida - It has always been my belief that people will somehow support any activity that they believe worthwhile without any government interference. It makes me angry when I read that a few are narrowing the choices for my children and myself even before we had become informed enough to protest. That's immoral, or unethical, if you prefer. As for my children, thank God there are some still speaking for their future even before I realized that there was in infringement on the last "right" we have--to choose what we eat....

(The 1982 Winter Yearbook, page 87)

David Ridge, Missouri - It is alarming to see what multinational corporations are doing to our crops through control of seed companies, fuel, fertilizers and food processing. They seem to stop at nothing to gain control....I am most interested in what you and others are doing to help save the old varieties....

(The 1982 Winter Yearbook, page 89)

Local Newspaper Articles - Several of our members have sent me articles that they've gotten their local newspapers to run about their gardening activities and also the Seed Savers Exchange. That not only helps the SSE, but it puts them in contact with like-minded individuals and gardeners who may be keeping unique vegetables. My sincere thanks to: Mrs. Edwin Beiswenger of Minnesota, Thomas Butterworth of Connecticut, Sally Dunmire of Pennsylvania, the late Phil Hewitt of Connecticut, Betty Emlen of California, Albert Beck of Missouri, Eldon Lansford of Tennessee, Robert Lobitz of Minnesota, Karl Urban of Oregon and Dan Driskill of Alberta, Canada.

(The 1982 Fall Harvest Edition, page 62)

Plumas Seed Savers - I really love The 1982 Winter Yearbook and have passed it around to gardening friends, producing some very positive results. I have quite a group interested in starting a local seed savers association. We are having our first meeting tonight (3/22/82)....I ran an ad in the local papers a few months ago for people saving seeds, and got no response. I was disappointed, but am by no means giving up. Now that spring is here and gardening is on people's minds, I plan to run more ads. Do you have any suggestions on how to contact old-timers who are saving seed?....(Notices in rural weekly newspapers are sometimes successful. The Ernest Strubbe interview in this issue is actually the end result of such an ad, so you see what can happen. Also try local garden clubs, talking to growers at farmers markets, notices in coops, check with elderly relatives who are gardeners and also their friends, family reunions, local garden writers, even the younger relatives of folks you may meet in nursing homes. But most of all, when you see what is obviously the garden of a master gardener, just stop and talk to them. They're not going to bite. If they're like most of the folks in the hills around here, they're starved for some conversation....Locating gardeners who are keeping heirloom vegetable varieties is hard, but it just takes persistence, and the rewards are incredible - Kent.)

From further correspondence, 10/2/82: Plumas Seed Savers had its second meeting last night. There wasn't a great turnout, but really a lot of interest. We're all going to compile a list of everything we have to trade. Another meeting is planned for late November to trade lists and decide what we want to try. We'd be really proud to have you run an article on our local group. Surely it would help inspire others to form similar organizations....(Plumas Seed Savers, c/o Cindy Robinson, PO Box 1917, Quincy, CA 95971)

William Kelley, Box 509, Quincy, CA 95971 - I was loaned a copy of your Winter Yearbook and am so grateful that you have started the Seed Savers Exchange to deal with this catastrophic trend in western agriculture. A group of local gardeners in our mountainous country here in California are very interested in establishing a "Regional Bank" to serve the needs of our own local growers as well as provide security in the years to come. I think becoming members of your Exchange would be very valuable for us and hopefully for the Exchange as well. I have read and understand the terms by which you authorize membership and commend you for your straightforwardness....

Albert Beck, Culver-Stockton College, Canton, MO 63435 - In late February about 20 of us like-minded souls gathered for our "First Annual" Seed Exchange....Just wanted you to know that we're spreading the good news....

(The 1982 Fall Harvest Edition, pages 65-66)

Sue Holstrom, Wisconsin - Our house burned to the ground April 3. I lost all of my own seed in the fire and also some Fleener Top Set onions. I'd like to point out the wisdom of not putting all one's eggs in one basket, which is what this network is all about. From now on I'll rent a freezer locker and keep half my seed there. Everything in my freezer cooked, the heat was so intense. Luckily the well in the barn is fine so I'll be able to water the garden....No lives were lost in the fire, and we're already re-building....Being able to propagate the heirloom seeds this summer will be the highlight of my summer....

(The 1982 Fall Harvest Edition, page 118)

Dollie Cameron, Kansas - I am very interested in the Seed Savers Exchange. A friend of ours died two years ago. He raised the best-tasting watermelons and cantaloupes. He brought the seed with him from Tennessee over 60 years ago. As far as I can find out, no one kept any seeds. What a horrible waste....

(The 1982 Fall Harvest Edition, page 119)

Jack Finn, Arkansas - I have just read your very enlightening and interesting article in the January/February 1982 issue of The Mother Earth News....You have cleared up for my wife and me quite a few facts that we have wondered about in regard to the seeds that are now unavailable. We are native Arkansans and well remember growing much better vegetables than from the hybrids that we now have to purchase....

Mrs. Wlliam Roth, Oklahoma - I am old enough to be your mother or possibly even your grandmother, but I am learning so much. What more can I say than a great big Thank You....

(The 1982 Fall Harvest Edition, page 120)

Enormous Vegetables by Secret Formula? - Cabbages three feet wide and up to sixty pounds each. Collard greens up to five feet long. Onions ten pounds each. These are vegetables grown by Mexican farmer Jose Carmen Garcia in a village 260 miles northwest of Mexico City. Without fertilizer. Some of the vegetables are now in a freezer of San Diego Police Department information officer Bill Robinson. There are photographs of Garcia and more of the huge vegetables.

Garcia, 50, says that another Mexican, a stranger, gave him a sketch in 1947 of strange symbols making up a formula for growing the vegetables. The stranger said he "was held captive by tall, fair humanoids beneath a nearby volcano." He memorized the formula before he escaped. He told Garcia to concentrate on the symbols. The formula would become clear to him and he would realize the secret of the planting. Garcia concentrated. He has grown the huge vegetables ever since.

Officials of Mexico's Agriculture Ministry have investigated Garcia's farming methods with no luck in duplicating or explaining them.

(Reprinted with permission from the Miami Herald.)

(The 1982 Fall Harvest Edition, page 120)

Laurel Horton, North Carolina - I just looked up your article in the June 1981 issue of Organic Gardening. I am enclosing some seeds for you. The beans are called Snow Bean and were given by Vega Thompson of Traphill, NC (Wilkes County) to John E. "Frail" Jones of Moravian Falls, NC. She said they were an old variety. I saw them in Frail's garden last year. They're pole beans and put out an incredible number of pods on little tendril-like things from ground level to the top of the vine. They taste like a half-runner....

The corn was developed by Frail Jones from a no-cob variety called McNeil's Special and two varieties of sweet corn. He was seeking large kernels, small cob, good flavor and good freezing ability. As you can guess from the kernels, there are both yellow and white on the same ear....

Frail died in April. His daughter is a close friend and I was given these seeds when I visited her last weekend. I can think of no better way to honor his memory than to send you these seeds and then to go out and plant some in my garden. Call the corn "Frail Jones Legacy"....

Tony Lincoln, Self Reliance Seed Co., Tasmania, Australia - I am writing with an exploratory offer of help for your work, with which we at Self Reliance Seed

Co. and very many others in this country agree with all our hearts. We in Australia are blessed with a range of climate that can accommodate anything that grows and a very widespread, vigorous and organized body of gardeners. We have been organizing these people into a network of growers through the many Organic Growers Societies and Permaculture Associations, thus forming a source of organically grown seed of the greatest diversity, which we consider to be the best defense against the effects of P.V.R. legislation, now closing in on us here....We have had a bad setback here which has put us out of business. But the strength of support for what we are doing, which I am sure will be familiar to you, is enough to rescue the work if not the company....I don't know in what way we can be of most help to you. My strongest feeling is that the more widespread our sources of seeds and their storage are, the safer we are. We have access to a large freezer here for long-term storage (with a very reliable electricity supply) and we have a strong, reliable body of good gardeners. Let me know if you can make use of it and us....

(The 1982 Fall Harvest Edition, page 121)

Mrs. Dortha Carey, Indiana - I read the article enclosed in our local newspaper and thought maybe you would enjoy adding this Brown Head lettuce seed to your heirloom collection....My grandmother gave this seed to my mother, who gave seed to me years ago. I have raised this type of lettuce for years. I am 76 years old, had garden last year, but don't plan to have this year....This seed I think came from Germany. I feel it is time to pass this on to someone who will keep it preserved and enjoy growing it as I have....

(The 1982 Fall Harvest Edition, page 123)

C. L. Doerr, 1545 Gowan Dr., Memphis, TN 38127 - I am looking for a book that was co-authored by the former director of research at the Desert Seed Co. of El Centro, CA. It is the only complete book ever written on the onion--Onions and Their Allies, 1943. It was mentioned in the March 1981 issue of the Readers Digest on page 17.

(The 1984 Fall Harvest Edition, page 32)

Ken Tansey, Ohio - In your interview in The Mother Earth News, you mentioned your interest in developing a youth-oriented movement....It seems to me that a preservation garden project would be particularly suited to the 4-H program. As a club or individual project, they could grow those vegetables best suited and traditionally grown in their particular area. Younger people also have access to and the trust of a great many older folks who might be more willing to share a prized seed with an interested youth than with someone older or someone they didn't know. Even if only one club in each county in a state such as Ohio participated, there would be 88 gardens here....

(The 1984 Fall Harvest Edition, page 35)

Correspondence

WHEALY

CORRESPONDENCE FROM SSE MEMBERS

From The 1978 Seed Savers Exchange --

The late Dr. John Wyche, OK — I live on the very top of a rock mountain and have moved hundreds and hundreds of tons of rock, soil, leaves, sawdust, pine needles, loam and manure. I have 22 terrace gardens with rock walls around my house. Some are large, many very narrow and quite long, some quite small....I used to own Cole Bros. Circus and have many show people for friends and patients. Carson & Barnes Circus winters here and the owner is a close friend. I have exclusive rights on the elephant manure and picking rights on the cat cages. Believe me, elephants turn out a lot of guano. I scatter the lion and tiger waste around my gardens to keep out rabbits and coons....(p. 12)

Napoleon and Tina Montoya, Alberta, Canada — We can't describe to you the pleasure of exchanging seed with so many kind and interesting people, and the lovely garden and new eating experiences we enjoyed as a result. Thank all of you and especially you, Kent....(p. 15)

Joy Horton, Ecuador —....I had a very unfortunate experience while up in the mountains, was accidentally poisoned and slipped a disk in my back the same day....I have terminated my formal studies and apply my energy full-time to my work. This includes introducing new varieties of fruits, vegetables and cover crops for experimentation at varying altitudes. Progress is slow, but very sure. Steve Spangler of Exotica Seed Co. in Los Angeles will be leaving here in several days with about 200 pounds of seeds, cuttings and grafting wood from many fruits and vegetables here....We discovered a new fruit unidentifiable in books. A small tree, with long pointed yellow tomato-type fruit. Sweet fragrance, with flavor of very ripe peaches. Very excited about this....Just shipped 50,000 coffee beans for planting....Loji province is like a crumpled-up piece of paper, so every step over means either a step up or down....Climates are not determined by north/south position as in U. S., but rather by altitude. We grow apples at 2720 meters, avocados at 2620 meters, and in San Pedro, which is an hour hike down the canyon and 1700 meters altitude, they grow sugar cane, papayas, coffee, bananas, oranges, etc....August 29-September 4 I attended the 25th Congress for the Society of American Horticulturists. There were 40 people from the U. S. and others from Costa Rica, Mexico, Chile, Colombia, Venezuela, Iran, Brazil, Italy, Japan, etc....(p. 17)

Soula Moore, Ecuador — I want fruit, herbs and vegetables which come true from seed, including natural wild species of North American woods, fencerows, etc. My interest is in propagating those species that can naturalize themselves on mountainsides here, as have "Blackcaps" that John brought here to Ecuador in 1962. I'm especially interested in those having a high calcium content, as the sub-tropical climate and overly-acid and overly-sweet fruit of this region tend to demineralize the body, causing many of the local natives to lose teeth to decay by 25 years of age....I want to extend my personal gratitude to you for organizing the Seed Savers Exchange. I've hoped for some time to find others who believe in the importance of saving and sharing the seeds of our garden paradise....(p. 17)

Howard Jones, Guatemala —....We're a missionary training center which teaches agricultural principles and supporting oneself by gardening....We just finished cutting nine acres of barley and triticale by hand. A machine came and threshed it and we got about 4,000 pounds of grain. There is a certain point where hand labor just doesn't make it. But tonight I saw a friend who is managing a 200-acre farm and has 155 acres of wheat. He just got done weeding it all by hand. It took a month and a half with 16 men. Our project seems real small after that....(p. 17)

From The 1979 Seed Savers Exchange —

Freddie Pate, FL — Concerning problems starting New Zealand spinach, my mother used to chop seed tops and plant in late fall. Some started in fall, like wheat. Rest came up early, and frosts and freezes didn't seem to hurt it....A good way to keep tomatoes from cracking is to add two tablespoons Epsom salts to planting hole. This is an old trick, and I know it works. (p. 5)

George B. Teall, MI — There is much misunderstanding, confusion and various names for multiplier onions, or potato onions....multiply as a cluster in the ground....there is a yellow outer skin variety and also a white skin variety which is milder....Egyptian onions or top onions or walking onions multiply by producing a cluster of small bulbs on a seed stalk....there is a red variety of these....Shallots multiply in the ground, do not have a seed stem....French shallots have reddish outer skin and the flesh has pinkish-blue streaks of color all through the bulb....flavor is different and milder than garlic....Dutch shallots have a yellow outer skin....Many people who have trouble eating onions can tolerate shallots in their food....(The catalog of Garden Way Publishing, Charlotte, VT 05445 contains many gardening books, including a 32-page booklet called All the Onions and How to Grow Them. It covers all the members of the onion family including tree onions, potato onions, Welsh onions, shallots, rocambole, leeks, garlic chives, garlic, elephant garlic, chives, scallions and common onions, and gives illustrations, descriptions, cultural information and sources for each - Kent.) (p. 9)

Leroy J. Schmidbauer, NJ — Thanks to the information request for the history of the names of tomatoes in the 1978 Seed Savers Exchange, I now have the dates of introduction of about 1,500 tomatoes....earliest records of tomatoes seem to be an early 1700s gardening ad in a Massachusetts newspaper and also the Thomas Jefferson gardening diary....literature seems to suggest that after the Spaniards took it to Europe (Spain), it wandered into France, then to Italy, eventually to England, then to the Northern Colonies (Pennsylvania and Boston)....Pennsylvania Dutch moon lore: Plant tomato seeds two days before the new moon for healthy and vigorous plants. (p. 11)

Wilmer Swope, OH — We must develop corn varieties or cultivars which are not bound by the limitations of hybridizing, are not tied to chemical fertilizers for high yields and are insect and disease resistant....The American Indian was a master in the use of "visual selection" for improved types or cultivars. The pecan and corn are excellent examples of "eyeballing" for certain traits and their intensification....My selection standards include: elimination of stalks which show smut susceptibility or insect damage, elimination of stalk tassels or stalks which cannot compete in a vigorous way in a VERY THICK stand of corn, ear height which is not too high to invite wind damage and not too low to invite animal damage, ears which hang and point away from the stalk for good moisture and insurance against mold, high sugar content and a kernel which is square and deep and especially full (square) at the cob end, giving the greatest shelled weight for an acre of corn....Two corn varieties--Reids and Krug--form most of the genetic material in American hybrid corn. I believe that Lancaster Sure Crop, developed by Mennonite deacon Isaac Hershey of Paradise, PA, is a far better genetic parent than either of these two corns....I have found a corn cultivar from Guatemala which keeps vigor and height even after a very heavy dose of radiation (20,000 rontgens). This cultivar has a much sweeter stalk sap than U. S. corns. Increase is necessary before I can offer it....(pp. 13-14)

Perile Casey, TN —....I hope people who order naranjilla from South America know that they are different than our garden tomatoes. They take up to 30 days to germinate, take 20 months to fruit, and are not frost hardy....Clean "wet seeds" by working them across a cotton cloth with a table knife until all juice and pulp are absorbed....(p. 16)

Carl Wilson, TX — I have decided to place my Nueces River ranch acreage, near Uvalde, TX, into a Land Trust for perpetuity, never to be sold. I am developing it into a Wild Foods Sanctuary which I will call Walnut Acres. It will be a Wildlife Sanctuary as well, along the same lines as the National Wildlife Federation. I am sending you a list of the wild edibles that I have already planted there (It is very extensive - Kent). I hope to develop this into a co-op type gardening project, four other families besides my own. Anything that you or fellow members can do to help me promote the Walnut Acres Wild Foods Sanctuary will be appreciated....I raise timber wolf pups and cross-bred wolf/dog pups for sale. I now have two female timber wolves, my male stud wolf Lobo, and female Elkhounds, German Shepherds, Huskys and Malamutes ....If a stranger goes into Lobo's pen, he'll go into his house. But if you move toward his house, Lobo'll have lunch....Crush a few Buckeye nuts, put them in a Bull Duram sack and throw it into a hole in the river. In a short time all of the fish will float to the surface and you can pick out what you want just like in the supermarket. In a few minutes the others will revive and swim off....You mentioned in your Companion Planting Guide last year that tomatoes are stunted by black walnuts. Few plants will do as well near a black walnut tree as they will 100+ feet away, due to a toxic exudation of the leaves and roots. My dad used black walnut trees as livestock shade trees because they repel insects. My grandfather kept his hammock strung between two of them for the same reason....I am retired and almost 100% back to nature, depending on wild food foraging and a small garden for my food....(p. 17)

Clint Sherril, VA — In Agriculture by Steiner, there is a good discussion about plants and the genetic vitality of their seed as affected by varying agricultural practices. In Dialogue with a Dirt Farmer there is more information about the importance of vital seed....(p. 18)

Wes H. Brown, Manitoba — If anyone knows of a good organic method of controlling the flea beetle, I sure would like to know it. Someone told me that nasturtiums would work, but they ate them, too. Hate to turn to chemicals, but I'm almost there....The amaranth I am offering is from Rodale's Organic Gardening Experiment, grows to over 6' tall and seems to be a good companion plant for pole beans. Both did better together than alone....If you have a spot in your garden where water tends to lie, that is a good place to put a compost pit, as the extra water will help rot the compost faster. The compost acts like a big blotter, will absorb excess water during a heavy rainfall, then slowly release it to the plants during a dry spell. Squash, climbing cucumbers and others that like moisture, but not wet feet, will do well beside it....(p. 19)

Mrs. Virginia Lindo, Jamaica — You should mention Rare Fruit Council, International, Inc., 3280 South Miami Ave., Miami, FL 33129, which is a non-profit organization dedicated to the introduction and growing of rare tropical fruits. They publish a monthly report of their activities, which is absolutely worth the $10 yearly subscription fee....(p. 21)

Howard Jones, Guatemala — At the very top of the mountain behind the Institute is a group of about 10 houses, and the people grow vegetables to eat and sell. There we watched a lady pick up a basket of radishes that weighed about 100 pounds, put it on her head and start walking for the highway which is about an hour away. There she would get on the bus and for 50 cents go to Guatemala City to sell the radishes. She hoped to get $4 for them, then pay 25 cents to return home and walk back up the mountain. That $4 had to pay for her seed, labor and transportation. That is what we are up against here. There is no market for organic produce, because nobody pays any attention to what it even means. Cows' milk here has 90 times the DDT allowable in the U. S. ....the big battle of the year was over oxen or a tractor. One can't imagine the ebb and flow of the controversy as it raged among the brothers that have the say. But thank God we have decided to get the oxen....Yesterday we harvested the mulberry tree and tonight had a mulberry pie fiesta. We have found that about 21 different kinds of fruits will grow here and have dedicated ourselves to planting a hundred of each. We are at 7,000 feet, so we can marginally grow apples and pears and a local peach, plus all sorts of exotic things that you will have to come and try sometime....(p. 21)

William H. Lambert, Honduras — I am a U. S. citizen and have been a legal resident of Honduras for the past eleven years....very few vegetable gardens here....most islanders grow coconuts, bananas, papaya, pineapple, cassava, cocoa, mangoes and corn. Most of these tropical plants are usually not affected by innumerable leaf-cutting ants, fine salt spray, seasonal droughts and poor soil due to leaching by torrential rains during the rainy season....I have corresponded with Howard Jones in Guatemala advising him of a medicinal plant grown here for treatment of diabetes. (p. 25)

From The 1980 Seed Savers Exchange —

Stephen Facciola, CA — I am working on a project for the California Rare Fruit Growers at the Quail Botanical Gardens. The Gardens, which is located near the Pacific Ocean and is basically frost-free, contains a unique collection of sub-tropical and near tropical fruits and nuts. Much of this collection was donated by the pioneering members of the California Rare Fruit Growers. The current project aims to restore and expand the original planting. Therefore, we would like to hear from people living in the warm climate areas who might have seeds of unusual fruits and nuts to exchange. (p. 4)

Ms. Gregg S. Everhart, CA — I'm a full-time gardener, having studied with Alan Chadwick (master horticulturist who developed the BioDynamic/French Intensive system) at the Round Valley Garden and presently helping build a garden for an alternative health care group called Commonweal....15 acres of intensive hand cultivation....seven gardeners at present....We plan to fulfill Chadwick's grand vision of a Garden....we have cool, foggy summers, difficult for ripening seed....I seek to become a source of coastal acclimatized seed especially some Japanese and French "heirloom" varieties that have often been more carefully selected for quality of flavor and nutrition....you might publicize Tree Crop Nurseries, c/o Dan Crebbin, Rt. 1, Box 44B, Covelo, CA 95428 which specializes in old and rare varieties of apples, pears, other fruit and nut trees, hedgerow plants, roses, bee plants and herbs--all organic by former Chadwick apprentice....(p. 5)

Dave Hattem, CA — I am working for the Department of Food and Agriculture, State of California, and I am studying landscaping and alternative farming. I live in Chico, CA where an amazing resource is being wasted and "converted" (destroyed). I am referring to one of the original plant introduction stations, started by the Federal government to experiment with new and different varieties of agricultural and ornamental plants....has been taken over by Dept. of Ag. Forestry Div....the new administrators have ripped out most of the existing orchards, but there are several outstanding varieties left....the area surrounding Chico is an excellent place to gather seed....much diversity....I am sure in one year I can collect seed from 40 or more fruit and vegetable species (several varieties of each)....interested in corresponding with other groups about trading ornamental plant seed....(p. 7)

Howard Cory, IA —....There used to be a seed company, Oscar H. Will & Co., Box 600, 6th St. at Front Ave., Bismarck, ND. They haven't sent a catalog since 1959, so must be out of business. They had a lot of Indian seeds--corns, a few beans, pumpkins and squash. I tried about all of them over the years, but didn't bother to keep seed....they obtained the Great Northern bean from "Son of a Star," a Hidatsa Indian, over 80 years ago. They also had a Hidatsa red bean....it would be a shame if those seeds are lost, as they could well be since no one had much thought of saving things like that a few years ago....maybe the local Chamber of Commerce would know what became of the Company....(p. 10)

M. Ross, MA — I wish to swap herb seeds and old timey remedies and recipes ....I'm writing a book on organic gardening and old-time methods, home remedies and recipes. Have collected lots of hundreds-of-years-old herbal information. Would enjoy letters and tape cassettes from folks interested in sharing information. Will answer mail anywhere in the world and any questions I can....I teach so-called handicapped persons on crutches or in wheelchairs (stroke victims, paralysis, ill from "long living," blown up in 'Nam, birth defects, etc.) how to grow all their own food with little work. It CAN be done. Also teach the same to blind persons by cassette "talking letter." Anyone interested please contact me....(p. 11)

J. T. Wickey, MO — How can moles be kept out of fresh-planted peanuts? ....Baum Nursery, Rt. 2, New Fairfield, CT 06810 offers about 100 antique apple varieties, one of which is supposed to be 700 years old from Switzerland. Also Roxbury Russet, St. Lawrence, Wolf River, Early Harvest, Fireside, Melrose, etc. They are $5.50 each plus $3.00 shipping per order. (p. 13)

Leroy J. Schmidbauer, NJ — This year I am working with the Burpee Long Keeping tomato test. The project ends when the last stored green tomato turns red....should be sometime in early February....my wife is trying to collect as many tomato jam, jelly, preserve, conserve and butter recipes as she can find....a commercial source for the Stone tomato is: Ferry-Morse Seed Co., Fulton, KY 42041 or Mountain View, CA 94042....I have been working for some time with Dr. George Larke on tomato projects....He is doing an extensive study on tomato seed storage....He has a tomato seed bank and has nearly every variety of tomato....(p. 15)

Perile Casey, TN — Wouldn't it be great if a Convention could be planned so that members and interested people could meet and get to know each other in person? Gardeners are a special people....(In the summer of 1981 the Whealys are hoping to open our house and 30 acres of woods and meadows to camping for a weekend convention - Kent)....What is the cost of listing items for sale through the SSE? I am in contact with many Indians, Cajuns, Amish and poor mountain people who could only afford to participate in that way....(There is no charge. I would much rather see old varieties listed for a price, than not listed at all - Kent)....Should anyone list the Mexican plant Jicama, remember to advise the public that the seed is poisonous....Kent, if you think the Guinea Bean is beautiful, you should grow the luffa or the balsam pear or bitter melon....I read a few years ago that Japanese Beetles love delphiniums (seeds are poisonous), will eat it readily, and die....soapy water will kill cabbage worms in seconds....I read with concern Lawrence D. Hill's letter about the "Common Catalogue." I truly expect commercial interests here to attempt legislation that would outlaw the Seed Savers Exchange....for no other reason than it is taking sales away from commercial sources. Time will tell. I hope I'm wrong....(p. 19)

Ted Telsch, TX — I get about 50 varieties of tomatoes planted each spring and Wow! Shipped out on job assignment five years straight--Indonesia, Saudi, Tokyo, Holland and what have you....cannot keep seed in supply, only enough to hold viable seed each year....Kent, am sending out copies of your letter to friends and associates also interested in vegetable seeds and "heirloom varieties"....please finish building your house and try to go fulltime as an author....(p. 19)

Mrs. B. J. Martin, TX — When tomatoes bloom and refuse to put on fruit, take a strong weed and give vines a good beating. It works. (That'll teach 'em - Kent.) (p. 19)

Enid W. Rummel, WA — I envy those who live in warmer climates and can grow all those dry beans and corn. We are limited in varieties here. Our climate is much like England....England's plant patenting legislation is very foolish. Their catalogues used to be a delight and now are disappointing....(p. 21)

Robert Martin, Sweden — Sorry I cannot help you with any seeds from here as this area is quickly dying out. That is why your work is greatly appreciated ....here seed is mostly imported from southern countries, food and all. This is mostly steel country....I've made contact with Russia, Ukraine and hope maybe to get some heirloom seeds from there....I could use seeds for all vegetables suited to a cool to cold rainy summer....corn, pumpkins, squashes, cucumbers.... I would like to get some of Lowdens raspberries. He is up in Canada and is very old now....(p. 24)

Mr. & Mrs. Don Peterson, West Indies — We are in the vegetable and fruit business here in Nevis....We have just put in over 100 citrus trees, mainly valencia and navel oranges. Gardening goes on all year long here if you can get water. Normally the dry season is from April to September....We can supply all sorts of tropical vegetable and flowering tree seed....We have found a local okra here that is much superior to anything that we could buy in the States. I sent a sample to Park Seed Co., but never heard anything back from them. The stuff grows about 15 feet tall, but has extremely tender pods. We have some perennial bushes over two years old that keep right on producing. We must pick every day to get this to happen....The birds are our worst pests and I need some seed of the toughest-skinned tomato I can get....I can also supply some seeds of some rather exotic flowering trees....We are trying to save more and more of our own seed and are having bad luck with beans. The dried beans we save to eat (pinto, lima, pigeon peas, etc.) we treat in the oven and then put in jars or cans with tight lids. The seeds we save to plant don't get the heat treatment and the bugs, worms, etc. are ruining them. Does anybody have a neat way of solving this problem?....(p. 25)

Jim Johnson, WI — I have an interest in historic vegetables, fruits, grains, herbs and flowers. I have a file of 2,200 varieties that were introduced before 1900. I either grow these or know where to get the seed....Receiving the Seed Savers Exchange was like walking out of the wilderness. It's nice to know others are interested in historic varieties. I've been working alone and often get discouraged because thousands of old varieties still exist, but so many are being lost....I have researched approximately 450 vegetable varieties which were introduced before 1888; some go back to the 1500s....I would like to participate in your organization, because I believe very much in helping people to save our old varieties. But my correspondence load is tremendous already and I'm afraid it would bring my research to a halt....Few museums have historic gardens which are well researched and of authentic varieties. Williamsburg and Old World Wisconsin are the best, but Williamsburg I feel is more a fairyland than what was really of the times. Some museums make claims about authentic gardens, but I have yet to see any of their research that is even close to good....The American Seed Trade Association, 1030 15th St. NW, Executive Building 964, Washington, DC 20005 has a book of vegetable varieties that were introduced from about 1935 forward for $5....Well worth it. I'm not sure of the exact title, but they will know what you want if you send for it....Yes, there is a directory of outdoor museums, in fact a number of them. Contact any outdoor museum or your state historical society or a good library....My fruit booklet is about half done. I would guess it will contain about 1,200 heirloom varieties of all fruits. I won't get it done until late 1980, just too much work involved to do it right. I have located 400 apple varieties and have ended my search of this....(pp. 28-29)

Carl Wilson, TX — is an expert on wild foods and foraging. His knowledge is not from books, but from living his young years with his grandfather on a huge island called No Man's Land in the Red River between Texas and Oklahoma. Carl has had no luck getting a manuscript published on Foods from the Wilderness. He agreed to write a much shortened form of Foods from the Wilderness, five pages, which he will send anyone for $3.00 postpaid....(p. 29)

John H. Miller, VA — As an Amish-Mennonite person, I have had an interest in saving seeds for a long time already. I knew of an elderly lady in our circles who had saved her own garden peas for many,many years. After she became ill and later died, I was shocked to find out that none of her family realized what they had missed in not keeping the seed for further planting....our old-time neighbor Noah Yoder had developed a superior open-pollinated yellow seed corn for a long time also....that too was lost forever, when hybrid seeds took over. Often the way it goes. (p. 41)

From The 1981 Winter Yearbook —

Jim Bob Good, AR — We are working on a seed exchange program here....I have appealed through the county newspaper for time-tested seed stock to perpetuate....I have located a dozen varieties, but only a couple are truly heirloom types, either scarce or long acclimated to this area....There are more seeds around here in need of preservation. The problem is in locating their husbanders, mostly old-timers....We are in the grips of an incredible drought, some 80-plus years natives say it's the worst they have ever seen. Trees are dying by the thousands, pastures are brown, even old wild muscadines are dying....our seed saving program picked a bad year to begin....(p. 4)

Dr. Jesse Schwartz, CA — Was just listening to the radio....it seems the tide of madness is rising. I turned it off and decided to commune with what is alive in this country by writing you....The Seed Savers Exchange is an outstanding service. Future generations will be indebted to you....

I have just returned from a five-month sojourn in Mexico and Guatemala. The highlands in the state of Chiapas are clearly one of the most extraordinarily fertile and botanically rich places on the earth. There side by side with cedar, fir and pine, one can find cacti or bananas and mangoes growing. Every sort of fruit and vegetable can be grown from tropical fruits to apples, pears and peaches....Unfortunately the Indian peoples of this region are subjected to the full blast of the whirlwind of commerce. In Guatemala they are fighting for their land (and souls), and in Mexico commercial methods, fertilizers and seed are forced upon them. This great region is certainly a storehouse for the rarest and most wondrous treasures of seeds, and as elswhere, this precious heritage is being threatened....

I have brought back some samples of the beautiful beans grown there—glowing reds, yellows, blacks, purples, pinks, oranges and greys with various brown stripes. Nothing would please me more than to see these propagated....Also have brought back Guatemalan blue corn which grows to an unbelievable 12' with ears 8' off the ground (very vulnerable, I feel, to wind damage); might possibly be grown in the Deep South. The corn makes excellent, vivid blue tortillas....

Am moving to Oregon to set my ideas on local self-sufficiency into practice. We seek to establish a community, organic farm, orchard and school....We are organizing a two-week seminar in bio-dynamic/French intensive gardening this June at a new age community in southern Oregon. Anyone who is interested in attending, write me....There is an abiding holy living spirit in this country, down deep below the shopping malls and MX missiles. Your Exchange is a manifestation of the irrepressible....(pp. 4-5)

Stephen Facciola, CA — I recommend the book Vegetables in Southeast Asia by G. A. C. Herklots. Available for $10.50 from Allen and Unwin Inc., 9 Winchester Terrace, Winchester, MA 01890. Many of the plants described are sub-tropical (including chayote, jicama and winged bean), but many can be grown in temperate areas. Is especially good on oriental brassicas and cucurbits. (p. 5)

Bob Thompson, CA — Jojoba will thrive on barren dry soil and produce oil-rich nuts (50% oil) in three to five years. Foliage is great for animal browsing; the Southwest Indians used the nuts as food; modern man is replacing whale oil with it. The oil will not turn rancid in storage and is starting to be used for restoring hair, skin conditioner, cosmetics, you name it....I have been a subscriber for two years now and can't tell you how much I have enjoyed it.... Thank God for the job you are doing....you're at the top of my most admired people list....America has fewer and fewer people of your caliber with the guts to tackle a badly needed service and get it done....(p. 5)

Freddie Pate, IL — To Control Moles—use a cup of stock grade (cheaper) castor oil, pour in a squeeze quart bottle, add boiling water and squirt down middle of planting row, or use sprinkling can and sprinkle lightly over whole area. Till in lightly anytime before planting. I found where moles started into garden, they backed up and never returned. This doesn't hurt fishing worms, praying mantis or ladybugs. Organic Control of Flea Beetles—use Larkspur; the old-fashioned single-flower purple worked for me. The blooms are what kills them, so they should be started early so blooms would come when needed, which is early spring. I discovered this one summer when staying 100 miles away and gardening only on Saturdays. One weekend everything, especially tomatoes, were actually black with the beetles. Returning the next weekend intending to spray, I couldn't find even one beetle. I noticed the blooms on my larkspur were either eaten completely or gnawed badly and the ground under each bush was 3-4 deep with flea beetles. In Illinois I plant in late fall and they come up real early. Frosts and light freezes don't hurt them. Corn Borer and cantaloupe, watermelon, a large tomato and some Indian corn. Show them how, but let them plant, maybe help them mulch it so they won't have to weed, and then step back and watch - Kent.) (p. 12)

Mrs. Ed Beiswenger, MN — It makes me sick to realize that I too have lost varieties, not realizing that someday they might not be available....We paid a visit to the Hutterites and were surprised to learn that they had not handed down very many old seeds....Have received a letter from a man in Minnesota who has done some very good work on his own in preserving corn seed. He was familiar with the old Oscar Will Seed Co. and still has a few varieties from it. He sent me 23 different kinds of corn—all beautiful and many that he himself had perfected....(p. 14)

Ora Vanderpool, MO — I tried for years to find some seed company to promote the Grandma Easley beans, but they all ignored me. But they are now well on their way, thanks to SSE....(p. 15)

Evert Pettit, MO — We have gained some wonderful people as friends, thanks to you and them....(p. 15)

Virgil Johnson, MO — I was very pleased to find the bean I have been searching for for many years. I wish to thank you for this organization. (p. 15)

Bill Drake, NM — Three years ago I located a source of the original tobacco species utilized by all the major new world Indian civilizations--Nicotiana rustica—which is completely different from the commercial tobaccos. My source was an old Indian man here in northern New Mexico who has since died. For centuries, old men grew Punche' behind their houses and smoked it green. Nowadays the tradition has just about disappeared....(Many Indian peoples attribute healing properties and clarity of mind to Nicotiana rustica. Gary Paul Nabhan told me a story which I believe was told to him in Sonora: When one tribe realized that the last of this seed had been buried along with its owner, they broke a longstanding taboo and dug up the jar containing the seeds from beside him. They gave the seed to one of their people to increase. It was grown for a few years and then lost again - Kent.) (p. 17)

Lawrence D. Brown, NY — I had a chance this past season to test the difference in old-time varieties when I planted Stowell's Evergreen and compared it to Silver Queen (a 90-day hybrid sweet corn). Stowell's was said to be a 100-day corn, but came on 10 days earlier than the Silver Queen. Both grew to about 12 feet, but the Stowell's easily was the more productive....(We sometimes grow a hybrid cantaloupe called Supermarket, but never again. This year I paid 90 cents for a packet of only 22 seeds, of which only four grew. And these plants were easily outproduced by an Amish cantaloupe I was growing. A seed importer recently wrote me that many vegetable seeds are now being produced in the Orient and in some cases just aren't doing that well here. Such seed does not have to be identified as long as it is packaged in packets which are printed in the U. S....reports have shown that while hybrid corns may be markedly more productive, standard tomatoes and peppers may be little different from their hybrid counterparts....Dr. Ernest Kerr of the Simco (Ontario) Research Station is quoted as saying, "We've tested a thousand or so

Ear Worms—spray several times with Bacillus thuringensis (I used Thuricide) and red pepper or cayenne. Spray tassels and silks as soon as these appear, several times. This also keeps out birds, squirrels and coons. Nematodes or Eelworms—plant African type or smelly marigolds over garden. Chop stems, etc. into the ground after bloom. Spray made of marigolds keeps bugs off cabbages and beans. Peach Curl and Fruit Worm—spray with Thuricide and green soap, which can be purchased at some drugstores but may require a search. Diatomaceous Earth—swimming pool filter grade. I mix this in compost, and a little mixed in bottom of planting hole and sprinkled light down rows of potatoes, corn, onions, etc., keeps down cutworms, wireworms and onion maggot. Doesn't hurt fishing worms, ladybugs or praying mantis. (pp. 8-9)

Dan Eaton, IL — I had about 128 requests for seeds from people who are in SSE. That's good. I wish there were more....I have been doing a lot of work with corn alcohol and some coal research. A bushel of corn will produce about 2.5 gallons of 170-180-proof alcohol and 2# of corn oil. The grain product left after fermentation is about 20#, of which 7# is a high protein food additive for baby food and older people who have trouble digesting, 6# makes an excellent sausage filler or hamburger filler or corn chips, and about 7# for animal feed that is very high in protein and aids in animal digestion. You can give the alcohol away and still make a profit on the corn. It is too bad our politicians are all bought by the oil companies. (p. 9)

Catherine Watkins, IL —....the very existence of such dedicated and selfless people as yourself gives me some small measure of hope for the future....(Aw, shucks. - Kent) (p. 63)

Milan Rafayko, KY — I'm looking for a deep-purple tomato, variously known as Pruden's Purple, Potato Top or Peruvian Black. The fruit was so purple it looked black, about the color of the Black Beauty Eggplant fruit....I'm an amateur herbalist, so I spend some time each year gethering herbs and seeds of whatever is available each year. Mama Nature is unpredictable that way.... Kent, I'm enclosing a small donation. I wish to God it could be many times that, but I'm an old man (63) retired on a meager pension. Your work must go on. No one else is doing it. We can't afford to lose you, for God's sake....(p. 12)

Kathy Carabella, KY — Thank you for doing us all a great service. Even my four-year-old daughter thinks saving seed is important, and that's due largely to your exchange. (I always love to hear about children in the garden and don't think parents can give their children a greater gift. Last summer I was being interviewed about the SSE on live nationwide Canadian radio and was asked if my children liked to eat heirloom vegetables. I replied, "Not only do they like to eat them, what's more important is that they like to grow them." To me a garden is a magical place and anyone who plants seeds and watches them grow each day can't help but be awed. That's why gardeners are such special people. And our children are much more aware than we are because their minds aren't nearly so full and busy. If we are wise enough to expose them to this magical experience, it will change their outlook and enrich their lives forever. Try doing what we do--this summer work up a small area on the edge of your garden, rope it off and give your child a few seeds of Russian sunflowers, a large different hybrids, but none of these are equal to the best standard varieties of tomatoes"....Although the seed industry always promotes their hybrids as more vigorous, more productive and more resistant, all you can really count on is that they are more expensive and you can't save seed from them - Kent.) (pp. 17-18)

Michael Cannon, NY — The SSE is an instant cure for "Cabin Fever"....Last fall I was allowed the ultimate joy. Phil Tolli was taken ill in early summer and had to be hospitalized. In his absence his garden dwindled and was practically a total loss. He was overjoyed when I brought surplus vegetables from my garden, not for just the vegetables themselves, but for future seed. He is 87 and all but forgot that he had provided me with the Roma tomato seed he originally brought from Italy. He did not hold seed back and thought he had lost it forever. He is well recovered and already starting this year's seedbeds....(p. 18)

Jeana Sirabella and Thomas Crissinger, NY — We have Jimmy Nardello's Italian sweet pepper, grown in his family for over 100 years in the U. S. When Jimmy is told by admirers of the plants, "You have to give me some of that seed," he says yes and ignores the request. We feel very privileged to have him share with us. He's 83....Thank you for stringing us all together....There are some really wonderful people involved--helpful, honest, overly generous--great models to aspire to....(p. 18)

Dr. John Wyche, OK — Last summer was the driest ever recorded by my people since they came over the Trail of Tears in 1834. My seed crop was a complete failure, but I keep a three-year supply in my deep freezers....May the Rainbow always rest on your shoulder....(p. 20)

Richard Grazzini, PA — My doctoral research here at Penn State will be dealing with maize, looking into the mechanism of starch synthesis....Bruce Bugbee is still here at Penn State. He hopes to be finished by June. I'm impressed by the depth and breadth of his knowledge and interests....How about reminding the membership to label their seed packets. I've gotten unidentified seeds--no name, no label, no return address on the envelope. The postmark was so obscure that I couldn't even trace it through that....If you've access to a decent library, read Crop Genetic Resources edited by O. Frankel and J. G. Hawkes....also of interest would be Tropical Legumes: Resources for the Future and Underexploited Tropical Plants with Promising Economic Value....I made a rough count of the HAS listings in the 1979 SSE that weren't in the 1980—eighty-two people didn't list this year! We should do something to keep the old members participating....you may want to consider a NEEDS HELP category, if someone needs help in multiplying a source, due to infirmity or whatever....The SSE is a good pool of heirloom varieties. I'm glad to be part of it....(p. 21-22)

Brian Smith, SC — I am a vegetarian who attempts to grow edible plants, by organic methods, in the sandy, salty soil of a barrier island off the South Carolina coast. For the past few years I have carefully selected seeds in an attempt to select plants more suited to this environment and my needs....If any members have a seedling var. of Jerusalem artichoke, they should consider sharing this valuable seed. Virtually all cultivated Jerusalem artichoke in North America seems of one clone. It flowers, but produces no seed. Seed would have many advantages over the tuber method of propagation....The potential of the Seed Savers Exchange seems almost infinite. May you go from strength to strength. (pp. 22-23)

Perile Casey, TN — I planted an allium (a green garlic from Taiwan) at the base of peach trees. The borers have really "worked on" the trees without the garlic, but not on the ones with it....Formula for plant food from household items: 1 tsp. baking soda (phosphorus), 1 tsp. saltpeter (potassium), 1 tsp. Epsom salts (magnesium) and 1/2 tsp. liquid household ammonia (nitrogen). Water plants every two weeks for four weeks and then once a month....Mr. Withee reported in his newsletter that the winged bean was grown in New York with some success. If you can consider 30 ft. vines with no blooms, I too grew them with some success....This has been the most hostile growing season I can remember in 35 years of gardening....Dale's parents used to repel Mexican bean beetles by planting carrots right along with the beans....If the government would just "butt out," the type of people who originally built this country would solve its problems....(p. 23)

Robert Gene Daubert, TX — I recommend these books: Alternative Foods - A World Guide to Lesser-Known Edible Plants by James Sholto Douglas, Pelham Books, London, 1978, and The Winged Bean: A High Protein Crop for the Tropics and Underexploited Tropical Plants with Promising Economic Value, both available from National Technical Information Service, Springfield, VA 22161. (p. 24)

Douglas Law, Ontario —....In 1923 a Seeds Act was passed in Canada which listed all field crop varieties grown and sold in 1922. All varieties introduced after that date were required to be tested for merit. If they were good enough, they were licensed for sale. Other varieties may not be imported, advertised or sold, although there are exceptions for research and for farmers' own use....It appears that open-pollinated field corn has not been listed in the Seeds Variety Order under the Seeds Act since 1962....From my own research, it appears that few of the old varieties were preserved in the Plant Gene Resource Centre. Fortunately the USDA has preserved some similar strains that were grown in the U. S. and a few that were grown here. It appears that many Canadian strains have been lost nevertheless....As far as field corn is concerned, the plant breeder's rights legislation can have little effect on the preservation of valuable germplasm here. It has been largely eliminated from this country already....The market is largely dominated by a few large multinational seed companies....It is possible to experience some red tape in getting seed corn from the U. S., but if one can't bring it in oneself, it can be mailed. There does not appear to be any duty....Canada maintains the world oat collection....This past year we tried two open-pollinated varieties from the 1979 SSE Commercial section: Krug's 90-Day corn from Stodden Seed Farms, which makes a very good ensilage corn here; and Wapsie Valley Dent, from Steinbronn Seed Farm, a good corn also, either to pick or to fill silo....(p. 28)

I. M. Laszlo, Australia — Your present yearbook has been photostated many times here and makes its rounds....the great disadvantages here are that vegetable matter in this country is seldom "heirloom" and the Establishment, which lives by the rule that anything done outside the commercial channels is antisocial and therefore your and similar publications and ideas are against the existing order of things, therefore subversive....we have here a rather powerful movement....my aim, together with others in the Permaculture movement, is to save genetic material and to pass on those which we can obtain to others in perpetuity....the Permaculture movement was started by two university lecturers from the U. of Tasmania who have since gone on to establish a community based on the precepts of permaculture, which basically goes back to your Russell Smith and his ideas of tree crops....permanent agriculture which uses the same soil on three layers (ground crops if need be, but above them low food-producing shrubs, above them small trees and above them tall trees....this is using the same soil three times over and will be very important in the difficult times ahead....I have an M. A. in Economic History and what is in store for you makes me shudder....Australia has provided only one small item into mankind's food larder, the Macadamia nut, but there must have been many other seeds and food materials worth saving, as the aborigines of Australia ate and conserved them over nearly 40,000 years. However, with the virtual genocide of most of them, this knowledge has disappeared....(p. 29)

Mrs. Margaret Adkin, Western Australia (Seed Bank Manager, Organic Growers Association) -- We are an organization which is very interested in saving "heirloom"-type seeds for the future....We can offer to send you unusual seeds of your choice, in exchange for others which you may have which could interest us. Only food crops are available from us at the moment as this is the vital issue in the world today, as we see it....let us know how we may benefit each other....(p. 29)

Donald McClure, Self-Reliance Seed Co., Tasmania, Australia -- We are a young non-hybrid seed company run by the Tagari Community and are endeavoring to obtain the disappearing traditional varieties of useful agricultural plants. We are setting up a grower network to propagate up small samples of rare and endangered species and maintaining them while making them available to the public through our mail-order catalogue....We are also involved in fighting plant breeder's rights legislation which may soon come before Parliament....non-hybrid agricultural/gardening seed has become almost impossible to obtain in Australia and will get worse due to takeovers of the few independent non-hybrid seed companies by multinational corporations. The proposed legislation will have a terrible effect on the non-hybrid seed range available, so we hope to offer the widest possible selection of seeds and act as a gene bank....One of our community members, Bill Mollison, is in the U. S. right now arranging the publishing of the books Permaculture I and Permaculture II....(p. 30)

Joy (Horton) Hofmann, Ecuador --....Kent, you were right about me having some adjusting to do when I got back to the USA for the first time in five years. But after facing and conquering Miami International Airport, everything else was simple--the new sliding doors with electric eyes, fancy cars, push-button everything. I really enjoyed the clean streets, clean walls and especially the clean bathrooms with toilet paper. The dogs seemed so tame, the people so nice. Everything so efficient and orderly. The telephones work!....Last night we all rocked through an earthquake which registered 6.2 on the R. scale, centered in Columbia killing 300 there....In the garden everything is positive. The rains are back, making the green world explode with life....(p. 31)

Don and Carol Peterson, West Indies -- have a large, tender okra--a local variety superior to anything in the U. S., grows to 15 feet tall, extremely tender pods. We have some perennial bushes over two years old that keep right on producing. We must pick every day to get this to happen....We are in the vegetable and fruit business here in Nevis....Gardening goes on all year long here if you can get water. Normally the dry season is from April to September ....We continue to experiment....We solved the problem of birds eating the tomatoes by building a galvanized pipe frame and covering it with Ross bird net....We are the only producers of melons on this island and we can sell all we raise to the hotels. One of the gardener's donkeys got loose the other day and ate about twenty ripe cantaloupe. At $3.00 per pound I could have bought the donkey for less money....(p. 33)

Carl Wilson, TX — For my Nueces River wild foods sanctuary which I'm calling Walnut Acres, I would appreciate donations of the following: berries of all kinds, nuts of all kinds, wild plums and cherries of all kinds, grapes of all kinds, rose hips, chufa nut, May apple, May pop, red and black Haw, muscadine, persimmons--in fact anything of an edible nature collected from the wild. Even though I might have a variety of an item, I would also like different varieties from other parts of the country....Sure could use a couple of your people to share the Walnut Acres project with me. No money involved. I will furnish free mobile home site and all the equipment needed for gardening 30,000 sq. ft. Outside employment available....last week I was visited by a flock of over 100 wild turkeys. We have lots of deer, turkey, quail, squirrel, etc. and I have seen as many as six Golden Eagles at one time....(p. 37)

Jim Davis, MO — I am currently involved in the development of an insect control project attempting to assist small vegetable farmers in southwest Missouri....Prior to 1940 there was a great deal of research attempting to locate or breed resistant sweet corn varieties that showed very little earworm damage below the ear tip. Using these varieties farmers were able to quickly cut off the ear tip and be left with a marketable ear of sweet corn. Unfortunately, this research and the availability of these sweet corn varieties have decreased tremendously since farmers became dependent on pesticides for corn earworm control....(p. 40)

Wesley M. Bailey, OH — I cut mole damage about 90% by using 2 parts castor oil, 1 part liquid soap, equal amount of water, zip in blender, use 3 tablespoons to a gallon of water, apply with sprinkling can. (p. 41)

Shirley Ann Steiner, OH — wants a large Italian Plum tomato and a two-toned pink and yellow tomato which was an old farmer's who lived in the neighborhood who died about ten years ago. His relatives will not share his seeds with anyone. They were so good. (Ever think of making a midnight raid on a tomato patch? - Kent) (p. 41)

Mrs. Clarence Grewer, MO — Balsam liniment is made of the fruit or melon of the Balsam Pear. The melon will ripen to an orange color, then start splitting at the end. When it reaches this stage, take about three of these melons and slice them into a jar. A quart jar or a coffee jar which is slightly larger is what I use. Then fill the jar with rubbing alcohol. Seal and let sit a few weeks or until it turns a cream or light yellow before using....

It is used for pain. I have arthritis, mostly of the muscle. It is real good at relieving pain. Rub the liniment on and rub it in real good. It doesn't stain clothing and it isn't oily. It seems oily, but it isn't. There is no need to bandage to make it work. The lady that started me using it said she had a sore spot on her stomach and it stopped it. I can truthfully say it works, but it isn't a cure for arthritis. It sure helps me get around with a lot less pain and a lot less aspirin....I don't know of anyone eating the melon. For all I know it could be poison. Also it is hard to get to sprout. Probably because the seed is so hard. Soak the seed in water for two days before planting....(The 1981 Fall Harvest Edition, page 58)

From The 1982 Winter Yearbook —

Jim Taylor, AZ — I recently gave a talk at our Yuma Organic Garden Club on how very important it is to keep the old lines of seed going. It was so well received that the group took up a collection for the SSE....(p. 7)

Robert E. Johnson, CA — I have always wondered about the old seeds that were being lost, even in our own family, but didn't know what I could do. I'm glad that someone has figured out what to do and went ahead and did it! (p. 10)

Dr. Edson Andre Johnson, CA — My work is with the oft-ignored sciences of Biomagnetics (magnetic fields), Electroculture (electric fields) and Photobiology (colored spectrum light) and how they affect seed germination and stimulate (and even mutate) plants and plant growth. If anyone is working with similar studies or has any data on the use of any type of radiations (cobalt, x-ray, ultra violet) on seed or pollen or plants to effect mutations, let me hear from you....I would like to recommend "Magnetism and Its Effects on the Living System" by Albert Roy Davis and Walter C. Rawls, Jr., 1974, Exposition Press, which can be obtained from the Albert Roy Davis Research Laboratory, PO Box 655, Green Cove Springs, FL 32043....To try a simple experiment in the stimulation of seeds by magnetic fields, take some garden peas or pumpkin or corn seeds (hybrids don't react well; open-pollinated varieties react better; Hopi Blue corn really reacts). Use a compass to find magnetic north. Plant part of the seeds with their seed scars (hilums) aligned with magnetic north and plant part of the seeds normally as a control. The magnetically aligned seeds germinate faster and show better growth....(p. 10)

Dortha E. McKim, CA — I have Evergreen onions, which are winter onions that keep on multiplying, grow very slowly, aren't affected by heat or cold, never have any bottom onions so you use them only as green onions. My Grandma Besse came over from Holland when she was four years old on a whale boat with her mother. They brought these same onions with them. When she took cold, her mother would fry some of these onions and make a poultice and put it on her chest to break up the cold in the lungs....(p. 11)

Maya Starchild, CA —....We went to India on business last spring and got some interesting Indian and Thai vegetable seeds and a great book on Indian vegetables called Vegetables by B. Choudhury, printed in India....the seed situation in India and many Third World countries is getting bad fast....Not much time left! (p. 12)

Nancy R. Wilson, CA — I have enjoyed the Seed Savers Exchange and am especially interested in your concern over Seed Banks and legislation. I use it in my talks to Garden Clubs and have aroused a lot of interest in not only saving seeds but following the Plant Patenting and Protection issues....(p. 13)

David Savory, CO — For the past two years I have been employed by the Denver Botanic Gardens. I am in charge of the Demonstration Vegetable Garden as well as an ornamental vegetable garden (Indian corn, gourds, etc.)....I am particularly interested in the Cucurbitaceae family....I will soon be leaving Denver to work on a degree in Agronomy at CSU in Fort Collins. There is a good chance that I will be working part-time at the National Seed Storage Laboratory. It depends largely on the current "Reagonomic" situation. Possibly such a contact could be beneficial to the SSE....(p. 13)

Brian W. Cole, CT — There was an apple variety called Everlasting which was grown in Maine 80 years ago which kept in storage for two years, but I haven't been able to locate it....I am looking for a way to keep potatoes over two winters so a crop loss won't be so disastrous....I was surprised to learn that Indians at least as far north as Montreal raised muskmelons. I wonder what they tasted like?....I think we are neglecting the biennials....From The Garden (Journal of the Royal Horticultural Society, Cto. 1980, Vol. 105, Part 10)--"The problems start when we want to breed new characteristics into our varieties--characteristics which may have only existed in material which is now extinct! ....What if we want to breed ringspot resistance into brassicas grown in the southwest, where the disease is prevalent? The answer is we can't, for the best source of ringspot resistance, the Old Cornish cauliflower, disappeared 30 years ago when it was replaced by French material"....I read that the British are finding vitamin deficiencies in the patented (hybrid) varieties and are slightly alarmed....The National Seed Storage Laboratory wrote me that they don't have descriptions of their holdings programmed into their computer because they don't have enough funding or personnel to do the job....(pp. 13-14)

Tommy Simmons, FL — Keep up the good work. I'm enthused by the potential of this work for earthlings worldwide....(p. 14)

Harold Miller, GA — We are a working farm operated by and for a resident community of 17 mentally retarded persons and a staff of 10. We grow much of our own food and share your viewpoint regarding the importance of perpetuating heirloom seeds....(p. 15)

Don Cornelius, GA — I'm more and more disturbed at what I see agriculturally in this country....I am very interested in the idea of natural farming--i.e. not disturbing the topsoil. I recently finished Plowman's Folly by Edward Faulkner which makes many very good points....If you know of anyone using such techniques for large-scale grain production, I'd appreciate hearing about it....You are a soul whose need on this planet becomes more and more apparent every day. Hang in there....(p. 15)

Glenn Drowns, ID —....the way the seed catalogs are changing really scares me. The last time I ordered from Shumway was in 1977 and they had four varieties of Cushaw pumpkins. I was shocked to find out all of the varieties they have dropped....Twilley's seed catalog has page after page of new hybrids replacing the old standbys....in a few years people are going to wish they hadn't been so hasty in destroying part of our heritage....I won a total of 43 ribbons at the fair with my vegetables and received a special award for having the most first places. I even won a prize on some of my heirloom beans....I plan on planting four to six hills of each variety I get, depending on supply and how rare they are. I hope to have between 400-500 hills next year....I had a terrible time last year with pheasants and quail eating my plants and seeds. I found that by planting the seed inside a can with both ends cut out, they aren't able to scratch the roots and eat the seed....(p. 16)

Dan Eaton, IL — It seems like a lot of us old-timers are gradually slipping away. I haven't heard from Lyle Settle in ages....On account of poor health and being 79 years old, I am trying to sell my seed business. I have about 2,000 customers, and it's just too much for me to look after. I have about 40 different kinds of beans, some quite rare. So if someone wants to acquire a business and thinks they can really make a go of it, I have enough seed to plant about five acres....(p. 18)

George S. Poole, IL — I have Mescher lettuce, an excellent heirloom lettuce which I got from Mrs. Mescher, an 86-year-old retired school teacher. She has kept the variety since 1909 when she got the seed from a neighbor called Granny Wallace. Granny Wallace had received the variety from her granny in 1849, who in turn had kept the variety since the late 1700s....I am working on a Master's degree in plant breeding. I enjoy developing my own varieties of vegetables, as well as preserving old varieties well adapted to my own region and soils. (p. 18)

Dale Anderson, IN — I have so many pleasant memories from the Campout Convention. I have hesitated to mention it to other members, not knowing whether it will be a yearly event or not. (It sure will - Kent.) It's such a great experience. I can see where it could balloon to nearly unwieldy proportions. We held our End of the Garden Vegetable Soup Party, Saturday, September 27. In five years it has grown from about 25 to 80. It was a huge success and not hard to do....the key is the large iron kettle where the soup is cooked outdoors....(p. 19)

John Edgerton, IN — I am turning my 80 X 80 ft. garden of deep beds into space for "heirloom and endangered seeds." Please, anyone, particularly people in the Midwest, who cannot for any reason renew or keep your seeds--send them to me. I will grow and renew them for you. I will even help with postage if absolutely necessary....I am planning to run some tomato trials of my own on a 30 X 150 ft. garden spot that has been mulched according to Permaculture II guidelines (see Permaculture I and Permaculture II by Bill Mollison). I will be requesting seeds from quite a few of you folks. Anyone who would be willing to send me some of their seeds would save me a great deal of time. Please give me a brief description of both fruit and plant habit. At the end of the season, I will publish my results for any SSE member who will send me a SASE....My present plans are to devote a good portion of the acreage of this farm to the preservation of heirloom and endangered seeds, fruits, nuts, berries, grains and herbs that have some history of survival and adaptability in our Midwestern climate. This will be along the lines of Gary Nabhan's work in the Southwest and will place an emphasis on Amish, Native American and other heirloom/endangered stock that can survive our climate. If any of you have any suggestions/advice, it would be most welcome....I wrote Richard Grazzini re Self-Sufficient Seeds here. Also I told him I would be glad to provide a regional trial garden on site. I sincerely hope that organization finds fertile soil. I agree with you--a real diversity of regionally based seed companies would help guarantee a lasting and varied gene pool....(p. 20)

L. E. Hileman, IN — It's been fun and gratifying to participate in the SSE and I have developed a very interesting friendship, letter-wise, with a good number of SSE members. They are fine folks and deserve good treatment....Good luck, Kent, and please don't get discouraged even though things look monumental to you now. You have a lot of good people on your side trying to make your dream come true....(p. 20)

Dorotha J. Shortridge, IN — There seems to be something in the soil, climate, or what have you here that causes such bush beans as Coach Dog, Trout, Jacob's Cattle, Appaloosa, etc. to go pretty much to the solid color. I haven't noticed that in any of the pole beans, though. It may be that those varieties are just not as stable as some. (Ralph Stevenson wrote me about the very same thing recently. Many of his varieties did that, and he doesn't know what is causing it either. Could it be the extremely wet weather? Could it be that the acid rains are worse than we think and it's a pH change? - Kent) (p. 21)

Susan Littke, IA —....This year I started an insect collection. I can't tell you how much this helped my organic gardening and pest control. Spent more time looking for insects in the garden, learned to identify and hand-pick the bad ones before they got out of hand....Who cares how the neighbors look at you? Everyone should get a copy of Rodale's Color Handbook of Garden Insects. (p. 22)

Richard M. Libby, ME - Discovering the Seed Savers Exchange has been one of the most exciting things to happen to me in years....after receiving a copy of the 1981 Yearbook, I went to visit my grandmother, and I took some time to poke around in the basement of the old farmhouse where she lives. What I found there made me cringe. Off in a corner on a shelf were several boxes and jars of old seeds left there since my grandfather died several years ago. Much of the packaged seed had been attacked by mice and was lost. I imagine the rest are so old that they will never germinate, but I certainly intend to try. This loss occurred because no one knew that the seeds were there....(p. 25)

Darrell A. Rolerson, ME — I'd like to remind everyone who requests seeds from me that it costs no more to include a sample of your favorite seed, if you have plenty of it, whatever it may be: food, medicine, sacrament, heirloom vegetables especially. I love to explore the possibilities. I am especially interested in Big Beans. Anyone who grows unusually big beans, I would be grateful for a sample. Only one rule: the bean must be edible. I will report on the results of my search in next year's issue....Can anyone supply me with Damiana seeds and a watermelon or any melon that will thrive in Maine? ....Also everyone please send samples of all cucumber seeds, as I am experimenting with a collection of the old varieties....(p. 25)

O. J. Lougheed, MA — Soybeans do very well here, and I have noticed that the older varieties produce many more nodules (more nitrogen?) than most of the newer ones. I will follow this up....I've had an offer of land at 8,000 ft. in the southern Rockies for research on mountain agriculture and a seed storage bank. It's sunny and very dry there....I have found myself wondering how to do what you are doing with "vegetables" for "farm" crops ("field crops"). Corn, beans and squash, the pre-European native crops, seem to be transitional and of course you list them. But what about wheat, barley, oats, rye, millets, sorghums and soybeans? What about native grasses? Foreign grasses? Trees? Etc....Let me know how you see all of this fitting together....(p. 26)

Louis Marcelina, MA — If you know of anybody who needs some plants grown for the seed, I will grow them, save the seed and mail them to that person ....(Write first to make sure that he isn't swamped. And of course you shouldn't ask unless you absolutely can't multiply it yourself - Kent.)....I have two seed banks, each one has a hundred tubes each (coin tubes), plus suitcases of jars. Some go back to 1955....(p. 26)

Rev. and Mrs. C. Frank Morrow, MN — We're looking for the Stove Wood bean—resembles the Lupini bean but is up to 10 times larger with some seeds over 2". They used to grow them on heavy racks of hog wire. The pods were a sick green/yellow color growing up to 1.5" thick, 4" wide and 40" long. The pods were so large that you could put three or four fingers into the cut end of it. A little boy could use a half-foot of pod as a gauntlet covering his four-year-old wrist and forearm. The seeds latched onto the inside of the pod and hinged at about 3" apart. There were about 12 beans per pod and the beans were crowded in the pod....My father raised them in Tennessee, east of Chattanooga. But in 1923 we moved north of Osakis Lake, MN and he put out all his potted seeds, leaving none behind, and bingo—a frost got them but good. He never seemed able to get any more seed....(p. 29)

Ralph and Vee Du Chemin, MS — How to Kill Slugs and Snails—After a rainy spell it seems like an army of slugs appears out of the blue in Mississippi winter and spring gardens. Place a pan in your oven and drop all egg shells into it. It doesn't take much heat for them to get really dry. Then put the shells into a bread sack, roll them very fine and put each sack into a grocery bag until needed. Sprinkle this powder around the affected plants. The slugs and snails will become coated with the sharp fine chips and it will kill them....We feel sure we can offer more varieties next year because now—we believe! (p. 29)

Laurence Gladfelter, MO — To keep coons out of corn, whitewash the ears with wood ashes. Apply a thick mix with a brush....What a wonderful thing you've gotten going....(p. 30)

Willard W. Osburn, MO — When I read your story in Organic Gardening it seemed like one of my prayers may be answered....my parents and grandparents were growing Cutshorts when they came to this region around 1890....my mother cried when she lost her seed, as she was the last in our family to grow this wonderful variety....(p. 30)

Mary Williams, MO — Mix baby powder and sulphur half and half and apply to the body to repel ticks....hair, either human or animal, will repel deer. Tie the hair in a little bag and tie the bag in trees or whatever the deer are working on....red pepper or cinnamon will keep ants out of your kitchen....spearmint will keep weevils out of garden seed....(p. 31)

Anna L. Rouf, NE — I live by a softball park, and my garden is a conversation piece. I am a widow and I use a crutch to walk and work in the garden....People are envious of me and my garden, but they won't try it on their own, as they say it's too much work. But every time they see I've been out harvesting, they call and ask if I have any extra....(p. 32)

Leroy J. Schmidbauer, NJ — Try lovage with your crookneck squash or zucchini to bring out the flavor....Thank you for the outstanding job on your Harvest Edition. It is excellent and I will be reading and re-reading it all winter.... The "John Baer" tomato was introduced in 1908 and is the same as Bonny Best. This is according to Dr. Richard W. Robinson, New York State Agricultural Experiment Station, Geneva, NY 14456....We are trying to develop a tomato seed bank--trying some storage projects....Thanks for all your support and encouragement, Kent. Hold us together. (p. 33)

Ralph Dutcher, NY — The seed of White Kidney that I will be giving out has five beans to the pod. It has taken me eight years to bring this strain up from the four beans per pod that I bought from a seed house (all of these beans have been selected anywhere from 8 to 20 years). I have some special seed that has more than five beans per pod and I will send 50 seeds of that Special White Kidney to two different gardeners who are willing to develop it further. At seventy-four, I am still in good health and will continue with them, but I would hate to see that work lost. (p. 34)

Larry Kuhn, NY — I work for the New York State Department of Agriculture and Markets. I have an Associate Degree in Horticulture and am very interested in organic gardening and the intensive methods of gardening described by John Jeavons....Many people throw out their old garden hoses each year, and I've been busy collecting, repairing and customizing them for my garden. When they are laid next to the plant to be watered, I then put a hole in the hose with an icepick. The hole can be as big as the plant needs it to be with just a little bit more energy on my part. Then I bury the whole hose system under a layer of leaves that act as a mulch. The system works beautifully with hardly any wasted water....(p. 35)

Jeff Schineller, NY — I am 17 years old and have now been saving seed for my own use for five years....You are doing a truly great service by trying to save the remaining heirloom seed varieties and thus help to maintain the world's genetic base. (p. 35)

Roy E. Ensley, NC — I am 60 years old and have lived most of my time here in the Great Smoky Mountains. The Cherokee Indian Reservation is located in my county and a nearby county. Seeds in the older families around here have been passed back and forth for several generations between our families and also the Cherokees. It will take me a few years, but I am sure I can locate some good old-time varieties. (p. 35)

Clayton Knepley, OH —....crop failure here for limas and South American squash, but thanks to your regularly published admonition not to plant all seedstock in one growing season, I'll be able to start over next spring....(p. 37)

Tom Knoche, OH — My family isn't Jewish, but one of their ancient agricultural traditions that I practice is to only save seed from crown fruits--the first fruit to appear on the vine. I believe this may impart strength, especially in the long run, and may also help somewhat in keeping varieties pure....In answer to your question on how I keep squash varieties from mixing, I plant five gardens in five different locations around my 67 acres....(pp. 37-38)

Mrs. Thelma Sinift, OH — I am so glad there are other people who believe that "old-fashioned" vegetables are worth their weight in taste, beauty and diversity. (p. 39)

Gail E. Wagner, OH — For people interested in Indian gardening techniques, I highly recommend: Agriculture of the Hidatsa Indians, an Indian Interpretation by Gilbert L. Wilson, The University of Minnesota Studies in the Social Sciences, No. 9, Minneapolis, originally publ. 1917, republ. 1977 by J & L Reprint Co., 410 Wedgewood Dr., Lincoln, NE 68510, Reprints in Anthropology, Vol. 5; Corn Among the Indians of the Upper Missouri by George F. Will and George E. Hyee, originally publ. 1917, republ. 1964 by University of Nebraska Press, Lincoln; and William L. Brown and Edgar Anderson, 1947, "The Northern Flint Corns," Annals of the Missouri Botanical Garden 34(1):1-29....I am an ethnobotanist (an archeologist who studies the relationships between cultures and plants). Since 1978, I have been working toward replicating a 1200 A.D. Fort Ancient culture Indian garden at an archeological site under excavation by the Dayton Museum of Natural History....My goal is to morphologically replicate the sorts of plants grown by the local villagers at that time....I began the garden with the idea in mind that it would eventually form a permanent part of the reconstructed village, and that seed would be made available to the public....We pattern our gardens after the Indian ones described for the Hidatsa/Mandan by Gilbert L. Wilson in 1917. These gardens are very pleasing to the eye and actually seem to produce better crops with less effort than the "white man's gardens" around us. The book makes very good reading and I recommend it highly to all gardeners....Our Fort Ancient house reconstruction was a huge success. The house is 28' X 23' and 13' tall. The walls are lathing (Big Bluestem) and daub. It took an acre of 6-7' tall Sudan grass to thatch the roof. The bench that runs around three of the interior walls is equivalent to six double beds. We used all original post holes to build this house....I am leaving in a week for three months in India. I have been hired by an archeologist from the University of Pennsylvania to set up a flotation recovery system for a Chalcolithic (ca. 2000 B.C.) site in the Saurashtran peninsula northwest of Bombay....(p. 40)

Elderly Cherokee woman, OK —....Reading your books is like a tonic for my soul....(p. 41)

Donald A. Cook, OR — Shades of Orvil Redenbacher, Kent, one would expect to find somewhere in the Seed Savers Exchange just how much one has to dry ears of home-grown popcorn so it will be sure to pop. I have been told that drying to the exact optimum moisture content is critical to success--too little drying and the kernels will not pop, and too much drying also prevents them from popping. If there is a reliable home test for this problem, I would sure like to know it....There are so many jewels of hand-me-down lore, sacred seed saving methods and treasured secrets of gardening, that the nominal stipend per year is worth it for all these "gems" alone....I can think of no better service to all of mankind than for you to be able to continue to do your wonderful work. (p. 42)

Perile Bogle Casey, TN — Does anyone know how to go about getting a nutritional analysis of a particular bean? The Soldier bean is the only bean I've ever grown that retains that soldier figure through cooking. The spots on other beans disappear in cooking. I can't help but wonder if it is a "message" that it is a superior bean in food value....We have a little peach tree that has produced fruit without insects while several other trees are subject to insect damage even with spraying. This tree is not even bothered with borers. The fruit is a cling and delicious. It is not a commercial peach, but a seedling. Would any member know if this could be a really resistant variety, or is it just coincidental? (p. 46)

Darryl Jordan, Plenty Agricultural Project, TN — I have a White-Seeded Grain Amaranth (Amaranthus cruentus) that was collected from Mayan Indians in the Highlands of Guatemala. The seed is very small, but the yield is good. The stand is uniform and has withstood drought and deer damage pretty well in Tennessee. It is presently being cultivated by, at most, four families in the tiny village of Choatalun, near San Martin Jilotepeques, Chimaltenango, Guatemala. We have been encouraging other Indians in the area to cultivate the amaranth. A few of them were glad to get the seed and said that they had wanted to grow it before, but had no seed and didn't know where to get any. Though a major staple crop in that area before the coming of the Spanish, most locals do not even know that the seed is edible, though they do eat greens of any of the weedy amaranths that grow there....I would like to know how to get a quantity of seed of Chrysanthemum cinerariaefolium or Chr. coccineus for the purpose of making pyrethrum. Are the commercially available flower varieties effective for these species? (pp. 46-47)

Janet Aldhizer, TX — We've had some wierd things happen in our garden this spring. We planted four rows of Mandan Bride corn--two sprouted beautifully, not a sprout in the other two rows! And our winged beans and asparagus peas sprouted, grew to about four inches, and are now just standing still. (That may be a day-length problem. I tried to grow some Jack Beans that Virginia Lindo sent me from Jamaica. The big, rank, square vines literally shot out of the ground, grew about two feet tall and then just stopped for the rest of the summer. It seemed like they realized, "Wait a minute--this ain't Jamaica!" - Kent) Someone asked how to get Winged Beans to germinate. My husband scratched them with coarse sandpaper, then planted. That's all. (Gary Nabhan suggested something similar in his writings. He said if you had some terribly hard seeds, for example beans that have become too old or too dry, sometimes you can get them to germinate by taking the corner of a razor blade and making a tiny nick through the seedcoat to let water in. Other people have suggested doing similar things to Bitter Melon seed which is extremely hard - Kent.) You asked what Plum Grannys are. I'm pretty sure they are pomegranates. At least that's what our German landlady calls them. (I'm sure you are right. Several other people have confirmed that - Kent.) Also, you wanted to hear from anyone about pleurotus mushrooms. We've been raising them all winter on sterilized chopped Johnson Grass hay. They are delicious. Our last crop will be ready tomorrow (May 14), then we'll have spore for the Chinese Rice Paddy mushroom coming for the summer. They require temperatures of 90 degrees F. and above. We obtain our sport from Mushroom Specialties, 445 Vassar Ave., Berkeley, CA 94708. They have the "recipes," as they call them, to raise them. (p. 47)

Carl Wilson, TX — For growing both nuts and stone-seeded fruits from seed, the following procedure works very well for me: I use boiling water, freezing, cool storage at 40 degrees F. and then a tepid water soak for 24 hours just before planting. A brush cleaning will remove flesh, meat and other debris from stone seeds. Soft-shelled seeds such as acorns and almonds, and small stone seed fruits such and cherries and plums, are placed in boiling water for two minutes. Large stone seed fruits such as peaches and nectarines are placed in boiling water for three minutes. The boiling water will eliminate mildew, mold and insect eggs (weevils, etc.) in the shell or in the kernel, but will not damage viability of the seed. (Don't cook the seed, just the two or three minutes time is sufficient.) Remove the seeds from the boiling water and allow them to cool. Then place them in a plastic bag with identification and put in the freezer for two or three weeks. After this they are refrigerated at 40 degrees temperature until just before planting time. Take them out and let them come to room temperature, and then soak them in tepid water for 24 hours before planting. All seeds should be planted "eye" down with the point end up....Sure nice that you got some funding from Rodale--a good organization. I started with them before Robert's time, with his father J. I., a very knowledgeable man and ahead of his time in his visions....Kent, is your seed bank to be a checking account sort of thing or is it a long-term savings bank for posterity? (The way I see it the program has to do both. This last year I've begun to build up an active collection here that I will deal out yearly for multiplication to the Growers Network. Pure new seed from that multiplication will go into relatively long-term storage. Steve Starr and I were just ready to start on a storage project when the Rodale organization proposed building a $200,000 cool storage facility with the same goal. We are waiting to see what happens with their proposal--just how quickly it may materialize. Also, Dr. John Rahart has expressed a desire to set up a storage project in Wyoming. So there are several ways it can go and I am sure that within a matter of months I'll know which way it will be going - Kent.) (p. 49)

Mrs. Eileen Wheeler, UT — I have a garden hint obtained from an old

Japanese man: In sandy soil that doesn't hold water well, dig a furrow about four inches deep and let the water run in it to soak. Then scatter your small seeds in the bottom and cover lightly with compost and shredded leaves. Keep a close watch for sprouting, then remove the leaves. (p. 49)

Alberto Vasquez, VA — The two largest edible beans are Vicia faba and Canavalia gladiata. Certain selections of either could qualify as Stove Wood beans (the enormous lentil craved by one of your correspondents is probably Fava bean). Canavalia gladiata can have 3' pods 2" wide. I expect that Gulf States may be the best place to grow. Seems to require at least as much frost-free season as cotton. Minnesota seems out of the question--probably the reason Rev. Morrow's father lost his stock after moving from Tennessee. As a journalist, you may be able to track this down: In the 1940s and early 1950s, pulp magazines (Popular Mechanics and the like, even comic books and occasionally Sunday supplements) had ads for "Amazing Giant Bean--grows pods a yard long and thick as your thumb," etc. I saw these many times as a boy. No doubt such ads were the source of some cultivated stock, although unless quick-maturing varieties were developed, they would not mature for their purchasers. In addition to long seasons, the beans may only set fruit on short days, thereby requiring long, frost-free autumns....

I sent Ben Quisenberry about 2 oz. of currant tomato seed about 10 years ago, and my wife sent a jar of tomato marmalade out of the same with the whole little tomatoes visible in the transparent gel. Really freaked the old boy out. That started the correspondence and resulted in our being sent all sorts of tomato seed. I believe that as long as the currant tomato seed held out, he sent free packs of ten seeds each to all his customers as a premium....

Encourage people to keep notes on season length, cultural practices, etc., and have all this information funneled in to you for editing and publication. This could be a self-reinforcing process that all members could benefit from. If you are spread too thin, I'll do it gratis until it becomes burdensome, just to get the information. Think about this "Field Trial Report" project and give me your ideas. At the moment it is just a notion, but I think it has great promise....

I'm not sure I agree completely with your idea that one shouldn't use the yearbook as a big seed catalog. I do. I figure I've "paid my dues." I distribute heavily and am very willing to give out to all. I think as long as one is willing to be a seed source, that that entitles him to "browse" the listings. The secret is spreading the seeds around and the desire to save one's own seed....(I agree with you completely. I have no qualms about members using the Yearbook heavily. That's what it's here for. What I object to are people who come into the SSE for the first time, order everything that sounds good to them, fill their gardens with heirloom varieties instead of commercial ones, and all this before they even realize that they don't yet have the knowledge to keep anything pure. Then they'll write and ask me if I think there's any chance that the five heirloom watermelon varieties that they are growing side by side might possibly cross. Or do I think that they may have hurt all of their bean seed which they accidentally left in the oven overnight--the ones that aren't scorched, that is. Or do I think that the rare sweet corn, that so-and-so kept pure for 40 years, might possibly have crossed with the 400 acres of field corn that is right across the road? For some reason the seed that they grew doesn't look like what they received....Letters like that are really frustrating for me because that type of misuse really can't be controlled by any means. Of course, the vast majority of new people who come to us are very enthusiastic and willing to learn. All I can really do is publish as much material as possible about the correct techniques of seed saving and storage--focusing specifically on purity - Kent.)....

I am willing to help the membership with plant identification to the best of my ability. I also know several experts whose friendship I can exploit. I will need a detailed description of growth habits, nearest town and state, date collected and the name of the collector. Ideally a pressed specimen of the entire plant including flowers should be sent. Several flowers and leaves represents about the minimum. Bean seeds can often be identified to species, as can those of most cucurbits, but seeds alone are not suitable for identification in most cases....(pp. 52-53)

Dr. John Rahart, WY — The Myona tomato was handed down from generation to generation in Italy and finally brought to New York about 15 years ago by an old Italian man. A local greenhouseman noticed the tomato in the man's garden and was impressed and obtained some seed to grow for sale. Many people wanted to know the name of the tomato, so the greenhouseman went back and asked the old man what kind of tomatoes they were and was told "Myona." It took a full year for the greenhouseman to realize that the old man in his broken English was merely saying that they were "my own." By that time the name had already stuck....My dad learned the story when he bought some of the plants at the greenhouse. He gave some of the tomatoes he grew to my aunt, who has been canning for over 60 years. She later told him not to bother bringing any other varieties to her because the Myonas were the best canning tomatoes she had ever seen. That's quite a tribute! (p. 55)

Alex Caron, Ontario —....We have been pushing seed saving among our membership. Recently I was at a one-day conference put on by the Bio-Dynamic Gardening and Farming Association of Ontario. One of the sessions was on seed saving, and it looks like this group may join with my group in saving seed. I have been in touch with pretty well all the Canadian SSE members and have obtained seed from them. I love multiplying it, and some of these are offered in my listing this year....(Alex and I have traded some letters this year exploring the idea of his starting a Seed Exchange in Canada. It would be patterned after the SSE and would parallel our efforts down here. I really hope Alex goes for it. I have always been disappointed that I haven't been able to develop our Canadian membership more. But I don't know their publications like I do ours and there is always a certain amount of hassle with trades across the border—not being able to get certain items across, different money, different stamps, etc. I really value our Canadian members, but I also feel that it could be handled much better from up there. I'd like to see Alex build a strong Canadian Seed Savers network. Eventually we could exchange new additions to each total collection once a year and then the new Canadian items could be distributed through the Growers Network. Maybe in the future by working at it from both sides, we might be able to do what it takes to cut through some of the red tape and start exchanging items other than seeds. Some of those potato varieties sound great - Kent.)

Douglas Law, Ontario — The restrictions regarding the importing of seed corn into Canada have greatly multiplied since last year. This compounds the problem of few of the old varieties being saved here. All our hybrid corn has been developed from six or fewer O. P. varieties. Regulations make it hard for me to give at least one O. P. corn variety still sold in your country to the Plant Gene Resource Centre in Ottawa for long-term storage and preservation, even though they would like it and some other varieties also. The new government restrictions and red tape here (due mainly to Head Smut, also restrictions due to unlicensed varieties, not at all due to plant patenting) are just about impossible to deal with....The plant patenting has not yet been passed here (11/2/81) but the present legislation has the same effect as patenting for field crops. A great number of old corn strains and varieties particularly adapted to Canadian (and Ontario) conditions have been lost....I was given a list of some corn varieties held in storage here. Almost no varieties were stored from Ontario farmers, although some indigenous Quebec varieties have been preserved. The prolific dents of South West Ontario (four ears matured per stalk and even earlier than southern U. S. prolific corn) as well as varieties grown here by the early United Empire Loyalist settlers seem unavailable....I have been unable to locate the "early" strain of Bloody Butcher which was grown here 40 years ago. This variety was likely similar to that grown in New York State. I have some more "southerly" Bloody Butcher which should be useful when acclimatized....

There has been much Canadian red tape and restriction on the importation of seed corn due to outbreaks of Head Smut. The Government set these restrictions in place because some hybrids grown here have no resistance to Head Smut, just as some varieties a few years ago had no resistance to Southern Leaf Blight....A Canadian wishing to import seed corn should write for an application for a permit to import corn seed to: Agriculture Canada, Plant Products and Quarantine Division, Permit Office, Ottawa, Ontario, Canada K1A 0C6. A phytosanitary certificate will be required to certify the corn free of Head Smut. This is available from the USDA offices. I am working on a way to reduce the red tape U. S. suppliers of small samples would have to go through, but as yet I have not succeeded. I am also trying to get a copy of the Canadian Seeds Act, and a list of corn varieties sold here prior to 1923 (although it is illegal to sell these as seed now)....

This year we tried Henry Moore Yellow from Stodden's Seed Farms, PO Box 6, Trilla, IL 62469. This corn has the ability to put up a lot of ensilage per acre. Its very strong root system does not seem to be much troubled by (western) corn root worms which are starting to be a problem here, even with regular rotation. Krug's 90-Day corn also from Stodden's gave a good account of itself as an ensilage corn. I am selecting this variety for earliness. Wapsie Valley Dent from Steinbronn Seed Farm, Route 2, Fairbank, IA 50629 has large ears and appears to be high yielding for ensilage or to pick as a full season corn here....Results of Ontario Agricultural College research on ensilage corns are just now appearing in farm papers here. This is the first basic research on ensilage corn here in about 40 years. Results seem to indicate corn grain yields are not necessarily indicative of high yields of ensilage. No corn is presently licensed here on its merits as an ensilage corn, or on its food value, that I am aware of....I do not believe the hybrid grain corns grown in this area are the best that can be had for ensilage. This is the result of my own observations....I have not found any Canadian farmer save myself who has open-pollinated field corns....(pp. 57-58)

Roy Lent, Costa Rica — I'm going to start collecting and maintaining peppers, especially sweet peppers. Will exchange with others, especially in tropical countries. Am also interested in any unusual tropical vegetable varieties....After reading in the Harvest Edition about the plight of your small seed companies, I began to imagine what the future could hold for you people living up there in that "bastion of free enterprise." Undercover "seed-leggers" send out their seeds in packages made to appear as innocent packets of pornography or as packages of violently poisonous ag chemicals. Shoot-outs as clean-cut federal agents close in on the farms of snarling, unshaven, illegal seed producers. A Coast Guard cutter races to cut off the wild dash of a dastardly foreign seed smuggler in his fast cruiser....All of the above is quite dramatic and silly, but not that much sillier than what was described as going on right now....(p. 59)

Paul W. Jackson, Mexico — I have been told that our Tahitian squash is an old variety that was discarded by the seed companies. I can see why it might have been, because a 20-pound squash will have only about 300 seeds, which is a fraction of the production of most other squash. For us here, the production has been unbelievable. Some of the plants here forget that they are supposed to be annuals and live over. One plant in back of the house produced large, seedless squash all winter long--there were no male blossoms--and grew vines over 70' long. I had to tear it out because it was swamping my grapes, and the job took all day and I still missed some that is still growing. It keeps very well--we have had some keep for over seven months laying out in the open, and it gets sweeter as it gets older. We gave a lot of its seed away around here, and I know that it is going to make a significant difference in the diets of the local children....

You could do us a favor--we need seed of all kinds so that we can grow the plants here and see which do well under our growing methods and conditions and add to the food self-sufficiency of the people. Any of your members who have a little left over and would be willing to send some would earn our greatest gratitude. We have grown over 70 different varieties of beans, and some just don't produce here while others give enormous crops. The Indian Blue corn does very well, and we got a second crop with seed from the first crop this year. The variance in yield from different varieties has amazed me, and I have been growing things for many years....Our trial gardens are being run on a totally organic basis because they must provide an example for the poorest who cannot afford fertilizers and insecticides. The only way you can really understand what the people here are up against is to live the same way they do. People in the United States just can't realize the frustrations facing the extremely poor....

Kent, why don't you and your family take a little trip and come down to see us? Birth certificates are all that are necessary to get past the check point, and you would find it interesting here. You haven't lived until you see a Boojum tree, and the Boojum National Forest is just about 200 miles south of here, with a few trees around here in yards. It would give me great pleasure to have a chance to visit with you....(I'd love it too and maybe we can someday. Paul and I have been corresponding off and on for almost six years. I am not sure if Paul is still putting out his excellent little magazine called "Good and Wild" now that he is working full-time on his projects in Mexico. His organization has worked for years to raise money to plant food trees as street trees to help the children in extremely poverty-stricken areas. I can't think of anything more worthwhile. I hope everyone will support his project by sending him seeds of varieties that are likely to produce abundantly for him. Our First Class mail goes to Canada and Mexico too, so it won't cost you any more to send seed to Paul than it would to send it to someone in the U. S. A trial garden south of the border being operated by someone with Paul's experience is an exciting development - Kent.) (p. 60)

Lars-Olov Rosenstrom, Sweden — I am a biology teacher and am planning a garden at the school this summer for my students. I feel that a school garden could be used to perpetuate and spread many things including interest in plants and their preservation....We have many hundreds of local potato varieties. In the 19th century people were starving and had to eat the potatoes they were saving to plant. So they had to grow potatoes from seed, and in that way all this genetic variation came to light and many new forms were created....I would like to obtain two Chinese vegetables—Sugarroots (Sium sacharum) and Choro-Gi (Stachys tuberosa) which was brought to Europe about 1900 by a German ambassador to China. Also any corns or beans that will mature in 90 days or less in this climate....Some other gardeners and I hope to start a Seed Exchange here and do research on vegetables suited to our climate....(p. 60)

Karen Burnette, KY — I am trying to find an old potato known locally as White Sprout. My mother said they were so long and slender that people used to go to the patch, dig them and carry them in stacked up in their arms like stove-wood....I am an English teacher. Each year my 10th grade students do a research paper on something to do with their heritage. One topic which can be chosen is heirloom seeds....(p. 63)

John R. Burch, MO — How did we manage to go so far in hybridizing everything and get so far away from the fresh taste of many of the old-fashioned varieties?....Our vegetables have undergone much the same changes that our fruit trees have: fruit that's beautiful to look at but not much taste....(p. 64)

Maurice Eisenstadt, NY — Harry Saier used to carry a wonderful Japanese hot red chili called Yatsufusu. Hudson dropped it and no other oriental sellers know of it....I had a disagreeable experience with the Vilmorin company in France, which seems rapidly going the way of the companies here—fewer and fewer varieties every year. They had a F1 hybrid eggplant called Croisette which was the eggplant of my dreams—prolific, intense flavor, grew 7' tall. When my supply of seed was exhausted, I discovered those morons discontinued it. I wrote them a long letter (in French, no less) begging for a few seeds, since I was sure they had a stash. I got a two-line brushoff....A good source of Oriental seeds is Dr. Yu Farm, PO Box 290, College Park, MD 20740. He is a Korean farmer (formerly a Ph.D. food chemist!) who I got to know in connection with oriental mushroom growing. You should write for his catalog, which is fantastic....(pp. 65-66)

John D. Payne, NC — My grandaddy had some field peas that he got from an old man in Dothan, Alabama. Grandaddy planted them for years and then one season they did not make it and he lost the seed. When he went to the old man for more seed, he had died and none of his children knew anything about the peas except that they were the best they had ever eaten....when the peas were picked and shelled they were a deep dark red. The dry seeds were jet black with a small red eye. And when cooked they made a rich walnut brown juice and the peas were real dark....They were about the size of a Georgia Clay pea. The taste was so good I can't begin to tell you. My grandaddy and I don't know the name of the pea....(p. 66)

From The 1983 Winter Yearbook—

Jean Boone, AK — I'm very proud to be your first Alaskan Listed Member (yes, there are gardeners in Alaska!), but I'm concerned about the exchange of mail. At that time of year (February) it sometimes takes up to a month for me to receive mail out here in the bush....I have less than 95 frost-free days to grow vegetables. To compensate for permafrost soil and short season, I grow non-greenhouse vegetables in my unheated greenhouse. In a pair of 2' wide beds, 25' long each, I harvest 90 lbs. of tomatoes, 5 gallons worth of "dill pickles," and over 400 sweet bell peppers annually. I have just added on 12' of greenhouse for what I hope are more cucumbers and space for cantaloupe and squash....In addition, because of cold climate and permafrost soil, I grow my winter squash in above-ground containers which are what you warm-climate people call compost bins. Here, they are decomposition/soil-making bins. They have clear plastic lids, and the boxes are insulated. It works! Don't laugh!....I feel I have a contribution to make even if my growing capacity is small, my location remote, my season short and my methods unorthodox. I am careful at record-keeping and observation (taking notes). I feel very strongly about food—any serious subsistence bush homesteader does. You're sorta inclined to take serious stock of life when the temperature daily drops to -70 degrees F., as it did here in 1974 for over two weeks! (p. 49)

Donna DuMouchel, CA — I just want to offer a few delighted words of congratulations for the 1982 Winter Yearbook. I never expected anything on that scale. Certainly makes me appreciate what a tremendous undertaking "The Garden Seed Inventory" must be. Thank God for home computers!....I really don't have any special projects going this year except my son's first garden (he'll be two this summer). I'm tilling up a special section just for him. I'm hoping to give him an idea what on earth Mommy does out there all summer and to start the love of gardening in him early. (p. 54)

Oscar Jaitt, CA — To repel ants from your garden, grow Feverfew, a perennial herb with dainty daisy-like flowers. Also repels bees. (Someone else wrote that Feverfew also repelled mosquitos if you rubbed it on your skin - Kent.) ....Sowbugs and earwigs are especially attracted to corn cobs. Use them as bait. The best time to collect is at night. Would like to know of any other organic methods of dealing with sowbugs....Next summer I will try to grow Moon & Stars watermelon commercially. I would like to encourage other farmers to try growing the old-time vegetables. Some of them have a definite commercial market. And this is one excellent way to ensure their future existence. (p. 57)

Brainerd and Loraine Lundin, CA — We think the SSE is great for elderly people who have gardened all their lives....More power to you....The members are wonderful. We enjoyed hearing from all of them....We use Bay leaves to keep the bugs out of beans....(p. 58)

Rob Sobolewski, CA — I think it's a great idea to get as much financial (and other) support as quickly as possible, as I sense the whole organization is mushrooming rapidly! (See what you've started!) Isn't it amazing--that when there is a need, the people step forward and give of themselves. What you are doing is spiritually right or else you would not have received so much human assistance. (pp. 60-61)

Craig and Katie Watts, CA — We are troubled with snails, slugs, sowbugs (pillbugs) and earwigs and would appreciate any organic aids to controlling them. So far the best we've discovered for snails and slugs is handpicking at night and dropping into hot, salty water. Earwigs are very fond of well-sugared coffee and it does do them in quite well. Have no aids for the sow/pill bug, though, and we have millions. We're told they do no damage, but don't believe it. (p. 62)

Joseph V. Rosanova, HI — I believe that the SSE is the most important alternative lifestyle network at work in America today. Living and gardening in Hawaii makes dealing with seed companies even less desirable than it is for mainland gardeners. We tend not to get our spring catalogs until some of the seeds are sold out. Many orders get lost or damaged in the humid, slow-moving Post Offices (since the packages are seldom sent First Class). The fact that we have no real seasons gives us a bounteous 12-month production but also necessitates a year-round seed supply. Hybrids are worthless here since they are almost always specialized for an environment totally unlike ours....(p. 69)

Chet Fowler, IL — I live in the inner city of Chicago, but have been blessed with a bit of land....You're doing our nation a very great service. Please keep it up. (p. 79)

Charles E. Voigt, IL — What is the status of Self-Reliant Seeds' collection of antique grain varieties? Do you have contact with them, Kent? It would be a shame if their work in preserving the varieties was lost because the seed company folded. (O. J. Lougheed spent several days with my family this summer on his way to New Mexico where he is relocating. He was pulling a trailer with 600 lbs. of seed. He has a listing under New Mexico. I think we'll be hearing much more of him in the coming years and am certainly glad he is sharing his collection with all of us - Kent.) (p. 80)

Lester E. Hileman, IN — I've had many people write to me requesting either seed or information that has led to a writing friendship. That is a real pleasure, and just part of the rewards of participating in your program. (p. 85)

William T. Arthur, IA — In 1853 my great-great-grandmother Elizabeth Robertson Fleener, a widow with 11 children, brought her family from Indiana to the farm where I live. She brought the Fleener Top Set onions with her. They have been raised by her descendants since then. They can be planted in the spring or in the fall and will winter over. They will produce a white onion 3" in diameter or larger under good growing conditions....In 1983 I had almost 200 requests for this onion. That coupled with the poor 1983 harvest will prevent me from offering samples in the spring....The large onion sends up a stalk which bears small sets. These are planted the next year and will then produce the large onions. Occasionally large sets will produce sets the first year planted. Large onions will produce sets for more than one year. Each year a person should plant both sets and large onions. If you do not replant them in the fall, they need to be kept in a very cool place. I keep mine in a cave that stays just above freezing. (p. 86)

Howard Cory, IA — I hope your little seed boat stays afloat....I will send you an old Oscar Will catalog soon. (Howard is the only remaining Charter Member (1976) who has been with the Seed Savers Exchange every year since its founding - Kent.)

Darrell Rolerson, The Dharma Farm, ME — I want to continue cucumber trials, so anyone requesting seeds I encourage to enclose a few seeds of their own favorite variety. As my collection grows I will be able to offer a greater selection....The "Biggest Beans" which I have "discovered" (thanks to the generosity of Seed Savers) include Aztec White, Aztec Purple, Mexican Red Runner, Bearpaw (a Seneca Indian bean), African Butterbean and White Dutch Runner. All cross easily. All are extraordinary beans weighing on the average of 4 grams each (often more). These beans are "meat." I will send a collection containing a few of each for $3 postpaid. The proceeds will be used to renovate a seed display room which I am opening to the public for educational purposes. Donations of books for this project will be acknowledged with real appreciation ....There is a nice enthusiasm among the gardeners here on the island where I live, which is three miles off the coast of Maine. I get a lot of help in maintaining my rare seed collection....This year I am appealing to Seed Savers to send me old pea varieties, any purple-podded varieties, and snow peas and sugar snaps. The higher climbing, the more enthusiastic I am about growing them. I would like to climb a ladder to pick fresh peas....My search for rare seeds is directed especially toward the preservation of American Indian strains, especially corn, beans and squash—"The Three Sisters of the Earth." These three vegetables are most representative of the "dharma" (that which contains the true nature) of this continent—hence the name of my farm. Any Indian seeds which are sent to me will certainly be cherished. I believe that the Seed Savers Exchange is one of the best things happening in this world. (pp. 92-93)

Joyce B. (Risner) Fetterley, MI — I was hurt last spring and did not get to plant all the seeds I wanted to reoffer. I smashed my ankle. I'm trying to catalog all the old rare seeds I have propagated over 20 years of saving and hope to offer them next year. I have a very large collection not listed in the SSE. I don't want them lost....I can approximately 2,000 jars to give to poor and needy people. I have many lovely people helping me this year. We even put up more than usual and are still canning deer now. My seeds are used for the needy people, too, for food. (p. 97)

Treasa Iho, MI — So many of the SSE members were very generous in sample sizes and gave personal replies. That's gratifying! After I won the Upper Peninsula Homemaker of the Year Award last August, a local radio station (in Escanaba, Michigan) did a five-minute spot on the SSE because I had listed it as one of my hobbies. I heard from many people who never imagined that seeds have become and are becoming extinct. They thought that only happened to animal life. (p. 98)

Mrs. Ed Beiswenger, MN -- Running ads in newspapers helped to acquire more unknown seeds. Good luck, Kent. Many people appreciate you. (p. 100)

Jocelyn Beiswenger, MN — I'm age 13. I think it's really neat how people save seed! I would like to hear from some foreign members. (p. 100)

Micheal Gould, MN — I'm very thankful that someone took it upon themselves to establish an exchange like the SSE....Each year I get several hundred seed and nursery catalogs from U. S., Canada and overseas....I am seriously thinking about starting a small seed service to supply old-time and foreign open-pollinated varieties. If I could just find a way to bypass some of the costly and time-consuming regulations now in effect. I could easily keep 250 to 1000 varieties pure. I've always planned to start a seed business, but so far all the red tape has stopped me. (p. 101)

Elaine Grotjohn, MN — I have really enjoyed the SSE. We have had much pleasure and good eating from varieties obtained as a result of your hard work. (p. 101)

Paula Northerner, MT — I forage extensively in the mountains and foothills of Montana. If anyone wants herbs or wild plants, please send a description (including Latin name if possible) and POSTAGE and I will try to collect it during the season. Some I have cultivated. Many native plants of great use are also endangered due to introduced plants, herbicides, cattle grazing, etc. Let's save and improve these, too! (p. 108)

Peggie C. Leifeste, NJ — Everyone should read Gary Nabhan's book The Desert Smells Like Rain, North Point Press or from your library. It is about Papago Indian farmers today. You will never feel down about your garden problems after you read about Papago gardening conditions. It is very dry land farming, but they succeed. (p. 110)

Peter R. Hoover, NY — Yours is a controlled burnout, which is a pretty exciting and demanding way to live. (p. 114)

Paul J. Valente, NY — To grow leeks without a lot of fuss: we start them in the house and transplant them to the garden. As they grow, we just pile mulch around the stems (usually grass clippings). We get long, white leeks without burying them a foot deep....(p. 116)

Mrs. Lucina Cress, OH — I will include a recipe on request for stuffing peppers with cabbage and canning....I planted about 400 miniature pepper plants this year and picked about four bushels at every picking....Again this year our branch of the hospital stuffed miniature peppers. We canned over 830 pints for the Hospital Bazaar and they were sold out by 11 a.m. (Lucina sent me a pint of her stuffed red peppers. We had them on the Thanksgiving table at my folks' in Kansas. They were great! Thank you - Kent.) (p. 122)

John R. Parker, OH — It's great to be a part of a group which is dedicated to "saving" rather than "using" only. Good luck to us all! (p. 125)

Thelma Sinift, OH — I am so happy you are now tax-exempt. Your (our) organization is certainly needed. Thank you for letting me be a part of the future. (p. 129)

Donald A. Cook, OR — We wish to thank you for your recommendation of De Giorgi Company (1411 Third Street, Council Bluffs, IA 51502). Their catalog, "Gardener's and Florist's Price List," is very comprehensive. An ample amount of seeds for the money, and they sell seeds, not just beautiful photos. A buck for this catalog is worth every 30 cents of its deflated value. Their 1982 issue is packed with 112 pages of goodies—with many old-time, open-pollinated vegetables, especially melons and squash and pumpkins. Let's continue to patronize them so the conglomerates won't gobble them up. (p. 133)

Al Kaplar, Omniversity of the One, Peace Seeds, OR — From thin-layer chromatrography of tomato juice from different kinds, we know that some have as many as 6-20 amino acids essential for the building of proteins, others have only 2-3. Color is independent from nutrition. More data when we move from 1-dimensional to 2-dimensional to columns....The One World Family Planetary Plant Seed Collection currently has about 2500 kinds of seeds from 147/360 plant families....We have 91 kinds of tomatoes and 55 kinds of lettuce from this year's work. There is more bio-sand, silicon dioxide, in lettuce than any other common garden food. Cultivars selected for this molecular property would be useful....(p. 135)

Mark Boettner, PA — I was told that soaking mint leaves in water two days and spraying cabbage plants with it would repel bugs. I tried it and it does seem to work. But I've only tried it one year. (p. 136)

Henry B. Lillibridge, RI — When the colonists arrived in what is now Massachusetts, in 1620, they found the American Indians growing a white flint corn. This flint corn soon became the main food crop of the colonists. Our present-day Rhode Island Whitecap flint is the result of selection made from the original white flint obtained from the Narragansett Indians by the colonists....Since the advent of the newer hybrid dent corn varieties, Rhode Island white flint corn has all but disappeared form the farms of Rhode Island. A few growers still plant small amounts for johnnycake meal. It is highly prized for that purpose....It is difficult for a grower to maintain his seed stock of Rhode Island white flint because of the prevalence of hybrid dent varieties in the neighborhood. Corn is wind-pollinated, and it is not long before the white flint becomes a mixture of dent and flint and of yellow and white kernels. The grower must then try to find some source of seed where the white flint has been grown in sufficient isolation to maintain its characteristics. This has become a very difficult task....In order to maintain at least a small source of pure seed of this variety, the Rhode Island Experiment Station has grown a small plot of this variety in as isolated a location as possible about once in every three or four years. In this way, a small amount has been available for growers who have wished to start with a new seed stock....(pp. 139-140)

Donna E. Buckingham, SC — I'm a member of an organization called the Carolina Farm Stewardship Association (CFSA). As a learning experience for myself and others, I'm attempting to form a seed savers group within CFSA. Your name and resource materials are often mentioned during group discussions, and I thank you for your efforts. (p. 140)

Bernard Williams, SC — I am a Master Gardener. I have gardened for many years, but in the spring of 1982 the Horticulture Extension Office of Clemson University (Greenville County, SC) had a Master Gardening Course and accepted 20 volunteers from widely dispersed areas of the county. We took a three-month course on gardening, lawn care, shrubs, trees and vegetables. After we completed the course we were supposed to help anybody who was in trouble. If we couldn't diagnose a problem, we could take a specimen to the Horticulture Dept. and they would give us an answer. I have received more help on raising vegetables than I have given. Some of the old-timers have tremendous gardening experience and are willing to share. Everybody wants to give you slips or seeds of flowers and vegetables. I have received nothing but good feelings from my experience....(p. 140)

Eugene V. Boren, TX — Next year I hope to offer poppy seed of the type that is widely used by cooks of Czech and German extraction in central Texas to cook kolaches and butka. Regular ingestion of these pastries leads to a happy, long life—the happiness because they are so delicious, the longevity because one is afraid to die for fear the goodies may not be available in the hereafter. (p. 144)

Robert Gene Daubert, TX — I have been reading some of Luther Burbank's works....He developed a hybrid chestnut that produced nuts in less than two years from seed! I am enclosing a part of an article on Tigridias (Mexican Shell flowers). He states that they are tasty and nutritious. He also discusses several other plants that could be developed as a food source....I hope to develop a small peanut and shallot collection. (p. 144)

Mrs. L. R. Hayes, TX — Just got the 1982 Harvest Edition....Those corn pictures are works of art, even in black and white. They must really be something in color! Mr. Quisenberry reminds me of my uncle and an aunt, although they are two or three years younger. Some of the older generation really have the energy and a zest for living. (p. 144)

Jeff McCormack, Southern Exposure Seed Exchange, VA — WARNING: Seeds delivered to rural mailboxes during the heat of summer can cook in the mailbox, especially if it is a dark color. This summer I plan to put a temperature recorder inside my home mailbox just to see how hot it gets on a hot summer day. I've had one batch of seeds ruined because of this problem. I've since painted the mailbox white, but it still might become quite hot. I am now having all of my seeds sent to a box in the Post Office. We are warning our customers about this problem. (p. 149)

Terri L. Halstead, WA — Another mosquito control/repellent besides citronella is the herb Feverfew. While some people do occasionally have a dermatitis-type reaction from the oils released in the fresh leaves, our family has never had that trouble. Feverfew is an annual, but self-sows easily. We use it two ways: 1) rub fresh leaves on our skin to effectively repel the mosquitos we're notorious for, and 2) burn the dry leaves in a smudge-pot effect to rid the air of mosquitos--they scram!....You were looking for a white-striped purple eggplant also. I don't know if this is the variety you mean, but our local Asian markets (in Seattle) sell one in the summer. It's like the golden egg-shaped eggplant, but has wide white stripes on purple. I have used it and it's quite good. When I asked where it's from, I was told Thailand....I understand your "Warning--Some of You Are Killing Your Seeds," about the damage done by heating seeds in the oven. I also don't remember where I first heard it (but I knew it was false), but the latest place I saw it suggested was in Rodale's Organic Gardening. The issue for May of 1982 (pp. 55-56) had an article "Field Peas: The Three-Way Bean" by Barbara Pleasant. On page 56 she states, "After a 10 minute stay in a 200-degree oven to kill any insects, the peas are ready to be stored in jars." So the myth goes on!....Besides freezing some of my seed, I use diatomaceous earth (1 Tb. per quart of beans, mixed throughout to keep bean weevils, etc. out). Have you ever heard anything negative about this practice? (I'm sure it's fine to use on seed that you know will be planted. I wouldn't ever use it on any seed that might be used as food. It is actuallly a micro-sharp abrasive--kills insects by lacerating them--and some people are afraid its effects may be similar to asbestos - Kent.) (pp. 153-154)

William D. Ross, Good Seed Company, WA — Thanks to all of you fine seedy characters out there, especially those who sent Christmas cards. And a special thanks to Kent, for leaving us no excuse not to stop what we were doing and start swapping seeds....For provocative reading I recommend: Weeds and What They Tell Us by E. E. Pfeiffer, who was a skilled scientist as well as a masterful farmer, and The Future Is Abundant by the Tilth Association (2270 NW Irving, Portland, OR 97210), a fine overview and access book into sustainable agriculture, very comprehensive....The following tomato seed cleaning suggestions assume that you are familiar with the general procedure. First: cut the fruit the "wrong way," like you would to see the star on an apple. Squeeze the pulpy seed mix out, retaining the flesh and the skin for sauce, soup, catsup or whatever. Second: on your rinse cycle use four or more parts water to one part seed pulp. This facilitates pouring off the unwanted debris in just two rinsings. Any smaller ratio and the mixture is so dense that it's hard for the good seeds to sink through all of the pulp. Also, using a glass jar makes it easy to watch what you're pouring out. I pour pulp through a sieve over a bucket for double insurance to catch my blunders. Third: I recommend using thick brown craft paper (heavy-duty shopping bags are made out of this) to dry soaked seed on. I want to discourage you from using newspaper to dry your seeds on, as the ink is very dirty and the paper is so thin that it sometimes "melts" onto the seeds. When the seeds are dry, I spread the thick paper on a smooth surface and lift off the seeds with a spatula or putty knife....Kent, your organization is so engrossing that I stand accused by my family of ignoring business to play seed swap. And so, I propose a toast: to the Seed Savers Exchange--a poor man's Atari and mail order Pac-Man! (p. 157)

Steven J. Bacon, WI — I've been in potato farming all my life (23 years). My parents and grandparents also were potato farmers. In the past four years I've been growing my own seed plots. I recently grew about twenty different types of potatoes, though most of them are still pretty common. I would like to someday grow about 40-50 different types....I use a three-year rotation: potatoes/oats (clover and timothy)/maturing clover and timothy that is cut in late July and plowed in. This assures me of no volunteering....(p. 160)

Dr. John P. Rahart, WY — I would like to obtain Missouri Basin Indian crops of all types, especially squash, corn and dye plants or medicinal plants. Similarly, I'm looking for much the same from the Southwest Indians. Specifically I am looking for a light lemon yellow flour corn from Hopi--a variety of corn known as Lazy Corn which grows on the ground instead of upright in stance, any corn varieties where the stalk and leaves display pigmentation other than the normal greens and green-yellows, any short-day flour corns from anywhere, any short-day sweet corns and any odd color varieties of any garden crops. I am quite interested in any short-season or cool-adapted varieties of any crops not normally so. I'm also very interested in any crops or any information anyone may have on ultra-violet tolerant fruits, vegetables, grains, etc....Any money I receive from non-listed members for seed will be donated to the SSE to strengthen the organization. I hope that other members will consider doing the same. (p. 163)

Mrs. Sam Brown, Alberta — Did anyone try the pea and bean carousel made by setting two buggy wheels with an axle between them up on end? Use bindertwine up and down to each spoke for them to climb on and put a dish of water on top for a bird bath. Makes a beautiful showpiece....(p. 164)

Alex Caron, Ontario — (Alex is Vice President of the Canadian Organic Growers (C.O.G.) and a member of the Natural Farmers of Ontario....Alex and friends intend to set up an "Adopt-A-Seed Program" within the Canadian Organic Growers. They are currently building interest through their C.O.G. Newsletter which reaches over 500 farmers and gardeners all across Canada. They also hope to use Canadian publications and Canada's nation-wide radio network to locate both materials and interested growers. The organization they envision would have specific growers permanently maintaining specific varieties....I met with several of Alex's associates when I spoke at the 1982 Harvest Festival of the Federated Organic Clubs of Michigan last August. They are dedicated, high-caliber young people of excellent vision. After meeting with them, I know they will succeed....I am really excited about their new program, will do everything that I can to assist them, and am looking forward to working with such fine folks in the coming years - Kent.) (p. 166)

Lee Creighton, Ontario — I would enjoy corresponding with Mexican and Southwestern gardeners, especially those with various chiles, herbs and spices....Have you thought seriously about Dr. John Rahart's suggestion of developing a network of small volunteer seed banks? Certainly, you've had other members suggest the same thing and even offer their own time and finances to set up banks for their regions. I have a good, dry basement if you feel it would be of some help to you for me to be the storage facility for South-Central Ontario....I also intend to contact Alex Caron in King City, Ontario, to see if I could be of some service to him in his efforts to set up a Canadian SSE....Kent, if you weren't performing this service for the good of gardeners and posterity, I don't know who would be. Continuing what you set in motion must be becoming an awesome task....I admire you for the enormous amount of work you're putting into this very important project and I hope you're getting as much satisfaction out of it as your members are. (p. 167)

Douglas Law, Ontario — My seed selection and hanging of ears has almost taken 1 1/2 months of full-time work. Our climate this year meant I had to look for good and mature ears in the case of later corns. Hopefully this year's seed will do better in our tough conditions....Some of my varieties are in limited supply due to our poor corn growing conditions here this year....In a bad year like this last summer, one should remember that seed corn is damaged by frost at 12 degrees F. if the moisture in the corn is 20%. Corn at 15% moisture is undamaged to -5 degrees F.; and at 13-14% moisture, corn is generally safe. Heat in excess also damages corn seed....I am pleased that I have had the opportunity to visit o.p. corn growers in Eastern and Western Ontario. Mostly they are working with seed imported from the U. S. I sent them samples of some of my corns....On a trip to the other end of Ontario, I found that less than 10 years ago there was a very early strain of Bloody Butcher grown for ears, or to put more grain in the ensilage. It had fairly deep grain, considering its early maturity, and had long, slender ears. It was a very old variety and was not North West Dent. It appears to be lost, although I will look for it further. It is a different Bloody Butcher than my dad remembers. It does not appear to be maintained in any institutions here....I hope something can be done to help preserve the o.p. field corns still in existence and locate the few that might still not be in the collections of seed....Canadian Indian corns need to be located....They have not been able to pass Plant Patenting Legislation here yet, but it is in existence in all but name anyway....I am still trying to locate other Canadians who have saved o.p. field corn from when it was widely grown. (p. 168)

Sara Bezaire, Mexico — There are six songbird species here which eat the reproductive parts of the squash flowers. I am looking for squash varieties which will fruit even if unpollinated, as zucchini sometimes does....(Rich Grazzini told me he had noticed several of the heirloom squash are self-fertile. He was puzzled as to what use that could be to a gardener, but of course he had never encountered your problem - Kent.)....I am especially looking for black or red currants, Kiwi fruit seeds and any hardy leguminous plants to cut for composting. I had more than an acre of alfalfa, but it can't stand our rainy season. Red clover and timothy do better....I spend many nights in the garden protecting it from armadillos, who find more earthworms here than in the whole Sierra Madre. There are so many animals which appreciate my labor that I am always amazed to get a share of the harvest....The plants I am offering are organic and are very distinctive from any squash or peppers I know of, and my interest in offering them is for breeding. Both have very large gene pools....I work and live alone, an hour's walk from any road and several hours' bus ride to town....have many other excellent and ancient Zapotec vegetable varieties and fruits, but it is a two-day journey to get out to mail seed that will not go in a letter. (My seed is also handled by J. L. Hudson, Seedsman, Box 1058, Redwood City, CA 94064)....Regulations here have made it a real struggle to ship seeds out, but I will do my best....

Kent, you have drawn such wonderful people around you....Many people here speak of such sad news about what's happening in the U. S., but I can smile back at them because of you and all the members who have written me....am so glad you express your devotion to family. I thought that was only in Mexico. You would feel so at home here....My word to God--"oh unlimited energy seeking to manifest all that is good"--always comes when I think of you. God bless you. I know he does, but I tell him, too. There is such love radiating from your work....It is very late. My candle flickers....(p. 170)

Joy Horton Hofmann, Ecuador -- I feel so limited in what I can offer to the SSE. The corns I have are photo sensitive and I haven't heard of too much luck with them north of the Mexican border. The squash need a very long growing season and must be forced, as well as some of the beans. Most of the plants resulting from seeds I've received through the SSE refuse to go to seed. One of the brassicas received from Steve Facciola is 7' tall by 4' wide with an 8" trunk diameter and still shows no sign of bolting. My "tall green kale" is over my head and leaves are picked from Swiss chard for three years until the plant finally dies out....I wish there were some way to integrate the gene pool from down here to up there. There are many characteristics still found here that are probably in some part lost there, i.e. drought tolerant, wind resistant, virus resistant, etc. Unfortunately, what is happening there is happening here also, i.e.--60-day bush beans from the U. S. are a popular thing to plant. But unfortunately they are not resistant to much here and must be sprayed several times to guarantee a harvest. Local beans need no spraying (yet)....Compliments on your new coding system. There is no easy way to keep track of such a wide assortment of vegetables and people. (pp. 170-171)

Paul W. Jackson, Mexico -- Last year we planted the Tahitian squash in our weediest area. It killed virtually all of the weeds and there were many fewer this year. Its production here has been unbelievable. This year we planted it in an unused roadway where the burr bearing plants were thick as fleas and five to six feet tall the year before. We hoed several times until the squash got a good start, then irrigated well several times and just stood back. It grew so thick that all you could see were green squash leaves and yellow blossoms--no earth anywhere....I took the delegado and the two local police down to see them last evening. They just shook their heads in disbelief. I am going to harvest all of them when ripe and make a big pile. Then I will get the ejido to bring all of the people down when they have a meeting after harvest and give each member some seeds. I know it will make a significant difference in the diets of the local children....The local secondary school teachers have all been down, the agriculture class three times last year and the animal husbandry class once this year....

You should include us in your "Apprenticeship Programs" section. We would welcome having some interested people come and learn with us. We have no dogma beyond pushing the ideas of Permanent Agriculture and the use of indigenous food plants and trees. We are attempting to try all different ideas of plant culture that might be adaptable for the use of the poorest people. Also we are exploring methods of utilizing foods....There will be no charge for apprentices and they will be "working apprentices." If they feel they have learned enough to make a donation worthwhile--fine. Apprentices are appropriate here year round. We are planting and harvesting all the time....The growing grounds here are a "showing and learning center"....

The more I do and read and see, the more I realize the deficiency of our "expert" knowledge. Also, that many of the "fads" have incorporated within them some good ideas....My wife and I are going to try to get to your August Conference in Colorado. She calls me a hermit, because I rarely leave here....Please invite any of your members who are making vacation trips to stop and visit us. We are about 125 miles south of the border, on the main highway that runs from Tijuana to Cabo San Lucas, in the little town of Colonet....Don't write and say you are coming--just come. (pp. 171-172)

Lars-Olov Rosenstrom, Sweden --....you have succeeded in planting your idea in at least one other country besides America. On April 4, 1982 we founded an organization called "Sesame - Sweden." As many of you know, sesame is the key word in one of the stories in The Arabian Nights: Ali Baba says "sesame" and a wall opens in the cliff. We hope that our Society will likewise open up great treasures of plants for gardeners here. The full name of it is "Sesame--Society for the Growth of and Seed Making of Vegetables in Sweden"....The first season we only had eight members, but it will grow. It is now very popular here to grow things in the backyard, but few gardeners save seed. We intend to change that....The Society is organized as a closed membership with every member being responsible for certain plants. They guarantee that seeds from those plants will always be available for other members and for the future. We are also planning a seedbank to back up our efforts. We feel that people who are not willing to take this responsibility would only be a drag on our organization. So only people who are contributing to the survival of these treasures will have direct access to them.

We like your idea of a "wanted vegetables" section in the Yearbook and plan to include one in a small magazine the Society will print. I want to thank you for all ideas we get from you and all inspiration....The last member we got is growing potatoes and collecting old Swedish vegetables. It is so good to have him with us, as we were having a hard time locating any really old Swedish varieties. He grows 38 old potatoes and knows an old man who grows 103 others....

I have a story to tell you. Elaine Grotjohn sent me a Swedish Brown bean which her grandparents brought with them from Sweden in 1872 when they immigrated to America. To understand my feelings you would have to see the wealth we have now, but also know of the poverty that was. In the winter of 1869, 100,000 people died in this country from starvation. I thought of her grandparents who had left this country for a better life, not knowing where they were going. Certainly they did not know that their small beans would return alive 110 years later. I was deeply moved by this....

We are at 59 degrees latitude and have 100 frost-free days, but we have a maritime climate with a cold spring followed by cold rains....In Sweden we have only one seed company and they only carry one corn variety (Queen Anne), which produces well in 80 days. We also have only one dry cookiong bean in cultivation--Swedish Brown beans (so anything not brown is beautiful). But in the U. S. you have hundreds of corn and bean varieties in cultivation. Surely some of them must be suited to our climate....We are also very interested in obtaining any vegetables which were originally brought from the Nordic countries (Sweden, Norway, Finland and Denmark) by immigrants and then kept alive in America....(pp. 174-175)

Albert Paul Lopez, CA — is looking for Rosary pea (Abrus precatorius)--very beautiful, very small (.25") oval seed that is half fire engine red and half jet black, used to make rosaries; the seed is poisonous....(from literature Albert sent): a vine indigenous to the tropics around the world, has been introduced into Florida where it has escaped and grows wild....a woody perennial vine to 20' in length. It grows as a weed trailing on the ground in citrus groves or twining on fences in Florida....The fruit is a many-seeded pod to 1.5" long, which is covered with fine hairs. These seeds are oviform, 1/4 to 3/8 inches long, glossy bright scarlet or orange with a black marking at one end....also called rosary beans, jequirity, love beads and precatory beans. They hang on vines and grow in pods like ordinary garden peas, but there the resemblance ends. Rosary peas are used only for jewelry, beads, necklaces and bracelets. Being about .25" (6 millimetres) long and brightly colored--orange at one end and jet black on the other--they make attractive and safe jewelry unless the wearer perspires. Then the natural perspiration may cause some of the poisonous compound called abrin to be absorbed through the skin. Ordinarily the beans have a hard coating and could be swallowed with no ill effects, but if only one bean is chewed and swallowed, death is almost certain. Abrin of the Rosary peas and ricin of the castor beans are similar in their effects. Symptoms may not develop immediately on swallowing or may not even appear for several days, but following this lapse of time there's a loss of appetite, stomach upset, delirium and finally total collapse. There is no known antidote. (I had some sent to me a couple of years ago (for identification) with the name Magic Beans. They are incredibly beautiful. Until now I didn't know what they were....If you have any around or use them for jewelry, you would be wise to coat them with polyurethane varnish. If you have children around, you better be extremely careful, because they look delicious - Kent.) (pp. 179-180)

Mrs. Zelma M. Schulte, CA — My mother has her 85th birthday on the 4th of February. I would like to surprise her with the knowledge that I had located her long-sought Rice Pea that was lost years ago. It has a very pale seed. The dry hulls or shells are parchment thin and off-white. The seed is about the size of a small garden pea and perfectly round. It has a tiny black eye, or at least dark. We raised them in southern Missouri (Birch Tree to be exact) about 45 years ago. (p. 180)

James K. Short, CA — When I was a kid up in Washington State, there was a little town called Winlock. Across the tracks from the depot was a huge egg made of sheet metal (probably 10' X 18') and under that egg was a sign that read "Winlock Washington--Egg Capital of the World"....That might have been about 55 years ago, of course not now. But I remember those farmers and those huge long chicken houses everywhere you looked. Those farmers didn't have egg mash as we do today. But one thing those old farmers did do was raise kale for a food supplement. Especially in the winter months, they would hang stalks of kale ever so many feet....Kale does increase egg production; I've tried it. But I can't find the kind of kale they had. It grew about 3' high with a 1.5 to 2" stalk and big broad leaves (not like this dwarf in the stores now). It was sometimes called Chicken Kale. (pp. 180-181)

Kathy Stowe, CO — We have an overwhelming need for non-hybrid varieties of vegetables which will do well at high altitudes. We have the experience and want to get involved in a Growers Network for mountain folks throughout the Rockies. We live at 7,500 feet and are already successful in stocking a freezer and root cellar. Now all we have to do is become independent of the seed companies. We really need and believe in your efforts. (p. 182)

Milan Rafayko, KY — I'm not one of those gardeners (usually neophytes) who go in for "novelties," but Pruden's Purple tomato appeals to me because I know it was the work (no doubt long and disappointment-filled) of an old-time gardener (Mrs. Pruden) and one from my own bailiwick, too....It is the "Last of the Mohicans" of the 19th century French violet-hued tomatoes. Who knows what unknown and potentially valuable (in the future) qualties lie in the genes of this plant? We know that any purple fruit is rich in iron. I feel that it is extremely important to get this plant resurrected from near extinction. (pp. 186-187)

Dave Christensen, MT — I had a valuable experience that would be good to tell people about....I have some beans (second generation from those found in an Anasazi Indian burial site in Utah, carbon-dated 900-1000 years old). I got three weak seeds and planted them. In our brief 90-day mountain climate, at first frost in September the pods were in the edible stage, the seeds very small and green--weeks from maturity. I pulled up one plant, the slowest one which I had no hope for, and brought it into the house to study the plant form. Meanwhile, we only had a few light frosts during the next two weeks. The leaves were mostly damaged, but we did not have a hard freeze, so the bean seeds continued to grow until they looked about 75% developed. At this stage I picked one pod, removed the seeds and put them in the house to dry. We had a solid freeze that night and all the beans in the garden were totally destroyed. The beans I took out of the pods in the house shrank and wrinkled, were useless and did not germinate....However, the very slow plant which I uprooted and brought in the house had dried. I opened the dry pods and found every pod to contain smooth, shiny, normal-colored beans. Some had actually taken enough nourishment from the plant that they developed into 50% of their normal size. But even the pods that appeared empty had miniature beans in them, about 1/10 normal size. Even the smallest of these germinated. Every single one!....So, if you have a crop that is rare, or if your season is too short, you can stil get fertile seed if you let the pods dry on the plants before opening. (p. 190)

Lyle L. Settle, NM — Around 1925, someone stopped at our farm near Ina, Jefferson County, Illinois and had apparently been eating the Alma Gem melon. It was so good that they had saved a few of the remaining seeds and gave them to my father. I recall my dad saying, "Surely this must be the best melon in the world." We grew them for several years. I spent the summer of 1931 in the hospital and each week my parents brought some of these melons. My folks moved to town that summer, and the next year there were no seeds ....this Alma Gem melon was about 7" in diameter and perfectly globular. It was rough-finished, but not sectioned. The meat was green turning to orange around the seed cavity. It was sweeter than a honeydew. The melons seemed to grow in pairs about 10 to 12" apart. Both seemed to ripen simultaneously. When the stem cracked off, they were at their peak flavor. There is not a melon on the market today that can compare to it....I have researched this melon over the years, telling the story wherever I could. The nearest description I have come to is that of the Cannonball melon, reportedly grown at one time near Enid, Oklahoma....

What I would like to develop (no success yet) is a White Aztec bean with red blooms. This would cause a great increase in production due to pollination. Who do you think could do it?....Phil Germain sent me a sample of the largest Aztec beans I have seen so far. His went 243 to the pound based on a 15-bean sample. The best I ever had was 315, though I've heard rumors of some that reportedly went around 200 without having to be selected....Dan Eaton of East St. Louis and a friend in Ohio grew my Aztec beans successfully, though both did grow in partial shade....Aztec beans will do better if you plant them where they will be shaded for an hour or two in the morning <u>and</u> the evening....(p. 192)

<u>Mrs. Marjorie B. Mendenhall, OR</u> — is looking for Mexican Marigold (Tagetes minuta), grows to 12" with tiny pale yellow star flowers and aromatic foliage. It is a half-hardy annual originating in Latin America, but distributed pretty much all over the world, notably in Australia where it is known as "Stinking Roger." In Rhodesia it is used as an insecticide and repellent and is also reported to be an effective nematacide and fungicide....Henry Doubleday researchers in Britain have discovered that T. minuta also has root exudates which inhibit and/or destroy certain weeds with starchy roots. (That's where I want to start with it. I have no less than five varieties of morning glory and/or bindweed on my place, and anyone familiar with those pests in a mild-winter climate such as western Oregon knows how tenacious they are!)....<u>Gardening Without Poisons</u> by Beatrice Trum Hunter, 1964, Berkley Windhover Books, New York, should be read and/or owned by all gardeners, organic or otherwise, for its invaluable suggestions. <u>The Organic Gardener</u> by Catherine Osgood Foster is another fine volume of "answers." (p. 195)

<u>John S. Baylor, PA</u> — I just can't tell you how I have enjoyed the SSE. I think it's great! It has given a whole new dimension to my gardening. I love to garden, but this is really a thrill....(p. 195)

<u>Wilford Herman, SD</u> — Until about five years ago, Henry Field's Seed and Nursery Co. listed seed of a white-fleshed watermelon. It was an icebox type, rarely growing more than six inches in diameter. It was early ripening, sweet, juicy and delicious....We were negligent, since we depended upon Field's for seed every year. When the variety was discontinued in their catalog, we

immediately sent the company a sizeable check asking for any stray packages of old seed that might be scattered around in one of their seedhouses. Our check was returned because no seed was available....If we can just get a few viable seeds, we'll never run out of seed of the white-fleshed watermelon again. Also, we'll be more than willing to supply your group with seed of uncontaminated purity. (p. 197)

Patrick Hughes, Henry Doubleday Research Association, England -- I am looking for any old English lettuce varieties mentioned in Vilmorin's The Vegetable Garden or any other old sources....Lettuce seed is, of course, very easy to produce with a decent degree of purity. I find it strange that more people do not save their own....(I too have found that curious. Why are there such an incredible number of heirloom beans and so very few heirloom peas and lettuce? It must have something to do with a greater fascination in the seeds than with the plants themselves - Kent)....

I have just heard from our Plant Variety Office that they are going to allow new varieties to use old names. The whole idea seems absolutely crazy to me....As regards the naming of new varieties, the laws are really very clear on this. All names must be submitted to the Department of Agriculture for approval, and the Department will not accept any name that has been in recent or common use. The trouble is that this is a subjective judgment, and only this year we have had a case where a bean was renamed Robroy. As I understand it, names are not supposed to reflect the name of the seed company that had right to the variety. In this case the seed company is called Robinsons and the name has been used before and is listed in Beans of New York and elsewhere. This is an obvious breach of the spirit of the legislation, but, as the people who granted this name are also the guardians of the legislation, there is little we can do. (The trouble is that a lot of the good names have already been used and the modern ones just don't have the same appeal. Many sound more like short bursts of flatulence than vegetables.) Obviously the situation is unacceptable as it stands and pressure will have to be brought to bear....

One of the problems I suffer from is seed companies saying that they are maintaining a variety, when in fact they are doing nothing of the sort. I recently tried to order Loo's Tennis Ball lettuce, which we had hoped to maintain. The seed company which had it listed in their catalog now tells me that they haven't carried it for some time....Loo's Tennis Ball lettuce may exist somewhere in America. If you come across it, please let me know....

I am becoming very interested in Trout beans. It seems that in this country they change color! That is, they take on a "Dove" pattern when grown in our climate. I have had beans from various sources, and every time the beans change. Have you ever experienced this? Have any of your members ever heard of this? It breaks the rules of bean genetics. I would like to hear from anyone who has seen a change in the pattern (not the color) of a Trout bean after an exceptionally cool summer. Obviously this could be due to cross-pollination, and so I would also like to acquire "pure bred" Trout beans that have been grown in isolation for a few years. If any member of the Seed Savers Exchange has got such a stock, I would be grateful if they would get in touch. (Within just the last two years I have had several people write me that their Trout or Jacob's Cattle beans were changing toward their darker color and becoming more solid. No one seems to know why this is happening. I even asked John Withee and he doesn't know either....One lady in her Member's Listing this year said that she grew the same bean in three different areas of her garden having low and medium and high humus and fertility. She got three different colors of beans; didn't say which was the darkest - Kent)....(Any organic gardener should seriously consider membership in the Henry Doubleday Research Assn. They maintain the best organic advice service in England and their quarterly newsletter describing their ongoing experiments is fascinating and first-rate. The Association will send you brochures containing descriptions and prices for their many publications, guides and experimental results. They maintain The International Fruit and Vegetable Library which catalogs varieties and their known sources. They are in the process of setting up several Vegetable Sanctuaries on the grounds of old mansions across England. They are desperately trying to locate funding to collect peasant varieties in Greece while they still exist - Kent) (pp. 202-203)

From The 1984 Winter Yearbook—

Bob Thompson, CA — We had a terrible gardening year. A late spring held up planting and then the wind blew from the north almost constantly, drying up water as fast as it could be applied. Most folks said to hell with it and let everything die. People should really appreciate the drought tolerance of some of the older varieties now....You and your work, Kent, are greatly appreciated by all of us. But it is time we honored some unsung heroes as well. I would like to dedicate 1984 to the Family of Kent Whealy: Diane, Aaron, Amy, Tracy and Carrie. What wonderful people you are and how lucky your dad and husband is to have you. Believe me, you are not taken for granted and we all love you very much. (p. 74)

Gary Staley, FL — My thanks to all the members who sent me seeds this year. A special thanks to Faxon Stinnett who sent me many seeds. Also to Will Bonsall, who went to a lot of trouble to get some very good information for me. I wish Richard Grazzini and Kent Whealy had more time to correspond....A suggestion for the SSE: Send out a monthly newsletter to let us know Kent is still alive and what's going on. I'm afraid a lot of members may drop out because they only receive mailings twice a year. I know that sometimes I think to myself, "Maybe Kent got run over by his tractor." (p. 81)

Bruce W. Hancock, GA — I recently called on the head of USDA Vegetable Breeding Station in Charleston, SC. Much sweet corn work is done at this station. I learned that 30 days after pollination, corn pulled and dried on the cob gave 65-75% germination. At 40 days there was 85-95% germination. In fact, later gathering gave no better germination. In our climate we have trouble curing sweet corn seed. I shall try cutting the whole stalk 30 days after pollination and dry the shocks under cover to see what happens. This information about corn germination should be of great help to those in short-season areas....I think we should try to plant at least 500 seeds (about 2 1/2 cups) for field corns, about 2 cups for sweet corns and 1 1/2 cups for popcorns. Then select

40 to 50 ears from the very best plants to ensure having enough seeds for two crops. We planted some corns from Ernest Strubbe that had been stored in jars in his basement for up to 11 years with excellent germination. Here we plan to store (in jars in the freezer) seed dried to 8-10% moisture....

Small lots of corn were grown in my garden under irrigation. All were hand-pollinated. We have a bee here that is twice as large as a honeybee and a third the size of a bumblebee. We think it is a worker bumblebee. It played heck with us by cutting holes in the paper bags over the corn tassels. The holes were 1/4 to 1/3 inch wide and 1 to 1 1/2 inches long. I don't think they caused any trouble with carrying pollen to silks because it only worked the tassels. The hole was always up the side of the sack. With squash it was a different story. After spending many hours removing open-pollinated squashes, and then many more hours taping closed the male and female flowers, it was a heartbreaker to find one to three bees in each female flower below the tape, but never at the base of the flower. We did not find a single male flower cut....

Larry Sallee sent me soil from the fields where the different corns were grown. This was added to the row where the corn was grown here. I could not be more pleased with the yield we got from all the flour corns. A very old primitive brown corn, each stalk having at least two and some with three long (12-14") ears, was really a find. I am sorry, we can spare none of these this year. We had been told not to expect these corns to do too well for two or three years. By having soil from the field where it grew, we also had the microbes from those fields, too, and we feel this is very important. The role soil microbes play in plant nutrition is a very new field of study and is very important. We use only "chicken house composted" fertilizer on everything grown here so we can keep these microbes alive and functioning. Heavy usage of chemical fertilizers kills these microbes. (pp. 83-84)

Albert Beck, IL — Back from sabbatical in Italy. The gardens there on the Tuscany mountainsides were quite small but very productive. In early April the seedlings were protected in the ground by planting each in a slight depression and covering them with small scraps of broken window pane glass. The olive groves were beautiful, as were the vineyards. (p. 96)

John B. Edgerton, IN — This will be the third and final year of my tomato trials. Once again I would like to look at any of the large paste tomatoes, any large orange or yellow or red or pink tomatoes and any of the compact dwarf types....I am interested in corresponding with others interested in regional trial gardens. (Count me in - Kent.) (p. 103)

John Hartman, IN — Had a house break-in and it has taken weeks to sort out the mess. They took my garden log, of all things, since it was with legal documents and other valuables. For that reason I do not have the days until harvest on my listing. At least they didn't take the seeds!....Donna Kline says you're moving to Iowa. Good luck. Hope we can have another campout, if it already is not too big for you. (Yes, we are moving to Decorah, Iowa in March. I've been told there is a 4-H Camp near there on the Upper Iowa River that can be rented, so maybe that's where it will be. We definitely plan to hold the campout each year, wherever we are. Sure hope you can make it again - Kent.) (p.

105)

William T. Arthur Jr., IA — You've done an excellent job on the Yearbook and developing the SSE. I especially like the coding at the front of the Yearbook. It makes it so easy to find all members offering a certain item. The SSE is going to save quite a few heirloom varieties and get more people interested in raising them. I had about 400 item requests last year and about 100 this year. With the new onions I'll probably have quite a few more. I've met a lot of nice people through the SSE, especially some others who are primarily interested in the allium family....Thanks for starting and guiding SSE. It's going to mean a lot to future plant breeders. (p. 107)

Will Bonsall, Scatterseed Project, ME — Last year, Warren Schultz of Organic Gardening did an extensive interview on the Scatterseed Project, which he said would appear sometime this summer. I've been looking forward to it, as it would give some useful publicity to the SSE, which I emphasized....Kent, two years ago, I grew out some things for the Growers Network. Some varieties yielded scant, and I further increased them this year with greater success. Would you like me to send more of those samples? Or do you only want samples from whoever's growing it that year? I don't know how you're handling it....(I've been asked this exact question dozens of times by Growers Network participants. If anyone needs to grow a plant a second or even a third year, that's great. You can often save things that would have been lost by doing exactly that. As long as the material is eventually returned (along with its original number), the Growers Network collection remains available - Kent.) (p. 115)

Darrell Rolerson, The Dharma Farm, ME — I am interested in developing a native agriculture, or an agriculture which is indigenous. So I am interested in communicating with anyone on this subject. I would especially like to hear from veterans of this way of growing....Hopi agriculture is one exciting example. For 10,000 years the Hopis have been developing their wonderful seeds on the mesas above the Arizona desert. These same seeds seem to be at home here on this island off the Maine coast. They are surprisingly adaptable and they are exemplary of the plants which I love to see growing in my garden. They are extremely hardy, require a very minimum of culture, produce an early bounty and they continue producing! Next year I hope to be offering a Hopi watermelon....It seems to me that the Seed Exchange gets more exciting each year....In about two weeks (mid-November) I will be traveling to Brazil and will be there until April. It is a trip I have dreamed of for the past five years. So my seeds won't be mailed until May 1. Maybe I will discover some exciting ones in Brazil....Does anyone have traditional recipes for beans like Leather Britches and Lime-Speckled Cutshort, etc.? (p. 124)

Louise Bastable, North Shore Community Gardens, MA —....I am tracking down seed varieties that were introduced by the J. J. H. Gregory Seed Co., one of the largest U. S. seed houses in the mid-1800s, and growing them in a garden on the original site. (p. 130)

Elaine Reidy, MA — Any baked bean can be made "gasless" by adding about one teaspoon of baking soda at the end of the boiling period. (Do this over the sink, as it foams up.) Then rinse thoroughly, and proceed as usual for baking. My uncle Al gave me this "secret" he learned years ago in the Maine woods while working with men whose food staple was beans. (p. 133)

James P. Tjepkema, MN — I am interested in integrated crop pest management and intermediate technology and would like to work on improving variety selection in crops, especially soybeans and particularly soybean cyst-resistant varieties....SSE should consider encouraging members to send some particularly rare varieties to U. S. Plant Introduction Stations for preservation, if these varieties aren't in their collections....I sometimes do consulting in pest management and am interested in intermediate technology. Also I am a trained plant nematologist. (p. 145)

Ted M. Gibbs, MO — Does anyone know anything about Vine okra? It looks more like a gourd than okra. Yet it is not like any gourd that I can find or anything else as far as that goes....(It's Luffa Sponge (Luffa cylindrica) which is neither okra nor gourd. If dried and peeled, makes a good sponge or scrubber. People call it okra because of the ridged fruits - Kent.) Some of my West Indies okra plants started producing branches immediately above the leaf nodes in late September. The leaves have not fallen. The apical tip is not damaged. They are becoming very dense. Why?....My Star of David okra is a single stem plant. No branches like other varieties. This plant is thought to have come from Israel. Any background on this type of okra? (p. 149)

Gary R. Toms, MO — I am compiling information on varieties of vegetables which were popular in western Missouri before the Civil War. Will share information in response to specific requests. (p. 153)

Dave Christensen, MT — I live in a climate that is too cold to mature dry beans and corn, but am developing quick-growing new varieties. I'm looking for very fast-maturing corn (to use in breeding)....(p. 155)

Steve McComber, NY — These (Tonawanda Seneca flint corn, Mohawk Round Nose flour corn, Mohawk Long Ears flour corn) are some varieties of our corn, the Iroquois of Upper New York State and Canada. I've been maintaining as much as I can. These native varieties have been in our families for hundreds of years and also with the help of the dedicated Seed Savers, it will continue....Many varieties have been lost by my people also....Thank you Kent ....Keep up the great work. (p. 167)

Carol Bishop, OH — A good source of pre-packaged silica gel is from your local drugstore. Prescription drugs come packaged with small containers of silica gel to keep the drugs dry. Pharmacists end up throwing these little containers away. I received a large bag of these pre-packaged little containers free for the asking from my pharmacist. (p. 173)

R. G. Centofanti, OH — I have enjoyed being involved with the SSE. I have also learned a lot about saving seed and better ways of sending it. You all are teaching me how to be better organized, too. I wish someone would get a flower SSE going. Thanks, Kent, for getting this together. May the Lord continue to bless you and everyone involved with SSE. (p. 174)

L. Arlie Gray, OK — From all the evidence I have accumulated, I have come to the conclusion that Cowhorn, Longhorn and Texas Longhorn okra are all the same varieties. My brother gave me some seed three years ago, but he was unable to find what it was called. After growing it one year, I got some Cowhorn seed from TX/HA/L and also from a Texan who had advertised his Longhorn seed in the Oklahoma Stockman-Rancher. All of the plants were exactly like mine. Also, I discovered last season that a neighbor was growing Cowhorn okra which was exactly like mine. According to articles in the Texas Gardener Members of the Houston Texas Garden Club, several members grew varieties called Cowhorn and Longhorn which were identical. Also, this magazine listed the following seed company as selling Longhorn okra at $1.50 per packet: Green Horizons, 500-T Thompson Drive, Kerville, TX 78028. (p. 184)

Al Kaplar, Omniversity of the One, OR — We have obtained high quality amino acid analyses of three tomato varieties (the colorless juice) and for asparagus tincture and soybean miso. The results are remarkable. In Benewah red tomato, there are 15 of 15 amino acids needed for human protein synthesis in the juice. We analyzed for 15 of the needed 20. In mM (large) amounts are aspartic and glutamic acids, tyrosine (a major surprise), serine and histidine. There is 13-fold less histidine in Sunray (an orange tomato, "low acid") than in Benewah (red fruit, "high acid"). The allergic reaction to tomatoes is probably histidine, not acid. Histidine is converted to histamine, the material in bee stings that gives the allergic, inflamed reaction. By analyzing the bean seeds for the amino acids in their cotyledonary leaf proteins, one can begin selecting bean varieties based on needed amino acids. Then by fermenting the beans as the orientals do soybeans, one obtains free amino acids in the miso. Thus miso soup is amino acid soup from beans. Fava beans, too, make miso!....

As OR/OM/1, we distributed 81 samples of seeds to 43 folks for the 15 kinds listed in the 1983 Winter Yearbook. Most requests were for the giant lentil. May we all see continued growth and appreciation for the gene-pool and Life itself....Our Food System--Local, State and Nation, Cornucopia Project, Rodale Press, publishers of Organic Gardening, have supported an enquiry into the state of our food system. It is a revelation worth investigation. Manuals for implementing local Cornucopia Projects are available for nominal costs. Analysis of food systems by state is available for more and more states. Application to local townships and communities is needed....(Peace Seeds, 1130 Tetherow Road, Williams, OR 97544, puts out a catalog of their "One World Planetary Plant Seed Collection." Send a long SASE to receive a copy. It contains many unique varieties and some heirlooms - Kent.) (pp. 189-190)

Tim Peters, OR — Of special interest to me are winter hardy and overwintering vegetables. Anyone doing work with these overwintering vegetables to genetically improve them, please contact me and let me know what you are doing. Share your seeds with me and if I have anything you may want, I'll share with you....If any Seed Savers live where your temperatures drop down to -10 and -20 degrees F., please keep an eye on your cabbage, kale, brussels sprouts, turnips, rutabagas and carrots. If any plants survive, write me in time so that seed can be saved from these unusual individuals (your survivors). (p. 193)

Mrs. L. R. Hayes, TX — I've noticed more and more older varieties in the seed catalogs. Even Gurneys had Cowhorn okra last year....I think everyone understands about your not being able to answer all letters....I enjoy your comments in the books....(Thanks for your understanding, Joyce. I have two big boxes of unanswered letters now, and it really does bother me. I just hope no one thinks I'm ignoring them. But I can only do so much and the most important thing is that the books get out - Kent.) (p. 204)

Jeanine Mankins, TX -- I have a Perpetual Food Garden--food trees, self-sowing perennials and annuals. I have stopped disturbing the soil. I plant in permanent raised beds which nobody steps into. I mulch and water (not much rain) and something grows year-round here in the semi-tropics. The only pests are nematodes, but I have read (in the Houston Garden Book) that nematodes dislike rich, healthy soil, and I have noticed a decline in their damage over the years. There are plenty of birds and possums, but I plant enough to share. Correspondence with other (especially organic) gardeners in the semi-tropics would be comforting....I have two books to recommend: A Garden Book for Houston and the Gulf Coast, River Oaks Garden Club, Pacesetter Press (Div. of Gulf Publishing Co.), Houston, TX, 1975 (4th printing 1982)--valued for its year-round planting schedule for the sub-tropics. This material is rare, in my experience. Also Ruth Stout's No Work Garden Book; my copy has disappeared so I have no bibliography for it, but it has been of major importance to me over the years. (pp. 205-206)

Dorothy Smith, TX — To all those country folks who feel like I do about our few acres: "Happy the man whose wish and care, a few paternal acres abound, Content to breathe his native air, In his own ground. Whose herds with milk, whose fields with bread, whose flocks supply him with attire, Whose trees in summer yield him shade, in winter fire." Benjamin Franklin. (p. 207)

Sara Bezaire, Mexico —....I get many requests for corn and beans, both of which are prohibited from entering the U. S., and Mexico prohibits exporting of corn. I respect their regulations since we have severe weevil problems here....I am concerned that people may have sent me money for seed they never received, since not one of the requests I sent off enclosing dollars was answered. But almost all of the ones enclosing seed to exchange did get through, both ways. If anyone who sent money and didn't get seed will write again, I will be happy to send it....I actually prefer to send my seed in exchange for stamps. It works much better, and letters with 20-cent stamps reach me as fast as those with more postage....Various SSE readers have sent or requested seed of tropicals. Please mention that it is very cool here at 7500 feet and things like cucumbers and eggplants won't grow. The climate is quite difficult for many vegetables in the rainy season. I plant chilacayote just outside the garden edges because it unerringly sends long roots to the more fertile soil and consumes much of the water in beds of lettuce, carrots, etc. which can't stand our rains. I also let it grow high in the trees, but during our strong winds of November it is hazardous with 30-pound squashes occasionally falling from 20 feet....Thank you most profoundly for your wonderful work and the SSE family....(p. 238)

Manifesto

(Editor's note: The following section was actually written as a grant proposal nearly six years ago. Having never seen a grant proposal, it turned out to be much too broad in scope, although it did get funded. It is a rather powerful manifesto describing current problems, what needed to be done at the time, and our goals as an organization. It is interesting to examine what our thinking was then and how the SSE has evolved to meet those challenges.)

DEDICATED TO
Baptist John Ott (June 24, 1894–January 5, 1974)
Helena Hackman Ott (April 16, 1897–December 29, 1982)

(Grandpa Ott's Morning Glory, Near St. Lucas, Iowa in 1972)

When someone you love dies, some folks say they "passed away." But you'll hear others say that they've "passed on." What a tremendous difference!

Their possessions, no matter how meager, are divided and passed down in our families as priceless heirlooms. Some of us gardeners treasure their seeds the most, because that is one small part of their lives that is still alive. As these seeds are passed down through each new generation, they become even more sacred.

If you have ever enjoyed the vivid storytelling, the ornery humor, or the magic sparkle of a Grandpa or a Grandma Ott, then you are very lucky. If you have some of their seeds, treasure them....and pass them on!

GRANT PROPOSAL

There is a world-wide crisis that few people know about, but that may very well determine how hungry the world is in the future. The scientific community and laymen around the world are voicing increasing alarm about the genetic wipe-out of our food crops and their ancestors. Two specific areas of this crisis which have received almost no consideration are the extinction of both "heirloom" vegetable varieties and also the vegetable varieties currently being dropped from seed catalogs.

Because the United States is a nation of immigrants, we have been blessed with the largest and most excellent collection of food crops ever bestowed on any nation. From literally every corner of the world, gardeners brought with them the best of their vegetable varieties and proceeded to acclimate them to varying regional conditions across the U. S. This natural ethnic wealth of seeds is greater than could ever possibly have been achieved through plant exploration and introduction. This incredible diversity of varieties made America the agricultural giant that she is today. But much of this wealth is being lost forever and we must act immediately to save what remains.

I. Heirloom Vegetable Varieties Becoming Extinct

I have worked for the last five years to save this vanishing vegetable heritage. Many of these living heirlooms have been kept alive in the same family for over 150 years, always being passed from generation to generation. But life today in the U. S. has become so mobile that the majority of families move every few years. This separation of the generations necessitates that at this point in time, we must help with this handing-down process or thousands of heirloom vegetable varieties will be lost during the next generation. Across our land, elderly gardeners are keeping unique vegetable varieties that for lifetimes have been bred up and adapted to local conditions, pests and diseases. But when these master gardeners pass away, unless they have found younger gardeners to keep replanting their seeds, that outstanding variety becomes extinct. Not only will future generations never enjoy it, but we have lost forever irreplaceable genetic characteristics which may be desperately needed in breeding future food crops.

How dangerous is this situation? Voices in the scientific community are very clear:

"The genetic diversity of our food crops is a national wealth which the varied racial and ethnic members of our nation have brought from their homelands. It is unthinkable not to preserve and maintain these and other reserves of genetic diversity that still exist for future plant breeding needs." (Dr. Garrison Wilkes, Professor of Biology, University of Massachusetts)

"Genetic wipe out....might well be tomorrow's greatest single problem—the problem of feeding the human species in the midst of the technological wealth it has created." (Dr. Erna Bennett, Crop Ecology and Genetic Resources Branch of the Food and Agricultural Organization of the United Nations)

"These resources stand between us and catastrophic starvation on a scale we cannot imagine. In a very real sense, the future of the human race rides on these materials....The line between abundance and disaster is becoming thinner and thinner, and the public is unaware and unconcerned. Must we wait for disaster to be real before we are heard? Will people listen only after it is too late?" (Dr. Jack Harlan, Professor of Plant Genetics, University of Illinois)

II. Vegetable Varieties Being Dropped from Catalogs Also Dying Out

A similar crisis has been developing for some time within the seed industry and only recently has reached epidemic proportions. Each year hundreds of vegetable varieties are dropped from seed catalogs, not because they aren't delicious and unique, but because it is only profitable for the large seed companies to stock the varieties which sell the most. When each is dropped from commercial availability, unless an individual decides to keep that variety alive, it becomes extinct.

I have seen lists of vegetable varieties that were available at the turn of this century and I was amazed. The names of bean varieties available then covered six standard-sized pages of single spaced typing, pea varieties covered four similar pages, onion varieties covered two and a half pages, etc. It is estimated that less than 20% of these varieties have survived until today.

Lawrence D. Hills, Director of the Henry Doubleday Research Association in England, sums up the situation:

"....Several thousand vegetable varieties have vanished from the catalogues in the past twenty years because few firms can afford to list anything that sells fewer than 500 packets a year....The position would not be so serious if all deleted varieties were automatically consigned to a 'living museum of vegetable varieties' for possible use in the future. No one knows what sorts of qualities plant breeders will be looking for in 10 years' time, let alone 100 years...."

Although many of the vegetable varieties that were dropped during the first half of this century had been superseded by superior varieties, that has not been the case for at least the last 20 years. Far from being obsolete or inferior, the varieties being dropped today are literally the cream of the crop(s). Each is the result of millions of years of natural selection, thousands of years of human selection and usually about a decade of intensive plant breeding and testing. Only the very best make it to the catalogs and each is unique and irreplaceable. But they are being allowed to die out, <u>due to the economics of the situation,</u> with no systematic effort being made by government agencies or lay organizations to keep them alive or store them. We are allowing the breeding materials for the food crops of the future to be lost forever. We must stop this useless and short-sighted destruction.

III. The Seed Savers Exchange

In 1975 an elderly friend and gardener gave me seed of three heirloom varieties that his family had brought over from Bavaria four generations before--a large pink potato-leaf tomato, a delicious and very prolific pole bean, and a purple morning glory with a red star in its center. The old man didn't make it through that winter and I realized that it was up to me to keep his seeds alive. Realizing, too, how much must be lost each year in this way, I decided to do something about it. I have a degree in Journalism, so I started playing with the media and the mail.

That winter through letters to gardening magazines, I contacted and traded seeds with six other heirloom seed savers. One of the six died that spring, but by then several of us were growing the Bird Egg bean that her grandmother brought to Missouri in the 1880s. During 1976 there were 29 of us, and I poured more energy into contacting other heirloom seed keepers. I became the focal point for all correspondence and communications. By 1977 there were 140 of us and I started mimeographing a yearbook containing each member's name, address and lists of the varieties they had to trade and lists of the varieties they had lost and were trying to find. In 1978 the network had grown to 238 members. I researched saving vegetable seeds and added a Seed Saving Guide to the yearbook to teach others how to save vegetable seeds and keep their varieties pure and disease-free. In 1979 there were 274 of us and the yearbook I write each winter was turning into a book. And during 1980 there were 328 of us participating in this unique network called the Seed Savers Exchange.

During the past five years approximately 600 different members have offered an estimated 3,000 heirloom or unusual vegetable varieties to over 9,000 interested gardeners. I would roughly guess, estimating conservatively, that 150,000 plantings have been made of vegetable varieties that aren't in any seed catalog and in many cases were on the edge of extinction. It is hard even for me to imagine the impact this kind of an exchange is having. Varieties that would have been lost due to the deaths of members are now available from many sources. Many members write expressing deep gratitude for finding varieties they had been trying to locate for up to 40 years. Our efforts have been praised in articles by over 100 magazines or newsletters, and national wire service articles have been carried by hundreds of newspapers around the country.

I am very proud of the Seed Savers Exchange and what I have been able to achieve with just my spare time and no financing. But this year will be a critical turning point for the Seed Savers Exchange. I have been desperately striving to build the SSE to the point where it would provide a meager living for my family, so that I could work full time with it. In this respect I have only been about one-third successful. SSE has hit a sort of growth plateau entirely due to my lack of time. The avalanche of correspondence generated by the Exchange is tremendous. While working another job, it is all that I can do to keep up with it by working late into each night. I no longer have time to pursue the opportunities for growth as they present themselves. It would take less than two years to achieve permanent self-support for the SSE, but unless I can spend full time with it now, there's no way.

It frustrates me that the SSE is just a shadow of what it could be, especially because events are taking place right now which are accelerating the deaths of vegetable varieties. Many gardeners keeping heirloom vegetable varieties are very old and their seeds will be lost within years. The weather is becoming very unstable and I have always feared a growing season that would be so short or so dry that it would wipe out many of the remaining heirloom varieties. With great concern I watched this summer's heat wave burn up everything in its path from Texas to Iowa. I'll never know how many heirloom varieties were lost by gardeners I haven't been able to contact yet. I lecture my members each year to never plant all of any seed because they may lose it. This winter I'll see how many of them heeded my warnings.

During the last five years my efforts have saved hundreds of heirloom vegetable varieties from extinction. But for every one that is saved, I hear about two that were lost. Every year now I locate elderly bean or tomato collectors who are keeping literally hundreds of varieties, but often their health is failing and their collections will be lost within a few years. Seed hoarders, whether breeders or interested amateurs concentrating on a particular crop, can often amass impressive collections. But when they die, there is often no transfer of seeds or knowledge about their varieties. They haven't prepared anyone to carry on and to understand the significance of some varieties compared to others. The Seed Savers Exchange has given gardeners a forum in which to discuss and offer their heirloom and endangered vegetable varieties. It is really filling a void, so that we can all learn from each other and stop these irreplaceable resources from being lost.

Burt Berrier is a good example of what happens when an amateur collector dies. He passed away in January of 1978 at the age of 84 and was keeping over 450 varieties of beans that he had collected over a 50-year period. Burt was one of the first combine servicemen and traveled all over the West in a period when farmers were still growing and combining fields of dry beans. Many of his varieties were collected during this period. I had been corresponding with Burt for over a year. We were just beginning to make plans to transfer his collection to the membership of the Seed Savers Exchange.

Then I heard that he had passed away and that officials from the National Seed Storage Laboratory had picked up his collection. It is the policy of the Laboratory to make available only seed of specific varieties when there is no other known source. So I figured that we had simply lost access to Burt's collection. But the NSSL's collections are, for the most part, varieties and relatives of large-scale agricultural crops. It stores mostly the breeding done at the State Experimental Stations and is sometimes criticized for the redundancy of its collections. NSSL is terribly underfunded and understaffed and doesn't have much money for growing varieties that need multiplying. They store only five-pound samples, and since almost all of Burt's samples were smaller than that, they offered the entire collection to John Withee. John directs a network of several hundred bean collectors and growers called Wanigan Associates. Each year they multiply over 900 varieties of heirloom beans. Approximately 30% of Burt's 450 varieties had already died due to his decreasing ability to carry the load of his collection during his final years. Out of Burt's collection, only 180 varieties still survive in the membership of Wanigan Associates.

A story on the front page of the June 2, 1980 Los Angeles Times sums up the situation quite well:

"....It is the fear that one day there may not be much to switch to that keeps Withee and Kent Whealy and thousands of others around the country hoarding antique seeds and trading them back and forth. As these heirlooms are ceremoniously mailed around the country each winter, planted each spring and then harvested in the fall for the next mailing, seed savers provide an important genetic 'fail-safe'—a chain of small backup seed repositories for the big government center in Fort Collins, Colorado, where thousands of varieties are supposed to be saved for posterity. The backup is especially important because government repositories are so low on funds they cannot always do what they are supposed to do....'I've got a valuable gene bank here, the government tells me,' says Withee....'The reason is simple—people won't correspond with the government. I get 20 letters a day from people who write and tell me about their latest leg fracture and then send me a packet of the family's heirloom beans....'"

This clearly shows why government collections cannot perform our function and why our work is definitely not duplicating their efforts. Th type of networks I am developing are satisfying needs that government programs aren't fulfilling and reaching people not normally reached by them. Government collections are not open to individuals, while SSE and Wanigan Associates are. The government isn't making any systematic effort to locate heirloom varieties (Burt Berrier's collection was picked up by them at his earlier request) or any systematic effort to save varieties being dropped by commercial sources. It is very obvious that a laymen's exchange of seeds that works as a supplement to government programs is the best plan to pursue.

IV. Proposed Activities

A. John Withee and Wanigan Associates

John Withee wants me to take over his network of heirloom bean collectors. He retired about two years ago and hasn't had a spare minute to himself since because of his "hobby." I have agreed to attempt to start taking over his organization this winter. I hope to combine our two networks into an even stronger organization. This is a unique opportunity because, as I have shown, much is often lost when collections change hands, especially a collection of this size. But John and I each have a sizeable network of growers already established. Also I have a communications and seed exchange network already set up and working well. And most important, John and I have the time to make an orderly transfer with all knowledge of these varieties intact. The success of my attempt to take over Wanigan Associates is to a large extent dependent on the time this grant will give me.

B. Inventory of Vegetable Varieties Currently Available Commercially

Each year hundreds of vegetable varieties are dropped from seed catalogs and die out. Recent developments in the seed industry have escalated these losses tremendously. During the last few years, large multinational chemical and drug corporations have been buying up seed companies at an incredible rate. The resulting conglomerates then drop most of the regionally-adapted varieties formerly carried by their newly acquired seed companies and replace them almost exclusively with either hybrids or patented varieties. At the rate things are going, more varieties will be lost in the next few years than have been lost in the last 20 years. As the economy continues to decay, I believe that many small seed companies will unknowingly be putting out their last catalogs this winter. Massive bankruptcies of small seed companies would result in the wholesale destruction of thousands of excellent, locally adapted vegetable varieties. It is essential that we immediately inventory all vegetable varieties that are still commercially available. I have already compiled lists of nearly every seed company selling vegetable varieties in the United States and Canada. I intend to make an inventory indexed by vegetable variety name, each of which is followed by a coded list of the companies which handle that variety. This would clearly show the varieties that are in the most danger of being dropped. Although "The 1981 Inventory of Vegetable Varieties" would be a massive task to compile and write the first year, it would be relatively easy for me to update each year after that.

Such an inventory would be an invaluable tool for anyone interested in preserving vegetable varieties. Many of my members buy up varieties, if they know they are in danger of being dropped. But most gardeners have no idea that a favorite variety is in danger until it simply doesn't show up in their catalog one year and they are unable to find it from another source. Such an inventory would show what all is available, what varieties are endangered and, as the years go by, what varieties have been dropped. But its most important value would be in showing what varieties are endangered before they are dropped. Such endangered varieties could then be bought up and stored or multiplied and exchanged.

C. Heirloom Seed Bank

It was becoming apparent that I needed a way to store seeds for long periods of time to ensure that endangered vegetable varieties that I had already located would not be lost. I became increasingly fascinated with seed storage projects in which seeds are dried, canned and frozen. When seeds are stored this way, most varieties have a shelf life of 20 years or more before they must be multiplied or replaced. Most varieties of tomatoes are still in fine shape after 50 years of storing by this method. As the Seed Savers Exchange continued to grow, it was becoming increasingly necessary to have such a frozen collection of the wealth of germplasm which flows through the SSE each year.

Early in 1980 I was contacted by Steve Starr, a young botanist living near Columbia, Missouri. Because of deep personal concerns over genetic wipe-out, Steve had bought a canning machine which makes an airtight tin can. It's just like the machines used at the National Seed Storage Laboratory. He also has access to expensive seed drying equipment at the University of Missouri. He asked for help in locating varieties that I thought were the most endangered. In return he offered to dry, can and freeze collections for both of us.

We decided to concentrate on processing as many endangered collections as soon as possible. We bought several dozen varieties of cold-hardy and sub-artic tomatoes from the collection of Edward Lowden, age 90, of Ontario. We are trying to buy 30 tomato varieties from the collection of B. Quisenberry, age 92, of Ohio. I am working to obtain about 150 tomato varieties from the collection of Mrs. Gladys Gronke of Illinois. I have also contacted Ted Telsh who is keeping a collection of 150 tomato varieties in Texas. Two large bean collectors who are working through both the SSE and Wanigan Associates have already sent most of their large collections, and John Withee will send in several hundred other varieties this winter. We also bought 30 pepper varieties from Horticultural Enterprises in Texas. And Gary Paul Nabhan, SSE member and ethnobotanist from Arizona, is supplying specimens of many very rare desert-adapted traditional food crops of several Indian tribes from the Southwest and Sonora.

Right now Steve and I are storing the beginnings of these collections in chest freezers in our basements. When I publish plans for the Heirloom Seed Bank in "The 1981 Seed Savers Exchange" during February of 1981, a couple thousand heirloom varieties will be added to the collections immediately from SSE members. After "The 1981 Inventory of Vegetable Varieties" is published in the spring of 1981, I plan to purchase and store at least a thousand endangered commercial varieties.

Steve and I agree that we should develop twin identical collections to be stored at both his home near Columbia, Missouri and at my home near Princeton, Missouri. I will publicize the Heirloom Seed Bank through the media and in the SSE and will supply empty seed packets and forms to interested gardeners to document the history and physical characteristics of each heirloom variety. Two samples of seed of each variety will be sent to Steve and he will continue to do the drying, canning and freezing and I will pick up my half of the collection each winter. I will maintain the records needed for each variety, keep files on the collections and distribute seeds through SSE as necessary to renew them.

I view the Heirloom Seed Bank as insurance against any loss of varieties until I have built my organization to the point that we can multiply the entire collection each year. At that time I intend to build a new seed vault and store that season's new collection in a different location each year.

D. Growers Network

During the past two years it has become increasingly apparent that many individual gardeners are beginning to realize how desperate this need for preservation is and are willing to work to help. Hundreds of gardeners have written to me saying that they do not have any heirloom seeds to offer, but that they would gladly work to increase such seed stocks. This has been very gratifying because I have always felt that ultimately individuals must become involved and take the responsibility themselves for locating and growing endangered varieties.

I intend to develop a system of growers who are not yet keeping heirloom varieties, but are willing to multiply rare seed stocks. These growers would multiply out a certain number of varieties, return part of the seed to the Heirloom Seed Bank, and share their remaining seed by participating in SSE. These growers plus SSE growers plus Wanigan Associates growers would allow me to take over John Withee's heirloom bean network fairly easily. With such a

Growers Network, I could offer aid to my collectors, who are keeping hundreds of varieties and can no longer carry the load, or to any gardener who needs help in multiplying an heirloom variety because of infirmity or whatever.

Until now my efforts through the printed media have been only to locate heirloom varieties. From now on my messages will be geared just as much to developing a Growers Network. Both of these needs can be presented to gardeners in the same mailings and articles.

E. Heritage Gardens Network

I want to contact all Living History Farms, Outdoor Museums, Botanical Gardens, State Horticultural Societies, etc., who are already keeping "period gardens" as part of their projects. I have already been contacted by about a dozen directors of such gardens, because they have a very hard time locating appropriate vegetable varieties from the correct time period and ethnic origins. I can supply such seeds to them in return for them multiplying and making them available on a permanent basis.

By generating as much publicity as possible through planting ceremonies/-events, such gardens can easily be used as educational tools to educate local gardeners to the need for preservation. Such gardens can grow varieties entirely for seed and make possible a second wave of seed distribution to local gardeners who have become interested in historic varieties. This would be a living back-up for the Heirloom Seed Bank.

V. Synopsis

My goals are to locate all heirloom vegetable varieties that are still being kept, to keep alive all varieties that are being dropped from catalogs, and to locate varieties that have already been dropped but are still being kept by individuals. I know from my experience with the Seed Savers Exchange that there are more and better vegetable varieties being kept by individual backyard gardeners than are currently available commercially. Many are uniquely resistant to disease, insects, drought or cold and are also very well adapted to local growing conditions. They must be saved for future breeding purposes.

During a period when the government has drastically cut funds for public breeding programs at State Experimental Farms, I believe that the programs I have outlined can effectively increase the genetic diversity of our future food crops. It has cost hundreds of millions of dollars for the programs that bred the commercial varieties that are now becoming extinct. I am asking for precious little other than the time to save them.

This is a pilot project that demonstrates a way to coordinate a large network of seed exchanges, storage projects and growers networks. The real value is not only my project, but as an experiment with a functional structure from which others can learn.

(*The 1981 Winter Yearbook*, pages 44-49)